Selected Titles in This Series

177 **Satoru Igari,** Real analysis—with an introduction to wavelet theory, 1998

176 **L. M. Lerman and Ya. L. Umanskiy,** Four-dimensional integrable Hamiltonian systems with simple singular points (topological aspects), 1998

175 **S. K. Godunov,** Modern aspects of linear algebra, 1998

174 **Ya-Zhe Chen and Lan-Cheng Wu,** Second order elliptic equations and elliptic systems, 1998

173 **Yu. A. Davydov, M. A. Lifshits, and N. V. Smorodina,** Local properties of distributions of stochastic functionals, 1998

172 **Ya. G. Berkovich and E. M. Zhmud′,** Characters of finite groups. Part 1, 1998

171 **E. M. Landis,** Second order equations of elliptic and parabolic type, 1998

170 **Viktor Prasolov and Yuri Solovyev,** Elliptic functions and elliptic integrals, 1997

169 **S. K. Godunov,** Ordinary differential equations with constant coefficient, 1997

168 **Junjiro Noguchi,** Introduction to complex analysis, 1998

167 **Masaya Yamaguti, Masayoshi Hata, and Jun Kigami,** Mathematics of fractals, 1997

166 **Kenji Ueno,** An introduction to algebraic geometry, 1997

165 **V. V. Ishkhanov, B. B. Lur′e, and D. K. Faddeev,** The embedding problem in Galois theory, 1997

164 **E. I. Gordon,** Nonstandard methods in commutative harmonic analysis, 1997

163 **A. Ya. Dorogovtsev, D. S. Silvestrov, A. V. Skorokhod, and M. I. Yadrenko,** Probability theory: Collection of problems, 1997

162 **M. V. Boldin, G. I. Simonova, and Yu. N. Tyurin,** Sign-based methods in linear statistical models, 1997

161 **Michael Blank,** Discreteness and continuity in problems of chaotic dynamics, 1997

160 **V. G. Osmolovskiĭ,** Linear and nonlinear perturbations of the operator div, 1997

159 **S. Ya. Khavinson,** Best approximation by linear superpositions (approximate nomography), 1997

158 **Hideki Omori,** Infinite-dimensional Lie groups, 1997

157 **V. B. Kolmanovskiĭ and L. E. Shaĭkhet,** Control of systems with aftereffect, 1996

156 **V. N. Shevchenko,** Qualitative topics in integer linear programming, 1997

155 **Yu. Safarov and D. Vassiliev,** The asymptotic distribution of eigenvalues of partial differential operators, 1997

154 **V. V. Prasolov and A. B. Sossinsky,** Knots, links, braids and 3-manifolds. An introduction to the new invariants in low-dimensional topology, 1997

153 **S. Kh. Aranson, G. R. Belitsky, and E. V. Zhuzhoma,** Introduction to the qualitative theory of dynamical systems on surfaces, 1996

152 **R. S. Ismagilov,** Representations of infinite-dimensional groups, 1996

151 **S. Yu. Slavyanov,** Asymptotic solutions of the one-dimensional Schrödinger equation, 1996

150 **B. Ya. Levin,** Lectures on entire functions, 1996

149 **Takashi Sakai,** Riemannian geometry, 1996

148 **Vladimir I. Piterbarg,** Asymptotic methods in the theory of Gaussian processes and fields, 1996

147 **S. G. Gindikin and L. R. Volevich,** Mixed problem for partial differential equations with quasihomogeneous principal part, 1996

146 **L. Ya. Adrianova,** Introduction to linear systems of differential equations, 1995

145 **A. N. Andrianov and V. G. Zhuravlev,** Modular forms and Hecke operators, 1995

144 **O. V. Troshkin,** Nontraditional methods in mathematical hydrodynamics, 1995

143 **V. A. Malyshev and R. A. Minlos,** Linear infinite-particle operators, 1995

(*Continued in the back of this publication*)

Real Analysis—
With an Introduction
to Wavelet Theory

TRANSLATIONS OF
MATHEMATICAL
MONOGRAPHS

VOLUME 177

Satoru Igari

Real Analysis—With an Introduction to Wavelet Theory

American Mathematical Society
Providence, Rhode Island

実解析入門

JIKKAISEKI NYŪMON
(An Introduction to Real Analysis)

by Satoru Igari

Originally published in Japanese by Iwanami Shoten, Publishers, Tokyo, 1996

Translated from the Japanese by Satoru Igari

2000 *Mathematics Subject Classification.* Primary 26-01, 28Axx; Secondary 42-01, 46Fxx, 42C15.

ABSTRACT. This introduction to real analysis is based on the author's lectures at Tohoku University. It begins with a thorough discussion of Lebesgue measure and Lebesgue integration, and continues with the basics of distribution theory and Fourier analysis. The final chapter is devoted to wavelet theory.

Library of Congress Cataloging-in-Publication Data

Igari, S. (Satoru), 1936–
[Jikkaiseki nyūmon. English]
Real analysis : with an introduction to wavelet theory / Satoru Igari ; translated by Satoru Igari.
p. cm. — (Translations of mathematical monographs ; v. 177)
Includes bibliographical references and index.
ISBN 0-8218-0864-8
1. Mathematical analysis. I. Title. II. Series.
QA300.I3813 1998
515—dc21 98-7552
CIP

♾ The paper used in this book is acid-free and falls within the guidelines
established to ensure permanence and durability.
Visit the AMS home page at URL: `http://www.ams.org/`

10 9 8 7 6 5 4 3 2 04 03 02 01 00

Contents

Preface

The reader may be a little confused at the words "Real Analysis" in the title of the book. Indeed, real analysis, like harmonic analysis, has a lot of contents nowadays, and the author cannot definitively state what are the subjects that the real analysis deals with.

I suppose that real analysis meant, at first, the field which corresponded to complex analysis and aimed to analyse and synthesize sets and functions on the real axis or the plane. However, complex variable methods have been deeply incorporated into the Fourier analysis which has been one of the great backbones of real analysis. Also the problems, the methods, and the connections of real analysis to other fields have changed drastically. As a consequence, the objects dealt with by real analysis have been diversified and have not always been limited to sets or functions on the Euclidean plane.

Real analysis is based on the real numbers, and it is naturally involved with practical mathematics. On the other hand, it has taken on subject matter from set theory, harmonic analysis, integration theory, probability theory, complex analysis, the theory of partial differential equations, and so on, and has provided these theories with important ideas and basic concepts. Such relations continue even today.

This book, with these backgrounds as its setting, contains the basic matter of "real analysis".

It is natural, in view of the object of this book, to introduce the definition of the real numbers, which is done by just mentioning characteristic properties.

The statements about sets and plane topology will be kept to an irreducible minimum.

Measurability of sets is one of the fundamental notions of Lebesgue integration theory. To define measurability, we apply the Carathéodory condition, although this method is not always intuitive. One of the merits of the choice is that it enables us to unify the measures on the plane and on general sets. Moreover, the Carathéodory condition is used to simplify the proof of the theorem on representation of linear functionals by measures.

The existence of Haar measure and integration on topological groups are not covered. For them, the reader may refer to the bibliography in the back of the book.

The basic theory of distributions and Fourier analysis is covered in chapters 7 and 8.

The last chapter is allotted to an introduction to wavelet theory. The theory of wavelets was born at the beginning of the 1980's from a practical purpose and, coincidentaly, a purely mathematical motive. At present, the lucid theory is constructed by an effective application of Fourier analysis.

Wavelet theory is also important in Fourier analysis. Roughly speaking, it enables us to treat a function and its Fourier transform more easily at the same time. Wavelet theory is now a useful tool of real analysis, like distribution theory.

I would like to thank Professors Hitoshi Arai and Kazuya Tachizawa for useful remarks and making graphs.

I thank all the staff of Iwanami Shoten. Espcially I would like to thank Mr. Hideo Arai, Ms. Mamiko Hamamo and Mr. Uichi Yoshida for generous help in preparing the manuscript.

Satoru Igari
Sendai
1996

Preface to the English Edition

The first mathematical textbook which I read was "Measure Theory", by P. R. Halmos.

Even now I remember clearly the polished exposition, so neatly put in order.

I have benefited, since then, from many books published in Eurpopean languages, and I have had the opportunity to make deep friendships with excellent mathematicians. Especially, I have spent a delightful time with the professors and the students of the University of Wisconsin in Madison.

Integration theory, distribution theory, and Fourier analysis are still basic subjects for all students who study analysis. Also, the theory of wavelets is becoming a new basic. I hope that this book will give students good training in these basics.

I thank Professor Katsumi Nomizu, who gave useful advice in translation. I am full of gratitude to the kindness of Ralph Sizer, who read through the manuscript and gave me polite and pertinent comments, and correction of English.

Finally, I thank the American Mathematical Society for giving me the opportunity to publish this book in English, and the staff there, especially Christine M. Thivierge and Vickie Ancona, for their kind assistance.

Satoru Igari
Sendai
1997

CHAPTER 1

Euclidean Spaces and the Riemann Integral

In this first chapter we cover the necessary minimum basic knowledge of real analysis, which will be required in the following chapters. I recommend that readers who are familiar with most of this chapter skip to Chapter 2, and when necessary, come back for any needed parts of this material.

When we deal with various concepts of analysis such as convergence, continuity, differentiation, integration and so on, we need the definition of the real numbers.

What are real numbers ? Let us attempt to survey a way of constructing them.

First, the natural numbers are defined by the axiom of Peano. The integers and the rational numbers are deduced from the natural numbers.

One way to define the real numbers from the rational numbers is to introduce the "Dedekind cut". Then, the real numbers comprise all of the Dedekind cuts. Intuitively, a Dedekind cut is a cutting point when all the rational numbers, linearly ordered, are divided into two nonempty parts, the right-hand side A and the left-hand side B. Here we assume that A does not have the largest rational number in it. Symbolically, we denote such a cut by the pair (A, B).

As a consequence, the set of rational numbers is extended to a set which is closed under countably infinite operations.

This is a brief description of a constructive introduction to the real numbers. But in this book we do not adopt such a logical definition; rather we take, from the practical viewpoint, the course of picking up some properties of the set of real numbers (see § 1.1).

We have intrinsically the notion of length of a line segment and the notion of area of subsets in the plane. These are mathematically mappings from sets of real numbers or subsets of the plane to nonnegative real numbers, and the concepts of "length", "area", "integration", and so on, which we have learned in calculus, are strictly defined as Jordan measure or Riemann integration (see §§ 1.7 and 1.8).

1.1. Real numbers

1.1.1. Real numbers. The set of all natural numbers $1, 2, 3, \ldots$ will be denoted by $\mathbb{N}$, the set of all integers $\ldots, -2, -1, 0, 1, 2, \ldots$ by $\mathbb{Z}$, the set of all rational numbers $\dfrac{m}{n}$ (n, m are integers and $n \neq 0$) by $\mathbb{Q}$, and the set of all real numbers by $\mathbb{R}$.

The set of real numbers is characterized as a set in which there are binary operations " $+$ " and " $\cdot$ ", called addition and multiplication, respectively. It is ordered and, furthermore, complete. We state these properties as axioms.

AXIOM 1.1. $\mathbb{R}$ has the following properties (A1)-(A9), that is, $\mathbb{R}$ is a *commutative field* in the terminology of algebra.

(A1) *Associative law for addition*: $(a+b)+c = a+(b+c)$, for $a, b, c \in \mathbb{R}$.
(A2) *Additive identity*: There exists a unique element $0 \in \mathbb{R}$ such that $a + 0 = a$ for all $a \in \mathbb{R}$.
(A3) *Additive inverse*: For every a there exists a unique $b \in \mathbb{R}$ such that $a+b = 0$. This b is denoted by $-a$.
(A4) *Commutative law for adition*: $a + b = b + a$.
(A5) *Associative law for multiplication*: $(ab)c = a(bc)$.
(A6) *Multplicative identity*: There exists $1 \in \mathbb{R}$ such that $a1 = a$ for all $a \in \mathbb{R}$.
(A7) *Multiplicative inverse*: For every $a \neq 0$ there exists $b \in \mathbb{R}$ such that $ab = 1$. This b is denoted by a^{-1}.
(A8) *Distributive law*: $(a+b)c = ac + bc$.
(A9) *Commutative law for multiplication*: $ab = ba$.

We have taken the natural numbers as granted. However, they are given by Axiom 1.1. The symbol 1 in (A6) has special meaning. A natural number is the real number which is expressed by a finite sum $1+1+\cdots+1$ of 1's. For any natural number n, $n+1$ is also a natural number.

A real number k is said to be an integer if $k = 0$, or if k or $-k$ is a natural number.

A real number r is said to be rational if r can be written in the form mn^{-1}, where n and m are integers and $n \neq 0$.

AXIOM 1.2. $\mathbb{R}$ is an *ordered field.* That is, there is a subset P, called the positive numbers, with the following properties:
(Ord1) $P \cap (-P) = \emptyset$,
(Ord2) $P \cup \{0\} \cup (-P) = \mathbb{R}$,
(Ord3) if $a, b \in P$, then $a + b \in P$ and $ab \in P$,
where $-P = \{-x : x \in P\}$.

We write $a > b$ or $b < a$ if $a - b \in P$. Remark that $0 \notin P$ by (Ord1) and that if $a, b \in \mathbb{R}$, we have $a < b, b < a$, or $a = b$, and only one of these three relations occures by (Ord2).

For $x \in \mathbb{R}$ we define

$$|x| = \begin{cases} x, & x \geq 0, \\ -x, & x < 0, \end{cases}$$

and call $|x|$ the absolute value of x.

We remark that the rational numbers $\mathbb{Q}$ constitute an ordered field. The real numbers are distinguished from the rational numbers by the axiom of completeness.

To introduce the concept of completeness, we need a few preliminary definitions. Let E be a nonempty subset of $\mathbb{R}$. A real number b is said to be an *upper bound* of E if $x \leq b$ for all $x \in E$. If E has an upper bound, then E is said to be bounded above. Let B be the set of all upper bounds of E. An element c of B is called the *least upper bound* or *supremum* of E if $c \leq b$ for all $b \in B$. The least upper bound c is denoted by $\sup_{x \in E} x$ or $\sup E$. The least upper bound of a set does not always exist, but if exists, it is unique.

AXIOM 1.3 (The axiom of completeness). If a nonempty subset E of $\mathbb{R}$ is bounded above, then the supremum of E exists.

The terms lower bound, bounded below, greatest lower bound and infimum (inf) will be defined similarly.

If E is bounded both above and below, E is said to be *bounded.*

Since

$$-\sup E = \inf(-E),$$

Axiom 1.3 is equivalent to saying that

If a nonempty subset E of $\mathbb{R}$ is bounded below, then the infimum of E exists.

1.1.2. Completeness of the real numbers. We shall rephrase Axiom 1.3 of the completeness of the real numbers in a way that is more closely connected with the calculus.

DEFINITION 1.1. A sequence $\{a_j : j = 1, 2, \ldots\}$ in $\mathbb{R}$ is called a *Cauchy*[1] *sequence*, if for every $\varepsilon > 0$ there exists an integer $N > 0$ such that

$$|a_i - a_j| < \varepsilon \quad (i, j \geq N). \tag{1.1}$$

A sequence $\{a_j\}$ is called a *convergent sequence*, if there exists a real number a such that for every $\varepsilon > 0$ there exists an integer $N > 0$ such that

$$|a_j - a| < \varepsilon \quad (j \geq N). \tag{1.2}$$

(1.2) can also be written as

$$\lim_{j \to \infty} a_j = a \quad \text{or} \quad a_j \to a\,(j \to \infty).$$

Sometimes (1.1) and (1.2) are expressed as

$$\forall \varepsilon > 0 \ \exists N : \ |a_i - a_j| < \varepsilon \quad (i, j \geq N),$$

and

$$\forall \varepsilon > 0 \ \exists N : \ |a_j - a| < \varepsilon \quad (j \geq N),$$

respectively.

THEOREM 1.1 (The theorem of Cauchy). *Every Cauchy sequence of $\mathbb{R}$ is a convergent sequence.*

PROOF. Let $\{a_j\}$ be a Cauchy sequence. By definition, for every $\varepsilon > 0$ there exists an integer $N > 0$ such that (1.1) holds. Thus $|a_j| \leq |a_j - a_N| + |a_N| \leq \varepsilon + |a_N|$ $(j \geq N)$. This implies that the sequence $\{a_j\}$ is bounded. Therefore, by Axiom 1.3 the numbers $\alpha_k = \sup_{j \geq k} a_j$ exist and are bounded below. We have, again by Axiom 1.3, a real number

$$a = \inf_{k \geq 1} a_k = \inf_{k \geq 1} \sup_{j \geq k} a_j.$$

By (1.1), $a_i - a_j < \varepsilon\,(i, j \geq N)$. Thus, if $k \geq N$,

$$\sup_{i \geq k} a_i - a_j \leq \varepsilon \quad (j \geq N).$$

Take the infimum of the left-had side over $k \geq N$ and observe that

$$a = \inf_{k \geq 1} \sup_{i \geq k} a_i = \inf_{k \geq N} \sup_{i \geq k} a_i.$$

Then we have

$$a - a_j \leq \varepsilon \quad (j \geq N).$$

[1] A. L. Cauchy, 1789-1857. French mathematician; author of *Cours d'Analyse de L'École Polytechnique.*

Since $\inf_{j\geq k}(-a_j) = -\sup_{j\geq k} a_j$, in the same way, we have

$$a_i - \sup_{j\geq k} a_j \leq \varepsilon \quad (i \geq N)$$

for $k \geq N$. Take the supremum of the left-hand side with respect to $k \geq N$. Since

$$\sup_{k\geq N}(-\sup_{j\geq k} a_j) = -\inf_{k\geq N}\sup_{j\geq k} a_j = -a,$$

we get

$$a_j - a \leq \varepsilon \quad (j \geq N).$$

Thus

$$|a - a_j| \leq \varepsilon \quad (j \geq N),$$

which implies that $\{a_j\}$ is a convergent sequence. □

The following theorem is called the Axiom of Archimedes.

THEOREM 1.2 (Axiom of Archimedes). *For a given $a \in \mathbb{R}$ there exists an integer n such that $a < n$.*

PROOF. Set $E = \{k \in \mathbb{Z} : k \leq a\}$. Since E is a subset of $\mathbb{R}$ bounded above, $\sup E$ exists. Let $b = \sup E$. Since $b - \frac{1}{2}$ is less than the supremum of E, there exists $k \in E$ such that $b - \frac{1}{2} < k$. Since $b < k + \frac{1}{2} < k + 1$, $k + 1 \notin E$. By the definition of E, $a < k + 1$. Thus $n = k + 1$ has the property required. □

As a corollary of the Axiom of Archimedes we deduce the denseness of the rational numbers.

THEOREM 1.3. *For any real numbers a and b with $a < b$ there exists a rational number r such that $a < r < b$.*

PROOF. By the Axiom of Archimedes we find a natural number n such that $(b - a)^{-1} < n$. Thus $\frac{1}{n} < b - a$. Pick the least integer p such that $nb \leq p$. Then we have $nb \leq p < nb + 1$. Thus $a < b - \frac{1}{n} = (nb - 1)/n \leq (p - 1)/n < \frac{nb}{n} = b$. If $m = p - 1$, $a < \frac{m}{n} < b$. □

1.1.3. Countability of the rational numbers. If there exists a one-to-one correspndence between the sets A and B, we say that A and B have same cardinality and we write

$$\mathrm{card}(A) = \mathrm{card}(B).$$

If a set A is finite, $\mathrm{card}(A)$ indicates the number of elements of A. Denote $\mathrm{card}(\mathbb{N}) = \aleph_0$. A set A is said to be *countable* (or enumerable) if either A is finite or $\mathrm{card}(A) = \mathrm{card}(\mathbb{N}) = \aleph_0$. If A is neither finite nor countable, A is said to be uncountable, and we write $\mathrm{card}(A) > \aleph_0$.

Every countable set can be enumerated as $a_1, a_2, \ldots$, and conversely. Any subset of a countable set is countable, too.

THEOREM 1.4. *A countable union of countable sets is countable.*

PROOF. Let $A_m = \{a_{mn} : n = 1, 2, \ldots\}$ $(m = 1, 2, \ldots)$ be countable sets.

We enumerate the union by following the arrow in Figure 1.1. If a row $\{a_{m,n} : n = 1, 2, \ldots\}$ breaks finitely, skip it. Then all points of $\bigcup_{m=1}^{\infty} A_m$ can be enumerated. □

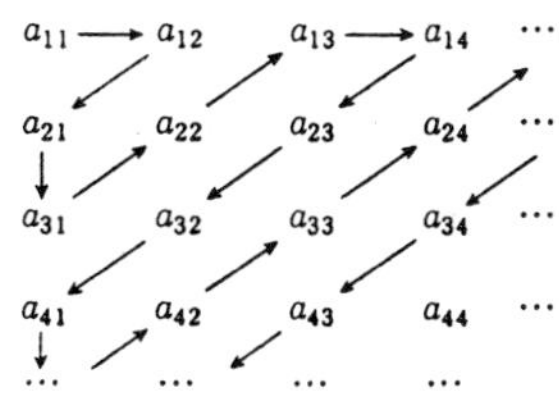

FIGURE 1.1. Enumeration

THEOREM 1.5. *The set $\mathbb{Q}$ of all rational numbers is countable.*

PROOF. For each $m \in \mathbb{N}$ the set $A_m = \{\dots, -\frac{2}{m}, -\frac{1}{m}, 0, \frac{1}{m}, \frac{2}{m}, \dots\}$ is countable, since

$$0, \frac{1}{m}, -\frac{1}{m}, \frac{2}{m}, -\frac{2}{m}, \dots$$

is enumerable. Thus by Theorem 1.4 $\mathbb{Q} = \bigcup_{m=1}^{\infty} A_m$ is countable. □

THEOREM 1.6. *The set of all real numbers $\mathbb{R}$ is uncountable.*

PROOF. It suffices to show that the interval $(0,1] = \{x : 0 < x \leq 1\}$ is not countable. Let $0 < a \leq 1$, and let $a = 0.a_1a_2\dots$ be a decimal expansion. If a has two representations, for example, such as $0.5000\dots = 0.4999\dots$, choose the one ending with zeros. Then every number is expressed uniquely.

Now assume that the set $(0,1]$ is countable and let

$$a_1 = 0.a_{11}a_{12}a_{13}a_{14}\dots,$$
$$a_2 = 0.a_{21}a_{22}a_{23}a_{24}\dots,$$
$$a_3 = 0.a_{31}a_{32}a_{33}a_{34}\dots,$$
$$\dots\dots\dots\dots\dots\dots\dots\dots$$

be an enumeration of $(0,1]$. Put $b_n = 2$ if $a_{nn} = 1$, and $b_n = 1$ if $a_{nn} \neq 1$. Define $b = 0.b_1b_2\dots b_n\dots$.

Then $0 < b < 1$ and b coincides with no a_n. Indeed, the n-th components of b and a_n are different. This implies that $\{a_n : n = 1, 2, \dots\}$ does not fill up $(0,1]$. □

If $\mathrm{card}(A) = \mathrm{card}(\mathbb{R})$, the set A is said to have the cardinality of the continuum, and we write $\mathrm{card}(A) = \mathfrak{c}$.

The set of all complex numbers is denoted by $\mathbb{C} = \{a + ib : a, b \in \mathbb{R}\}$.

1.2. Euclidean spaces

Let $\mathbb{R}^d$ be the set of all ordered d-tuples $x = (x_1, \dots, x_d)$ of real numbers.

Two operations, called addition and scalar multiplication, are defined for points $x = (x_1, \dots, x_d)$ and $y = (y_1, \dots, y_d)$ of $\mathbb{R}^d$ and for a real number a by

$$x + y = (x_1 + y_1, \dots, x_d + y_d), \tag{1.3}$$

$$ax = (ax_1, \dots, ax_d), \tag{1.4}$$

respectively. Thus if $x, y \in \mathbb{R}^d$ and $a, b \in \mathbb{R}$, then $ax + by \in \mathbb{R}^d$. We write $0 = (0, \dots, 0)$.

Furthermore, we define the inner product by

$$x \cdot y = x_1y_1 + \dots + x_dy_d$$

and the distance between x and 0 by

$$|x| = (x \cdot x)^{1/2}.$$

The space $\mathbb{R}^d$ with the above two operations (1.3) and (1.4), and the inner product, is called the *d-dimensional Euclidean space.*

For $x, y \in \mathbb{R}^d$ and $a \in \mathbb{R}$ we have

(i) $|x| \geq 0, \quad |x| = 0 \Leftrightarrow x = 0,$
(ii) $|ax| = |a||x|,$
(iii) $|x + y| \leq |x| + |y|$ (Minkowski's inequality),
(iv) $|x \cdot y| \leq |x||y|$ (Schwarz's inequality).

(i) and (ii) are obvious. We begin by proving Schwarz's inequality:

$$(x_1y_1 + \cdots + x_dy_d)^2 \leq (x_1^2 + \cdots + x_d^2)(y_1^2 + \cdots + y_d^2). \tag{1.5}$$

Note that $(tx_1 - y_1)^2 + \cdots + (tx_d - y_d)^2$ is a polynomial in t of order 2, and nonnegative. Thus the discriminant is nonpositive, which implies the inequality (1.5).

To prove Minkowski's inequality, consider the equation

$$\begin{aligned} |x + y|^2 &= (x_1 + y_1)^2 + \cdots + (x_d + y_d)^2 \\ &= (x_1^2 + \cdots + x_d^2) + (y_1^2 + \cdots + y_d^2) + 2(x_1y_1 + \cdots + x_dy_d) \\ &= |x|^2 + |y|^2 + 2x \cdot y. \end{aligned}$$

Apply Schwarz's inequality to the last term on the right-hand side. Then it follows that

$$|x + y|^2 \leq |x|^2 + |y|^2 + 2|x||y| \leq (|x| + |y|)^2,$$

which implies (iii).

DEFINITION 1.2. Let X be a set. A mapping ρ of $X \times X$ to $\mathbb{R}$ is called a *metric* (or distance function) if ρ satisfies the following three conditions:

(D1) $\rho(x, y) \geq 0 \quad (x, y \in X), \quad$ and $\quad \rho(x, y) = 0 \Leftrightarrow x = y,$
(D2) $\rho(x, y) = \rho(y, x),$
(D3) $\rho(x, y) \leq \rho(x, z) + \rho(z, y) \quad (x, y, z \in X).$

$\rho(x, y)$ is called the distance between x and y, and the space X with the metric ρ, which is often denoted by (X, ρ), is called a *metric space.*

The d-dimensional Euclidean space $\mathbb{R}^d$ is a metric space with the metric defined by

$$\rho(x, y) = |x - y| \quad \text{for } x, y \in \mathbb{R}^d.$$

DEFINITION 1.3. Let (X, ρ) be a metric space.

(i) A sequence $\{a_j\}$ in X is called a *Cauchy sequence* if

$$\rho(a_i, a_j) \to 0 \quad (i, j \to \infty).$$

(ii) A sequence $\{a_j\}$ in X is called a *convergent sequence* if there exists a point a in X such that

$$\rho(a_j, a) \to 0 \quad (j \to \infty).$$

The space (X, ρ) is called a *complete metric space* if every Cauchy sequence in X is convergent.

If $\{a_j\}$ is a convergent sequence, then it is a Cauchy sequence. Indeed, let $\rho(a_j, a) \to 0$ $(j \to \infty)$. Then we have

$$\rho(a_i, a_j) \leq \rho(a_i, a) + \rho(a, a_j) \to 0 \quad (j \to \infty).$$

THEOREM 1.7. *The Euclidean space $\mathbb{R}^d$ is a complete metric space.*

PROOF. Let $\{x^j, j = 1, 2, \dots\}$ be a sequence in $\mathbb{R}^d$. Let $x^j = (x_1^j, \dots, x_d^j)$. Then $\{x^j\}$ is a Cauchy sequence in $\mathbb{R}^d$ if and only if $\{x_i^j\}$ is a Cauchy sequence for each $i = 1, 2, \dots, d$. Thus the theorem follows from the Cauchy theorem 1.1. □

1.3. Metric spaces and topological spaces

1.3.1. Metric spaces. Let (X, ρ) be a metric space. For $x \in X$ and $r > 0$, the set

$$B(x, r) = \{y \in X \, : \rho(x, y) < r\}$$

is called the *open ball* with center at x and radius $r > 0$.

Let A be a subset of X. A point $x \in X$ is called an *accumulation point* of A if

$$(A - \{x\}) \cap B(x, r) \neq \emptyset \quad \text{for all } r > 0.$$

The set of all accumulation points of A is called the *derived set* and denoted by A'. The set

$$\overline{A} = A \cup A'$$

is called the *closure* of A. If $A \supset \overline{A}$, A is said to be *closed.*

A point x of A is called an *inner point* if there exists $r > 0$ such that $B(x, r) \subset A$. The set of all inner points of A is called the *interior* of A and denoted by $\operatorname{int}(A)$ or A^i.

If $A^i = A$, A is said to be *open.*

1.3.2. Topological spaces. The notion of open sets is introduced in the metric spaces. This idea is applied to define a topology in a wider class of spaces.

DEFINITION 1.4. Let X be a set and let $\mathcal{O}$ be a family of subsets which satisfies the following conditions:

(O1) $X \subset \mathcal{O}$ and $\emptyset \in \mathcal{O}$.
(O2) If $O_\alpha \in \mathcal{O}$ for all $\alpha \in I$, then $\bigcup_{\alpha \in I} O_\alpha \in \mathcal{O}$.
(O3) If $O_1, O_2 \in \mathcal{O}$, then $O_1 \cap O_2 \in \mathcal{O}$.

A set X with a family $\mathcal{O}$ satisfying these three properties is called a *topological space* and denoted by $(X, \mathcal{O})$.

When a family of open sets is presumed, we will denote a topological space simply by X.

To make clear the topology on X, the family $\mathcal{O}$ is occasionally denoted by $\mathcal{O}(X)$. Subsets in $\mathcal{O}(X)$ are called *open* sets.

For example, if $\mathcal{O}$ is the family of all subsets of X, $\mathcal{O}$ satisfies the conditions (O1), (O2) and (O3). The space $(X, \mathcal{O})$ is called a *discrete topological space.*

An important example of topological spaces is a metric space. Let (X, ρ) be a metric space. Define $\mathcal{O}$ to be the family of all subsets A in X such that $A^i = A$. Then $\mathcal{O}$ satisfies the conditions of open sets (see Problem 1.1 at the end of this chapter). Thus every metric space is a topological space with such a family of open sets $\mathcal{O} = \{A \subset X \, : A^i = A\}$.

In a topological space X, open subsets which contain x are called *neighborhoods* of x.

The space X is said to be a *Hausdorff space* if for any two distinct points $x, y \in X$, there exist neighborhoods V_x and V_y of x and y, respectively, which are mutually disjoint (see Figure 1.2).

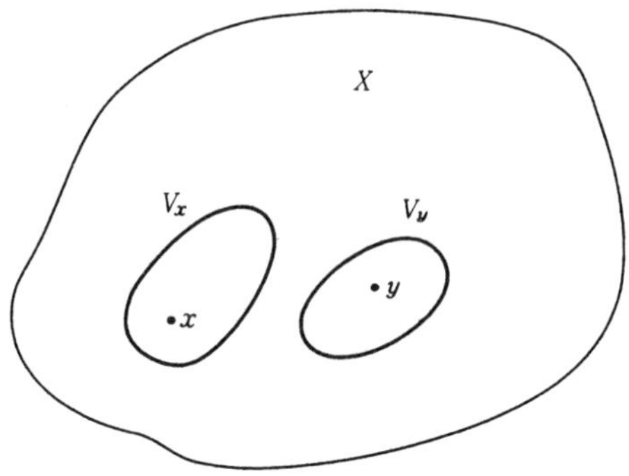

FIGURE 1.2. Hausdorff space

We give again the definitions on sets in the topological space $(X, \mathcal{O}(X))$. Let A be a subset of X. A point $x \in X$ is called an *accumulation point* of A if

$$(A - \{x\}) \cap V_x \neq \emptyset, \quad \text{for all neighborhoods } V_x \text{ of } x.$$

The set of all accumulation points of A is called the *derived set* and denoted by A'. The *closure* of a set A is defined by $\overline{A} = A \cup A'$ as in metric spaces, and A is said to be *closed* if $\overline{A} \subset A$.

A point x of A is called an *inner point* if there exists a neighborhood V_x of x such that $V_x \subset A$. The set of all inner points of A is called the *interior* of A and denoted by $\mathrm{int}(A)$ or A^i.

THEOREM 1.8. *In a topological space $(X, \mathcal{O}(X))$, a set O is open if and only if $O = O^i$.*

PROOF. Let O be an open set, that is, $O \in \mathcal{O}(X)$. If $O = \emptyset$, then the assertion is trivial. Suppose $O \neq \emptyset$. Since O is itself a neighborhood of every point contained in O, it follows that $O \subset O^i$. Thus $O = O^i$. Conversely, if $O = O^i$ and O is not empty, then for any $x \in O^i$ we can choose a neighborhood V_x of x such that $V_x \subset O$. Thus $O = \bigcup\{V_x : x \in O\}$, which is open by property (O2). Thus the theorem is proved. □

Let D be a subset of X. Then the family $D \cap \mathcal{O}(X) = \{D \cap O : O \in \mathcal{O}(X)\}$ satisfies conditions (O1), (O2) and (O3) with D in place of X, and it is called the *relative topology* of D.

By definition, a subset G of D is open in the relative topology if there exists an open set O in X such that $G = D \cap O$.

Therefore, a subset G of D is open in D if and only if for every $x \in G$ there exists a neighborhood V_x in X such that $V_x \cap D \subset G$. In fact, let $G = D \cap O$, where O is an open set in X. Then for every $x \in G = D \cap O$, the set O is a neighborhood of x such that $O \cap D \subset G$. Conversely, suppose that for every $x \in G$ there exists a neighborhood V_x such that $V_x \cap D \subset G$. Put $O = \bigcup\{V_x : x \in G\}$, where V_x runs over all neighborhoods of x in X such that $V_x \cap D \subset G$. Then O is open in X and $G = O \cap D$.

For a set A the set $\{x \in X : x \notin A\}$ is called the *complement* of A and denoted by A^c.

THEOREM 1.9. *A set A is closed, if and only if the complement A^c of A is open.*

PROOF. Let A be a closed set. If $A = X$, then $A^c = \emptyset$ which is open. Suppose $A \neq X$ and $x \notin A$. Since x is not an accumulation point of A, there is a neighborhood V_x of x such that $V_x \cap A = \emptyset$, that is, $V_x \subset A^c$. This implies that x is an inner point of A^c. Thus A^c is open.

Conversely, suppose that A^c is open and nonempty. Let $x \in A^c$. Since x is an inner point of A^c, there is a neighborhood V_x of x such that $V_x \subset A^c$ or, equivalently, $V_x \cap A = \emptyset$. Thus x is not an accumulation point of A. Thus $A^c \subset (A')^c$. Thus $A \supset A'$, which implies that A is closed. The case $A^c = \emptyset$ is obvious. □

By definition, a union of open sets is open. Therefore it follows from Theorem 1.9 that if the sets A_ι are closed for all $\iota \in I$, then $\bigcap_{\iota \in I} A_\iota$ is closed.

A set A is said to be *perfect* if $A \neq \emptyset$ and $A' = A$.

A subset B of A is said to be *dense* in A if $A \subset \overline{B}$.

A set A is said to be *dense in itself* if $A \subset A'$.

The set $A^s = A - A'$ is said to be the *isolated* set of A, and an element of it is called an isolated point.

The set $\partial A = \overline{A} - A^i$ is said to be the *boundary* of A.

1.3.3. Product topology. For two given sets A and B, $A \times B$ denotes the set of pairs $\{(a, b) : a \in A, b \in B\}$ and is called the *product set* of A and B.

We have, obviously, $(A_1 \times B_1) \cap (A_2 \times B_2) = (A_1 \cap A_2) \times (B_1 \cap B_2)$.

Let $(X, \mathcal{O}(X))$ and $(Y, \mathcal{O}(Y))$ be two topological spaces. Let $\mathcal{O}(X \times Y)$ be the family of sets in $X \times Y$ which consists of all (finite or infinite) unions of sets $U \times V$ where $U \in \mathcal{O}(X)$ and $V \in \mathcal{O}(Y)$. The family $\mathcal{O}(X \times Y)$ clearly satisfies conditions (O1) and (O2) of Definition 1.4. If $O_1 = \bigcup_{\alpha \in A} U_\alpha^1 \times V_\alpha^1$ and $O_2 = \bigcup_{\beta \in B} U_\beta^2 \times V_\beta^2$ are sets in $\mathcal{O}(X \times Y)$, then we have $O_1 \cap O_2 = \bigcup_{\alpha \in A, \beta \in B} (U_\alpha^1 \cap U_\beta^2) \times (V_\alpha^1 \cap V_\beta^2)$, which belongs to $\mathcal{O}(X \times Y)$. Thus the family $\mathcal{O}(X \times Y)$ introduces a topology on the product set $X \times Y$, which is called a *product topology*.

EXAMPLE 1.1. The Euclidean space $\mathbb{R}^d$ is a topological space with the Euclidean metric $\rho_d(x, y) = |x - y|$, where $x = (x_1, \dots, x_d)$ and $y = (y_1, \dots, y_d)$.

The product topology $\mathcal{O}(\mathbb{R} \times \mathbb{R})$ on $\mathbb{R} \times \mathbb{R}$ coincides with the metric topology $\mathcal{O}(\mathbb{R}^2)$ on $\mathbb{R}^2$. In fact, let $O \in \mathcal{O}(\mathbb{R}^2)$. If $x = (x_1, x_2) \in O$, then, since x is an inner point of O, there exists $r > 0$ such that $B_2(x, r) \subset O$, where $B_d(x, r)$ denotes an open ball in $\mathbb{R}^d$. We have $B_1(x_1, r/\sqrt{2}) \times B_1(x_2, r/\sqrt{2}) \subset B_2(x, r) \subset O$. Thus O is expressed as a union of the product sets of two open sets in $\mathbb{R}$. Therefore $\mathcal{O}(\mathbb{R}^2) \subset \mathcal{O}(\mathbb{R} \times \mathbb{R})$. Conversely, suppose that $O \in \mathcal{O}(\mathbb{R} \times \mathbb{R})$ and $x \in O$. Then there exist $r, s > 0$ such that $B_1(x_1, r) \times B_1(x_2, s) \subset \mathcal{O}(\mathbb{R} \times \mathbb{R})$, which implies that $B_2(x, t) \subset \mathcal{O}(\mathbb{R} \times \mathbb{R})$ with $t = \min(r, s)$. Thus we can conclude that $\mathcal{O}(\mathbb{R} \times \mathbb{R}) \subset \mathcal{O}(\mathbb{R}^2)$, and thus $\mathcal{O}(\mathbb{R} \times \mathbb{R}) = \mathcal{O}(\mathbb{R}^2)$.

1.3.4. Open intervals and closed intervals. Let $-\infty \leq a < b \leq \infty$. The sets

$$[a, b] = \{x \in \mathbb{R} : a \leq x \leq b\} \quad \text{and} \quad (a, b) = \{x \in \mathbb{R} : a < x < b\}$$

are called a closed interval and an open interval respectively. Every closed interval is closed in $\mathbb{R}$ and every open interval is open in $\mathbb{R}$.

The sets

$$[a,b) = \{x \in \mathbb{R} : a \leq x < b\} \quad \text{and} \quad (a,b] = \{x \in \mathbb{R} : a < x \leq b\}$$

are called a right half-open interval and a left half-open interval respectively.

Let $I_1, \dots, I_d$ be intervals of $\mathbb{R}$. The set I defined by

$$I = I_1 \times \cdots \times I_d = \{x = (x_1, \dots, x_d) : x_j \in I_j, j = 1, \dots, d\}$$

is called a d-dimensional interval. If all intervals $I_1, \dots, I_d$ are closed (open, right half-open, left half-open), then I is said to be closed (open, right half-open, left half-open).

EXAMPLE 1.2. For every nonempty open interval (a,b), there exists a rational number r such that $r \in (a,b)$. Thus the set $\mathbb{Q}$ is dense in $\mathbb{R}$.

On the other hand, $\operatorname{card}((a,b)) > \aleph_0$ and $\operatorname{card}(\mathbb{Q}) = \aleph_0$, which implies that any interval with nonempty interior contains an irrational number. Thus the set of irrational numbers is also dense in $\mathbb{R}$.

EXAMPLE 1.3 (The Cantor ternary set). We shall construct a subset of $\mathbb{R}$ called the *Cantor ternary set* in the following way (see Figure 1.3). This set has particlular properties, and furnishes various examples revealing interesting problms in analysis.

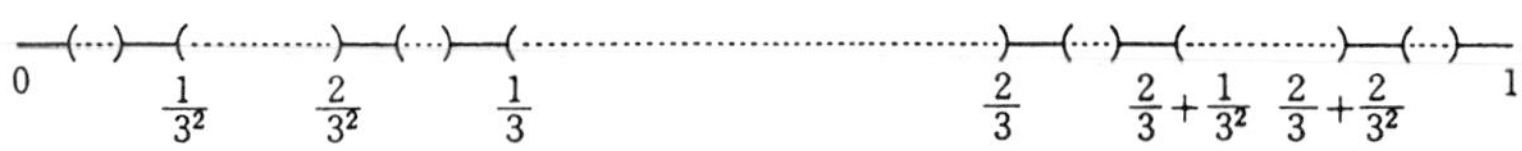

FIGURE 1.3. The Cantor ternary set

The first stage of construction is to remove from the closed interval $I = [0,1]$ the middle open interval $(1/3, 2/3)$ of length $1/3$. The remaining set $C_1 = I - (1/3, 2/3)$ consists of two closed intervals $[0, 1/3]$ and $[2/3, 1]$. Next remove from each of these closed intervals the middle open intervals, $(1/3^2, 2/3^2)$ and $(2/3 + 1/3^2, 2/3 + 2/3^2)$, respectively, of length $1/3^2$. The remaining set C_2 consists of 2^2 closed intervals of length $1/3^2$.

By the j-th construction we get the set C_j, which consists of 2^j closed intervals of length $1/3^j$. The Cantor[2] ternary set is defined as the infinite intersection $C = \bigcap_{j=1}^{\infty} C_j$.

Note that $I - C_j$ consists of $2^j - 1$ open intervals of the form

$$\left(\sum_{i=1}^{j} \frac{2\varepsilon_i^*}{3^i}, \sum_{i=1}^{j} \frac{2\varepsilon_i}{3^i} \right), \tag{1.6}$$

where ε_j and ε_j^* satisfy the following conditions: $\varepsilon_j = 0$ or 1 and $\varepsilon_j \neq 0$ for some $1 \leq j \leq k$. Let j_0 be the largest j such that $\varepsilon_j \neq 0$. Then $\varepsilon_j^* = \varepsilon_j$ for $j \neq j_0$, and $\varepsilon_{j_0}^* = 1/2$.

(i) We have the following expression:

$$C = \{x : x = \sum_{j=1}^{\infty} 2\varepsilon_j / 3^j, \varepsilon_j = 0, 1\}.$$

[2]G. Cantor, 1845-1918. Known for his contributions to set theory.

This implies that C consists of x's in $[0,1]$ whose triadic expansions do not contain 1. For example, $1/3$ is expanded as $1/3 = 1/3+0/3^2+0/3^3+\cdots$ but it has another expression $1/3 = \sum_{j=2}^{\infty} 2/3^j$. Thus $1/3 \in C$.

To get the expression of (i), observe that x belongs to $I - C = \bigcup_{j=1}^{\infty}(I - C_j)$ if and only if x is contained in an interval of the form (1.6) for some j. This implies that if $x = \sum_{j=1}^{\infty} \eta_j/3^j$ is a triadic expansion, there exists j such that $\eta_j = 1$.

(ii) C is a perfect set.

Indeed, since the sets C_j are closed, $C = \bigcap_{j=1}^{\infty} C_j$ is closed. Thus $C' \subset C$.

Conversely, if $x = \sum_{j=1}^{\infty} 2\varepsilon_j/3^j \in C$, then $(C - \{x\}) \cap B(x,r) \neq \emptyset$ for every $r > 0$. To prove this, put $x^k = \sum_{j=1}^{k} 2\varepsilon_j/3^j + \sum_{j=k+1}^{\infty} 2(1-\varepsilon_j)/3^j$. Then $x^k \neq x$, and $x^k \in C$ by (i). Since $x^k \to x$ as $k \to \infty$, we get the assertion. Thus $C \subset C'$. Therefore $C = C'$.

(iii) $\mathrm{card(C)} = \mathfrak{c}$.

In fact, the mapping $\psi : C \mapsto [0,1]$ defined by

$$\psi\Big(\sum_{j=1}^{\infty} 2\varepsilon_j/3^j\Big) = \sum_{j=1}^{\infty} \varepsilon_j/2^j$$

is surjective.

1.4. The structure of open sets and closed sets in Euclidean space

Open sets and closed sets in the Euclidean space $\mathbb{R}^d$ have particular properties, as shown below, which will be useful in various places.

THEOREM 1.10. *Every open set of $\mathbb{R}$ can be written as a countable union of mutually disjoint open intervals.*

Zorn's lemma is used in the proof (see Appendix A for Zorn's lemma and the terminology).

PROOF. Let O be a nonempty open set in $\mathbb{R}$. Let $\mathcal{I}$ be the collection of all open intervals contained in O; then $\mathcal{I}$ is a partially orderd set with the inclusion relation " $\subset$ ". If a subfamily $\{I_\alpha\}$ of $\mathcal{I}$ is a chain, that is, if it is a linearly ordered subfamily of $\mathcal{I}$, then $\{I_\alpha\}$ has an upper bound. Indeed, $\bigcup_\alpha I_\alpha$ is an open interval and contains all intervals $\{I_\alpha\}$. Thus by Zorn's lemma the family $\mathcal{I}$ contains at least a maximal interval.

Let $\mathcal{I}_o$ be the collection of all maximal intervals of $\mathcal{I}$. Then $\mathcal{I}_o$ consists of mutually disjoint countable open intervals, and $O = \bigcup\{J : J \in \mathcal{I}_o\}$. In fact, if x is a point in O, then x is contained in an interval I of $\mathcal{I}$ and I is contained in a maximal interval I_o. Thus $x \in I_o \subset \bigcup\{J : J \in \mathcal{I}_o\}$. Thus $O \subset \bigcup\{J : J \in \mathcal{I}_o\}$. The reverse inclusion is obvious.

If I and J are intervals in $\mathcal{I}_o$, then $I \cap J = \emptyset$ or $I = J$. For, if $I \cap J \neq \emptyset$, then $I \cup J$ is an open interval contained in O. Thus $I \cup J \subset I$ and $I \cup J \subset J$ by the maximality of I and J. This implies $I = J$. Therefore $\mathcal{I}_o$ is a family of mutually disjoint open intervals.

The final step is to show that $\mathcal{I}_o$ is countable. For each interval $I \in \mathcal{I}_o$ choose a rational point r_I in I. The mapping $\mathcal{I}_o \ni I \mapsto r_I \in \mathbb{Q}$ is one-to-one. Since the rational numbers are countable, $\mathcal{I}_o$ is countable. □

The above proof works only in the one-dimensional case. In general, open sets in $\mathbb{R}^d$ ($d > 1$) cannot be written as a countable union of mutually disjoint open intervals (see Problem 1.2).

Let $k \in \mathbb{Z}$. The d-dimensional cubes of the form

$$[j_1 2^{-k}, (j_1+1)2^{-k}) \times \cdots \times [j_d 2^{-k}, (j_d+1)2^{-k}), \quad (j_1, \dots, j_d) \in \mathbb{Z}^d$$

(see Figure 1.4), are called right half-open dyadic cubes of side length 2^{-k}, or simply *dyadic cubes*. Let $\mathcal{Q}_k$ be the family of all dyadic cubes of side length 2^{-k}, and put

$$\mathcal{Q} = \bigcup_{k=-\infty}^{\infty} \mathcal{Q}_k.$$

$\mathcal{Q}_k$ is countable, so the countable union $\mathcal{Q}$ is countable.

Note that if $I, J \in \mathcal{Q}$, then $I \cap J = \emptyset$, $I \subset J$, or $I \supset J$.

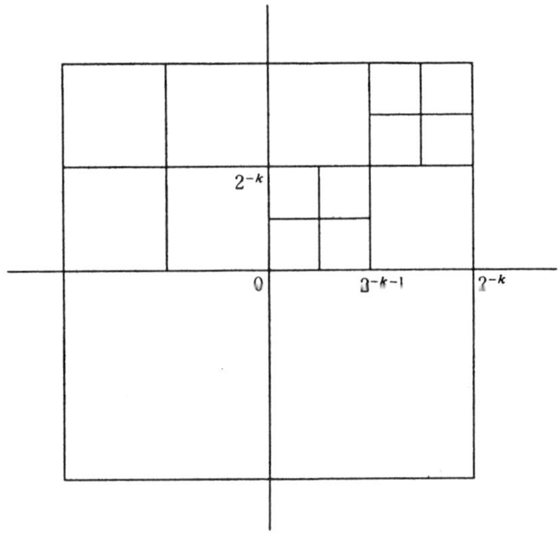

FIGURE 1.4. Dyadic cubes

THEOREM 1.11. *Every open set in $\mathbb{R}^d$ can be written as a countable union of disjoint right half-open dyadic cubes.*

PROOF. Let O be a nonempty open subset of $\mathbb{R}^d$. Let $\mathcal{I} = \{I \in \bigcup_{k \geq 0} \mathcal{Q}_k : I \subset O\}$. The family $\mathcal{I}$ consists of dyadic cubes of side length ≤ 1, and $\mathcal{I}$ is a partially ordered set with the inclusion relation. Let $\mathcal{I}_o$ be the collection of all maximal elements. Then $\mathcal{I}_o$ is a nonempty, countable family of mutually disjoint cubes.

To prove the theorem it suffices to show that $O \subset \bigcup\{J : J \in \mathcal{I}_o\}$, since the reverse inclusion is obvious. In fact, if $x \in O$, we can choose a small cube $I \in \mathcal{Q}$ so that $x \in I \subset O$. Since I is contained in a maximal cube in $\mathcal{I}$, $x \in \bigcup\{J : J \in \mathcal{I}_o\}$. □

Two subsets A and B are said to be nonoverlapping if A and B have no common inner points.

The closures of any two dyadic cubes either are nonoverlapping, or one of them includes the other.

THEOREM 1.12. *Every nonempty open set in $\mathbb{R}^d$ can be written as a countable union of nonoverlapping closed dyadic cubes.*

PROOF. Let O be a nonempty subset of $\mathbb{R}^d$, and put $\mathcal{J} = \{J \in \bigcup_{k\geq 0} \mathcal{Q}_k : \overline{J} \subset O\}$. Let $\mathcal{J}_o$ be the collection of all maximal elements of $\mathcal{J}$. Since cubes in $\mathcal{J}_o$ are mutually disjoint, the closures of cubes in $\mathcal{J}_o$ are mutually nonoverlapping. By the same way as in the proof of Theorem 1.11 we get $O = \bigcup\{\bar{J} : J \in \mathcal{J}_o\}$. □

DEFINITION 1.5. A family of sets $\mathcal{G}$ is said to be a *covering* of A if

$$A \subset \bigcup_{G\in\mathcal{G}} G = \bigcup\{G \in \mathcal{G}\}.$$

If all sets in $\mathcal{G}$ are open, then $\mathcal{G}$ is said to be an *open covering.*

If a subfamily $\mathcal{G}'$ of a covering $\mathcal{G}$ of A is also a covering of A, then $\mathcal{G}'$ is called a *subcovering* of $\mathcal{G}$.

EXAMPLE 1.4. Let $A = \{1, \frac{1}{2}, \frac{1}{3}, \dots\}$ and $I_n = (\frac{1}{n} - 2^{-n-1}, \frac{1}{n} + 2^{-n-1})$, $n = 1, 2, \dots$. The family $\mathcal{G} = \{I_n\}$ of open intervals covers the set A (see Figure 1.5), but it does not cover the set $A_0 = A \cup \{0\}$, since $0 \notin \bigcup_{n=1}^{\infty} I_n$. On the other hand, let I_0 be any open interval containing 0 and let $\mathcal{G}_0 = \mathcal{G} \cup \{I_0\}$. Then $\mathcal{G}_0$ is an open covering of A_0. Since I_0 is open, there exists an integer N such that $\frac{1}{n} \in I_0$ for all $n > N$. Thus the finite subfamily $\mathcal{G}_0' = \{I_0, I_1, \dots, I_N\}$ is a subcovering of A_0.

FIGURE 1.5. An open covering

LEMMA 1.1 (Lindelöf's theorem). *If $\mathcal{G}$ is an open covering of a subset A of $\mathbb{R}^d$, then $\mathcal{G}$ contains a countable subcovering of A.*

PROOF. By the assumption, $A \subset \bigcup\{G \in \mathcal{G}\}$. For $x \in A$ choose $O \in \mathcal{G}$ such that $x \in O$. Since x is an inner point of O, there exist a point $a_x \in \mathbb{R}^d$ whose components are all rational and a rational number $r_x > 0$ such that $x \in B(a_x, r_x) \subset O$. The collection of all such balls $\{B(a_x, r_x) : x \in A\}$ is countable (see Theorem 1.4).

For each ball $B(a_x, r_x)$ pick an open set $O' \in \mathcal{G}$ such that $B(a_x, r_x) \subset O'$. Then $\{O'\}$ is a countable subfamily of $\mathcal{G}$ which covers A. □

THEOREM 1.13 (The Cantor-Bendixson theorem). *Every closed subset of $\mathbb{R}^d$ either is countable or can be written as a union of a countable set and a perfect set.*

PROOF. Let A be a subset of $\mathbb{R}^d$. A point a is called a *condensation point,* if

$$\operatorname{card}(A \cap B(a, r)) > \aleph_0 \quad \text{for all } r > 0.$$

Let F be a closed set of $\mathbb{R}^d$. Let P be the set of all condensation points of F and put $N = F - P$. We shall prove that N is countable and P is perfect or empty.

In fact, since no point in N is a condensation point, for each $a \in N$ there exists $r_a > 0$ such that

$$\operatorname{card}(A \cap B(a, r_a)) \leq \aleph_0.$$

Since $\{B(a, r_a) : a \in N\}$ is an open covering of N, we can choose a countable subcovering $\{B(a_j, r_{a_j}) : j = 1, 2, \dots\}$ of N by the Lindelöf theorem. Thus

$$N = N \cap F \subset \bigcup_j B(a_j, r_{a_j}) \cap F.$$

The last union is a countable union of countable sets, so that N is countable.

Suppose that P is not empty. Let $a \in P$. For $r > 0$ we have $P \cap B(a, r) = F \cap B(a, r) - N \cap B(a, r)$. By definition the first term on the right-hand side is uncountable and the second term is countable. Thus $P \cap B(a, r)$ is uncountable for every $r > 0$, which implies that $a \in P'$. Thus $P \subset P'$.

Conversely, assume $a \notin P$. Since a is not a condensation point, there exists $r > 0$ such that

$$\operatorname{card}(F \cap B(a, r)) \le \aleph_0. \tag{1.7}$$

On the other hand, any point b of $P \cap B(a, r)$ is a condensation point. Thus

$$\operatorname{card}(F \cap B(b, s)) > \aleph_0 \quad \text{for all } s > 0. \tag{1.8}$$

If $s > 0$ is so small that $B(b, s) \subset B(a, r)$, then (1.7) and (1.8) are inconsistent, which implies that $P \cap B(a, r)$ is empty. Thus $B(a, r) \subset P^c$, that is, a is an inner point of P^c. Therefore P^c is open, or equivalently, P is closed. Thus $P' \subset P$.

Therefore we conclude that $P = P'$, and P is a perfect set. □

1.5. Covering theorems

DEFINITION 1.6. Let K be a subset of a Hausdorff topological space. K is said to be *compact* if every open covering of K has a finite subcovering.

Let K be a compact set. If $\{G_\iota : \iota \in I\}$ is a family of open sets and $K \subset \bigcup_{\iota \in I} G_\iota$, then there exists, by the definition, a finite set $\{\iota_1, \dots, \iota_N\}$ such that

$$A \subset \bigcup_{i=1}^{N} G_{\iota_i}.$$

We shall show later that any bounded closed subset of a Euclidean space is compact, and conversely.

DEFINITION 1.7. A family of sets $\mathcal{F}$ is said to have the *finite intersection property*, if every finite subfamily of $\mathcal{F}$ has nonempty intersection.

THEOREM 1.14. *A set K in a topological space is compact, if and only if we have $\bigcap_{F \in \mathcal{F}} K \cap F \neq \emptyset$ for any family $\mathcal{F}$ of closed sets such that $\{K \cap F : F \in \mathcal{F}\}$ has the finite intersection property.*

PROOF. Let $\mathcal{F}$ be a family of closed sets. The theorem follows from the definition of compactness and the following observations:

(a) $\mathcal{F}$ is a family of closed sets $\Leftrightarrow \mathcal{F}^c = \{F^c : F \in \mathcal{F}\}$ is a family of open sets.
(b) $\bigcap_{F \in \mathcal{F}} K \cap F = \emptyset \Leftrightarrow K \subset \bigcup_{F \in \mathcal{F}} F^c \Leftrightarrow \mathcal{F}^c$ is an open covering of K.
(c) $\{K \cap F : F \in \mathcal{F}\}$ does not have the finite intersection property
$\Leftrightarrow$ there exists $F_1, \dots, F_N \in \mathcal{F}$ such that $\bigcap_{j=1}^{N} K \cap F_j = \emptyset$
$\Leftrightarrow K \subset \bigcup_{j=1}^{N} F_j^c$
$\Leftrightarrow \mathcal{F}^c$ has a finite subcovering of K.

□

LEMMA 1.2. *Every compact subset of* $\mathbb{R}^d$ *is closed.*

PROOF. Suppose that A is not closed. This implies that there exists a point $a \notin A$ such that $A \cap B(a,r) \neq \emptyset$ for all $r > 0$. Thus the family of sets $\{A \cap \overline{B(a,r)} : r > 0\}$ has the finite intersection property. On the other hand, $\bigcap_{r>0} \overline{B(a,r)} \cap A = \{a\} \cap A = \emptyset$. Therefore A is not compact by Theorem 1.14. □

LEMMA 1.3. *Closed subsets of a compact set are compact.*

PROOF. Let K be a compact set and C be a closed subset of K. If $\{O_\iota\}$ is an open covering of C, then $\{O_\iota\} \cup \{C^c\}$ is an open covering of K. By the compactness of K there exists a finite subcovering $O_{\iota_1}, \dots, O_{\iota_N}, C^c$ of K. Here we add C^c to make the covering sure. Thus

$$C = K \cap C \subset (O_{\iota_1} \cup \dots \cup O_{\iota_N} \cup C^c) \cap C \subset O_{\iota_1} \cup \cdots \cup O_{\iota_N},$$

which implies that C has a finite subcovering of $\{O_\iota\}$. □

A subset of $\mathbb{R}^d$ is said to be *bounded* if it is included in a ball.

LEMMA 1.4. *Every closed bounded subset of* $\mathbb{R}^d$ *is compact.*

PROOF. Due to Lemma 1.3, it is sufficient to prove that a bounded closed interval I in $\mathbb{R}^d$ is compact. Let $\{O_\iota\}$ be an open covering of I. Suppose that $\{O_\iota\}$ does not have any finite subcovering of I. Let us divide I into 2^d congruent closed intervals whose side length is half of the corresponding side length of I. By our assumption there is at least one interval which does not have a finite subcovering. Denote it by I_1.

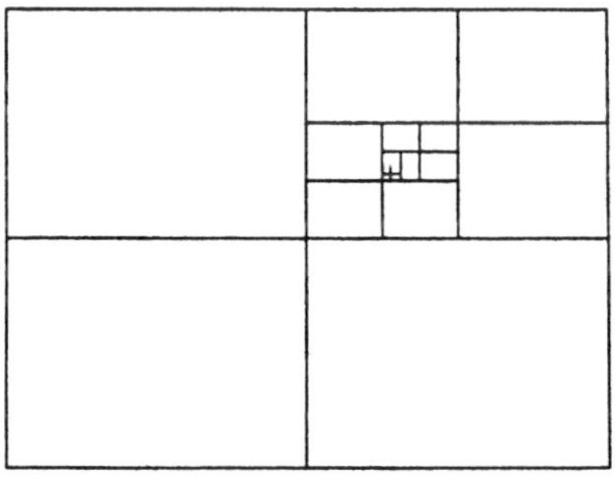

FIGURE 1.6. A sequence of decreasing closed intervals

Next, divide I_1 into 2^d congruent closed intervals whose side length is half of the corresponding side length of I_1. Then, by the same reasoning as before there is an interval I_2 which does not have a finite subcovering.

Continuing this process, we get a sequence $I_1 \supset I_2 \supset \cdots$ of closed intervals (see Figure 1.6). Every I_j does not have a finite subcovering, and I_j converges to a point a, which is contained in I since I is closed. By the assumption a is contained in some O_ι. Since I_j converges to a, we have $I_k \subset O_\iota$ for a large k, which contradicts the fact that I_k has no finite subcovering. □

THEOREM 1.15 (The Heine-Borel theorem). *A subset of* $\mathbb{R}^d$ *is compact, if and only if it is closed and bounded.*

PROOF. The "if" part is given in Lemma 1.4.

Let C be a compact set. Then C is closed by Lemma 1.2. Since the family of open balls $\{B(0,r) : r = 1, 2, \dots\}$ covers C, it has a finite subcovering, which implies that $C \subset B(0, R)$ for a large $r > 0$. Thus C is bounded. $\square$

1.6. Continuous functions

Let D be a subset of $\mathbb{R}^d$ and let f be a real valued function in D. Let $x \in D$. Then f is said to be *continuous* at x, if for each $\varepsilon > 0$ there exists $\delta > 0$ such that

$$|f(x) - f(y)| < \varepsilon \quad (y \in D, |x - y| < \delta). \tag{1.9}$$

When f is continuous at every point in D, f is said to be continuous in D.

In general, the number δ depends on x. When we can take a number δ independently of x (that is, there exists $\delta > 0$ such that (1.9) holds for all $x, y \in D$ such that $|x - y| < \delta$), f is said to be *uniformly continuous* in D.

THEOREM 1.16. *A real valued function f in a subset D of $\mathbb{R}^d$ is continuous in D, if and only if $f^{-1}(O) = \{x \in D : f(x) \in O\}$ is open in D in the relative topology for any open set O in $\mathbb{R}$.*

PROOF. Let f be a continuous function in D and let O be an open set of $\mathbb{R}$. If $x \in f^{-1}(O)$, then $f(x) \in O$. Thus we can choose $\delta > 0$ so that $f(y) \in O$ for all $y \in B(x, \delta) \cap D$. Thus $B(x, \delta) \cap D \subset f^{-1}(O)$, which implies that $f^{-1}(O)$ is open in D.

To prove the converse, let $x \in D$ and $\varepsilon > 0$. Then the set $f^{-1}(B(f(x), \varepsilon))$ is open in D by assumption and contains x. Therefore there exists $\delta > 0$ such that $B(x, \delta) \cap D \subset f^{-1}(B(f(x), \varepsilon))$, which is equivalent to saying that $|f(x) - f(y)| < \varepsilon$ for $y \in B(x, \varepsilon) \cap D$. Therefore f is continuous at x. $\square$

In particular, a real valued function in $\mathbb{R}^d$ is continuous if and only if

$$f^{-1}(O) \in \mathcal{O}(\mathbb{R}^d) \quad \text{for every } O \in \mathcal{O}(\mathbb{R}).$$

THEOREM 1.17. *Every continuous function on a compact set is uniformly continuous.*

PROOF. Let f be a continuous function on a compact set K of $\mathbb{R}^d$. Let $\varepsilon > 0$ be given. For every $x \in K$ there exists $\delta_x > 0$ such that

$$|f(x) - f(y)| < \varepsilon/2 \quad (y \in K \cap B(x, 2\delta_x)) \tag{1.10}$$

by the continuity of f. Since $\{B(x, \delta_x) : x \in K\}$ is an open covering of K, we can choose a finite subcovering $B(x_1, \delta_{x_1}), \dots, B(x_k, \delta_{x_k})$ of K. Put $\delta = \min(\delta_{x_1}, \dots, \delta_{x_k})$. Note that $\delta > 0$. The rest of the proof is to show that

$$|f(x) - f(y)| < \varepsilon \quad (x, y \in K, |x - y| < \delta),$$

which proves our theorem.

Indeed, for every $x \in K$ pick j so that $x \in B(x_j, \delta_j)$. If $y \in K$ and $|x - y| < \delta$, then $|x_j - y| \le |x_j - x| + |x - y| < \delta_{x_j} + \delta \le 2\delta_{x_j}$, which implies that $y \in B(x_j, 2\delta_{x_j})$. Thus

$$|f(x) - f(y)| \le |f(x) - f(x_j)| + |f(x_j) - f(y)| < \varepsilon/2 + \varepsilon/2 = \varepsilon.$$

$\square$

EXERCISE 1.1. Let f be a real valued function on $\mathbb{R}^d$. Suppose that $f(x_j) \to 0$ for every sequence $\{x_j\}$ such that $|x_j| \to \infty$. Then for a given $\varepsilon > 0$ there exists a compact set K such that

$$|f(x)| < \varepsilon \quad (x \notin K). \tag{1.11}$$

The converse is also true.

If f satisfies condition (1.11), f is said to *vanish at infinity*, and we denote it symbolically by $f(\infty) = 0$.

PROOF. If (1.11) were not true, there would exist $\varepsilon > 0$ such that for each $j > 0$ we would have $|f(x_j)| \geq \varepsilon$ for some x_j with $|x_j| \geq j$.

Conversely, if (1.11) holds, then for every $j = 1, 2, \dots$ there exists $R_j > 0$ such that $|f(x)| < 1/j$ $(|x| > R_j)$, from which our assertion follows. □

THEOREM 1.18. *The range of a real valued continuous function on a compact set is compact.*

Thus a real valued continuous function on $\mathbb{R}^d$ maps compact sets in $\mathbb{R}^d$ to compact sets in $\mathbb{R}$

PROOF. Let f be a continuous function in a compact set K. Let $\{O_\alpha\}$ be an open covering of $f(K)$. Then $\{f^{-1}(O_\alpha)\}$ is an open covering of K. More precisely, there exist open sets $\tilde{O}_\alpha$ of $\mathbb{R}^d$ such that $f^{-1}(O_\alpha) = \tilde{O}_\alpha \cap K$. Since $\{\tilde{O}_\alpha\}$ is an open covering of K, there exist a finite number of α's, say $\alpha_1, \dots, \alpha_k$, such that $K \subset \tilde{O}_{\alpha_1} \cup \dots \cup \tilde{O}_{\alpha_k}$. Thus $K \subset f^{-1}(O_{\alpha_1}) \cup \dots \cup f^{-1}(O_{\alpha_k})$, which implies that $f(K) \subset f(f^{-1}(O_{\alpha_1}) \cup \dots \cup f^{-1}(O_{\alpha_k})) \subset O_{\alpha_1} \cup \dots \cup O_{\alpha_k}$. Therefore $f(K)$ has a finite covering. □

COROLLARY 1.1. *Let f be a real valued continuous function on a compact set K. Then f is bounded on K, and f attains $\sup_{x\in K} f(x)$ and $\inf_{x \in K} f(x)$ on K.*

PROOF. $f(K)$ is a compact subset of $\mathbb{R}$ by Theorem 1.18. Since every compact set of $\mathbb{R}$ is bounded and closed, it contains the upper and the lower bounds. □

1.7. The Riemann integral

B. Riemann[3] gave a rigorous definition of integration for the first time in 1854. Although it might be inconceivable from today's ground gained by the Riemann integral, his purpose was to define Fourier coefficients, which will be explained later.

For a closed interval $I = [a_1, b_1] \times \dots \times [a_d, b_d]$ of $\mathbb{R}^d$, we denote the volume (area or length) $(b_1 - a_1) \times \dots \times (b_d - a_d)$ by $|I|$. Here we assume the volume $|I|$ is the same if the intervals $[a_j, b_j]$ are replaced by open or half-open intervals.

Let f be a real valued function in I. Let $\Delta = \{I_j : j = 1, \dots, k\}$ be a partition of I consisting of mutually nonoverlapping intervals. Note that $|I| = \sum_{j=1}^k |I_j|$. Let

$$\overline{f}_j = \sup_{x \in I_j} f(x) \quad \text{and} \quad \underline{f}_j = \inf_{x \in I_j} f(x)$$

[3] G. F. Bernhard Riemann, 1826-1866. Known for his studies of Abelian functions and especially his non-Euclidean geometry.

be the supremum and the infimum of f on I_j, respectively, and put

$$\overline{s}(f,\Delta) = \sum_{j=1}^{k} \overline{f}_j |I_j| \quad \text{and} \quad \underline{s}(f,\Delta) = \sum_{j=1}^{k} \underline{f}_j |I_j|.$$

From the definitions of $\overline{f}_j$ and $\underline{f}_j$ we see that $\overline{s}(f,\Delta)$ decreases when the partition Δ is refined into smaller pieces, and $\underline{s}(f,\Delta)$ increases.

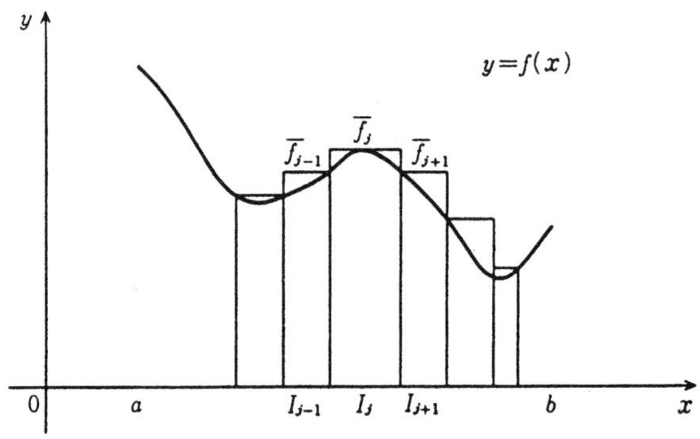

FIGURE 1.7. $\overline{s}(f,\Delta)$

The supremum and the infimum defined by

$$\overline{\int_I} f\,dx = \inf_{\Delta} \overline{s}(f,\Delta) \quad \text{and} \quad \underline{\int_I} f\,dx = \sup_{\Delta} \underline{s}(f,\Delta)$$

are called the *Darboux upper integral* of f on I and the *Darboux lower integral*, respectively.

When

$$\overline{\int_I} f\,dx = \underline{\int_I} f\,dx,$$

f is said to be *Riemann integrable* on I. The common value is called the *Riemann integral* of f on I and denoted as follows:

$$\mathcal{R}\int_I f\,dx \quad \text{or simply} \quad \int_I f\,dx.$$

For a complex valued function f we divide it into the real and imaginary parts, $f = u + iv$. If u and v are both Riemann integrable, f is said to be Riemann integrable, and the Riemann integral is defined by

$$\mathcal{R}\int_I f\,dx = \mathcal{R}\int_I u\,dx + i\mathcal{R}\int_I v\,dx.$$

THEOREM 1.19. *Every continuous function on a bounded closed interval is Riemann integrable.*

PROOF. Let I be a bounded closed interval and f be a continuous function on I. It suffices to prove the theorem for real valued functions dividing f into its real and imaginary parts.

Since I is compact, f is uniformly continuous in I by Theorem 1.17. Thus for $\varepsilon > 0$ we can choose a partition $\Delta = \{I_j : j = 1, \ldots, k\}$ of I so that

$$|f(x) - f(y)| < \varepsilon/|I| \quad (x, y \in I_j),$$

$j = 1, \ldots, k$. Then we have

$$\overline{f}_j - \underline{f}_j \leq \varepsilon/|I|.$$

Therefore, from the definition of the sums $\overline{s}(f, \Delta)$ and $\underline{s}(f, \Delta)$, it follows that

$$\overline{s}(f, \Delta) - \underline{s}(f, \Delta) = \sum_{j=1}^{k} (\overline{f}_j - \underline{f}_j)|I_j| \leq \sum_{j=1}^{k} (\varepsilon/|I|)|I_j| = \varepsilon.$$

Since $\varepsilon > 0$ is any given number, the Darboux upper integral and lower integral coincide. □

REMARK 1.1. There exists a Riemann integrable function which is not continuous. For example, a function is Riemann integrable if it is bounded and continuous in the bounded intervals $[a, b]$ except at a point c, $a < c < b$, which follows from a careful reading of the above proof.

REMARK 1.2. By Problem 1.8 every monotone function on a closed bounded interval is Riemann integrable.

We remark that the set of all points of discontinuity of a monotone function on $\mathbb{R}$ is countable.

The limit $h \to +0$ implies that h tends to 0 on the right, and for a function f we write $f(x + 0) = \lim_{h \to +0} f(x + h)$ if the right-hand side exists.

Let f be an increasing function. Then, for every x, $f(x + h)$ decreases and is bounded when h decreases to 0. Thus $f(x + 0)$ exists for all x. Let D be the set of all points of discontinuity of f. To every point x in D there corresponds the nonempty open interval $I_x = (f(x - 0), f(x + 0))$. Since f is monotone, it follows

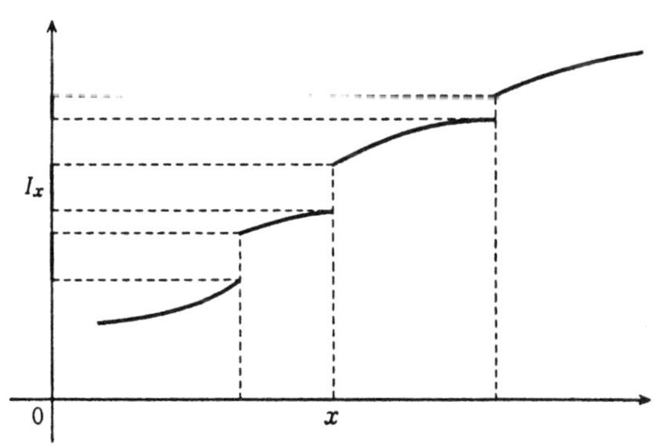

FIGURE 1.8. Monotone function

that the intervals $I_x, x \in D$, are mutually disjoint (see Figure 1.2), which enables us to choose a rational point r_x from each of them. Since the rational numbers r_x, $x \in D$, are mutually distinct, there is a one-to-one correspondence between D and $\{r_x : x \in D\}$. Therefore D is countable because the rational numbers are countable.

Ultimately we shall prove in §3.8 that a bounded function is Riemann integrable in a bounded interval if and only if it is continuous almost everywhere, that is, continuous except on a null set. "Almost everywhere" is one of the characteristic concepts of the theory of the Lebesgue integral, which we shall study in Chapter 2.

As a special case, it follows that a bounded function continuous except on a countable set is Riemann integrable in any bounded interval.

Let E be a set. The function defined by

$$\chi_E = \begin{cases} 1 & x \in E, \\ 0 & x \notin E \end{cases}$$

is called the *characteristic function* of the set E.

EXAMPLE 1.5. The characteristic function $\chi_{\mathbb{Q}}$ of the rational numbers is not Riemann integrable on any interval $[a, b], a < b$.

In fact, every interval with nonempty interior contains a rational number and an irrational number. Thus $(\overline{\chi}_{\mathbb{Q}})_j = 1$ and $(\underline{\chi}_{\mathbb{Q}})_j = 0$. Therefore for every partition Δ of $[a, b]$ we have $\overline{s}(\chi_{\mathbb{Q}}, \Delta) = 1$ and $\underline{s}(\chi_{\mathbb{Q}}, \Delta) = 0$.

We remark that $\chi_{\mathbb{Q}}$ is discontinuous everywhere.

1.8. Jordan measure

Let $I = [a_1, b_1) \times \cdots \times [a_d, b_d)$ be a half-open interval of $\mathbb{R}^d$. Denote

$$|I| = \prod_{j=1}^{d} (b_j - a_j)$$

and call it the measure of I. The measure $|I|$ coincides with the length, the area, and the volume for $d = 1, 2, 3$, respectively.

One answers to the question of how to define length, area, or volume for sets of more complicated shape is found by introducing the notion of Jordan measure, which agrees substantially with the idea of Riemann integration.

Let us recall the definition of the Riemann integral of a positive function f over the interval $I = [a, b]$. First, fix a partition $\{I_j\}$ of $I = [a, b]$. Consider two sums, the upper sum and the lower sum. The first is the sum of the areas of the minimal rectangles containing the graph of f on I_j and with the bases I_j. The second is the sum of the areas of the maximal rectangles which are contained in the region surrounded by the graph of f and with basis I_j. These two sums are considered as approximations of the area surrounded by the graph of f and the interval I.

Let us apply this process to bounded closed sets in $\mathbb{R}^d$. For a bounded set A in $\mathbb{R}^d$ define the *Jordan outer measure* by

$$(1.12) \qquad m_J{}^*(A) = \inf\{|I_1| + \cdots + |I_n| \,:\, A \subset \bigcup_{j=1}^{n} I_j\},$$

where $\{I_1, \ldots, I_n\}$ runs over all finite coverings of the right half-open intervals (see Figure 1.9).

If intervals I_i and I_j are overlapping, $I_i \cup I_j$ is written as a sum of nonoverlapping right half-open intervals. Thus we can assume that every covering of (1.12) consists of nonoverlapping intervals.

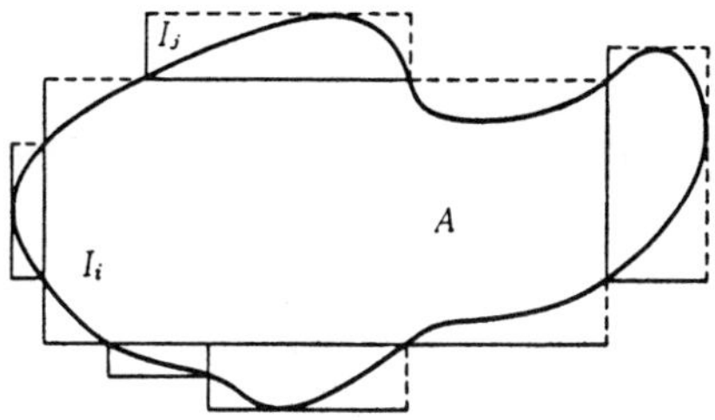

FIGURE 1.9. Jordan measure

Next, define the Jordan inner measure by

$$m_{J*}(A) = \sup\{|I_1| + \cdots + |I_n| : A \supset \bigcup_{j=1}^{n} I_j\},$$

where $\{I_1, \ldots, I_n\}$ runs over all coverings of nonoverlapping right half-open intervals.

When $m_J{}^*(A) = m_{J*}(A)$, the set A is called *Jordan measurable*, or mesurable in the sense of Jordan. The common value is denoted by $m_J(A)$ and called the *Jordan measure* of A.

It is not difficult to see that all bounded intervals are Jordan measurable.

EXAMPLE 1.6. Let A be the set of all rational numbers contained in $[0, 1]$. Then the 1-dimensional Jordan outer measure $m_J^*(A) = 1$ and the inner measure $m_{J*}(A) = 0$, so that A is not Jordan measurable.

We observe that

$$m_J{}^*(A) = \overline{\int} \chi_A dx, \quad m_{J*}(A) = \underline{\int} \chi_A dx$$

by the definitions of the upper and lower Darboux integrals.

1.9. Problems

1.1. Let X be a metric space and let $\mathcal{O} = \{A \subset X : A^i = A\}$. Prove that $\mathcal{O}$ satisfies conditions (O1), (O2), and (O3).

1.2. Prove that the open unit disc $B(0, 1)$ in the plane cannot be written as a countable union of mutually disjoint open intervals.

1.3. Let φ be a mapping of X to Y. If $A, A_\alpha \subset X$, and $B, B_\beta \subset Y$, then prove that

(i) $\varphi(\bigcup_\alpha A_\alpha) = \bigcup_\alpha \varphi(A_\alpha)$,
(ii) $\varphi(\bigcap_\alpha A_\alpha) \subset \bigcap_\alpha \varphi(A_\alpha)$,
(iii) $\varphi^{-1}(\bigcup_\beta B_\beta) = \bigcup_\beta \varphi^{-1}(B_\beta)$,
(iv) $\varphi^{-1}(B^c) = \varphi^{-1}(B)^c$.

1.4. Let A be a perfect set of $\mathbb{R}^d$. Prove that $\text{card}(A) = \mathfrak{c}$ (this is not easy).

1.5. Let $F_1 \supset F_2 \supset \ldots$ be a sequence of nonempty closed subsets of $\mathbb{R}^d$. If $\text{diam}(F_j) \to 0\ (j \to \infty)$, where $\text{diam}(F_j) = \sup\{|x - y| : x, y \in F_j\}$, then prove that $\bigcap_{j=1}^{\infty} F_j$ consists of one point.

1.6. Let $\{a_j\}$ be a bounded sequence in $\mathbb{R}^d$. Prove that $\{a_j\}$ has a convergent subsequence (the Bolzano-Weierstrass theorem).

1.7. Let f be a real valued continuous function on $\mathbb{R}^d$. If $f(\infty) = 0$, then prove that f is uniformly continuous.

1.8. Prove that a monotone function on a bounded closed interval $[a, b]$ is Riemann integrable.

1.9. Open sets are not necessarily Jordan measurable. Prove it in the following way. Let $\{r_j\}$ be an enumeration of $\mathbb{Q}$ and let $I_j = (r_j - 1/2^j, r_j + 1/2^j)$ be an open interval containing x. Then $O = \bigcup_{j=1}^{\infty} I_j$ is an open set. Check that

(i) $1 < m_{J*}(O) \leq 2$,

(ii) $m_J{}^*(O) = \infty$.

CHAPTER 2

Lebesgue Measure on Euclidean Spaces

Let us again attempt to think about the question: "What is the area of a plane set?" First, the area of a rectangle is defined as the product of its two side lengths. In the preceding chapter we introduced the Jordan measure to define the areas of more complicated sets.

Here, we require that the concept of the area—from now on we call it measure —be closed under countably infinite calculation operations. This mathematical requirement will allow us to take the limits of sequences of operations.

We summarize our demand more clearly as follows:

Let $\mathcal{M}$ be the collection of all sets in the plane for which the measures are defined. Then:

(i) $\mathcal{M}$ contains all rectangles, and the measure of a rectangle is equal to the product of its two side lengths.

(ii) The countable union of sets of $\mathcal{M}$ belongs to $\mathcal{M}$; that is, $\mathcal{M}$ is closed under countable union operations.

Furthermore,

(iii) The difference of two sets in $\mathcal{M}$ belongs to $\mathcal{M}$.

The smallest family of sets which satisfies conditions (i), (ii) and (iii) is called the Borel field (see §2.5 for the definition).

In addition, the notion "null" is introduced. The family of Lebesgue measurable sets is the smallest family of sets which satisfies conditions (i), (ii), (iii) and (iv):

(iv) $\mathcal{M}$ contains all null sets.

In this chapter, the concepts of Lebesgue measurablility and Lebesgue measure in Euclidean spaces are introduced; they are both closed under the countable union or the complementation operations of sets, and they enable us to define the Lebesgue integral, which is also closed under countable operations in a sense.

We define measurable sets as sets which satisfy the Carathéodory condition. In a later chapter this method will be applied to a more general situation.

An *extended real number* is a real number or one of the symbols $\pm\infty$. Between a real number a and $\pm\infty$ the following operations are defined:

$$a + (\pm\infty) = (\pm\infty) + a = \pm\infty,$$

$$a \cdot (\pm\infty) = (\pm\infty) \cdot a = \begin{cases} 0, & a = 0, \\ \pm\infty, & a > 0, \\ \mp\infty, & a < 0, \end{cases}$$

$$a/(\pm\infty) = 0.$$

Furthermore, we define

$$(\pm\infty) \cdot (\pm\infty) = \infty, \quad \text{and} \quad (\pm\infty) \cdot (\mp\infty) = -\infty.$$

2.1. Measure of intervals

DEFINITION 2.1. Let $\mathcal{R}$ be a family of subsets of a set X. $\mathcal{R}$ is said to be a *ring of sets*, if $\mathcal{R}$ satisfies the following conditions:

(i) If $E, F \in \mathcal{R}$, then $E - F \in \mathcal{R}$.
(ii) If $E, F \in \mathcal{R}$, then $E \cup F \in \mathcal{R}$.

Furthermore, if

(iii) $X \in \mathcal{R}$,

then $\mathcal{R}$ is said to be an *algebra of sets* or a *Boolean algebra*.

If $\mathcal{R}$ is an algebra of sets, then we have

(a) $E \in \mathcal{R}$ implies $E^c \in \mathcal{R}$,
(b) $E, F \in \mathcal{R}$ implies $E \cap F \in \mathcal{R}$.

In fact, if $E, F \in \mathcal{R}$, then we have $E^c = X - E \in \mathcal{R}$ and $E \cap F = (E^c \cup F^c)^c \in \mathcal{R}$.

Denote by $\mathcal{P}(X)$ the family of all subsets of X. Then the family $\mathcal{P}(X)$ is an algbra of sets.

Denote by $\mathcal{R}(\mathbb{R}^d)$ the family of all finite unions of bounded right half-open intervals in $\mathbb{R}^d$ (see Figure 2.1). Then the family $\mathcal{R}(\mathbb{R}^d)$ is a ring of sets.

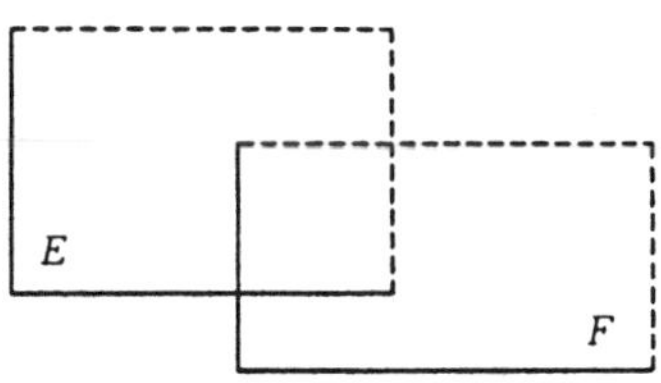

FIGURE 2.1. Union of two right half-open intervals

Indeed, the family $\mathcal{R}(\mathbb{R}^d)$ obviously satisfies condition (ii). To prove (i) it suffices to show that if E and F are bounded right half-open intervals, then the set $E - F$ is a finite union of right half-open intevals (see Figure 2.1).

For the right half-open interval $I = [a_1, b_1) \times \cdots \times [a_d, b_d)$, define the measure by

$$m(I) = \prod_{j=1}^{d} (b_j - a_j).$$

For $E \in \mathcal{R}(\mathbb{R}^d)$, express E as a finite union of mutually disjoint right half-open intervals $E = \bigcup_{j=1}^{k} I_j$ and define the measure by

$$m(E) = \sum_{j=1}^{k} m(I_j). \tag{2.1}$$

The left-hand side does not depend on the expression of E. In fact, let

$$E = \bigcup_{j=1}^{k} I_j = \bigcup_{p=1}^{q} J_p$$

be two expressions of E as the finite union of mutually disjoint right half-open intervals. Then $I_j = I_j \cap E = \bigcup_{p=1}^{q} I_j \cap J_p$, which means that $\{I_j \cap J_p : p = 1, 2, \ldots, q\}$ is a partition of I_j. Thus

$$m(I_j) = \sum_{p=1}^{q} m(I_j \cap J_p).$$

In the same way, we have $m(J_p) = \sum_{j=1}^{k} m(I_j \cap J_p)$. Therefore

$$\sum_{j=1}^{k} m(I_j) = \sum_{j=1}^{k}\sum_{p=1}^{q} m(I_j \cap J_p) = \sum_{p=1}^{q}\sum_{j=1}^{k} m(I_j \cap J_p) = \sum_{p=1}^{q} m(J_p).$$

In particular, if E and F are disjoint sets in $\mathcal{R}(\mathbb{R}^d)$, then we have

$$m(E \cup F) = m(E) + m(F). \tag{2.2}$$

DEFINITION 2.2. A mapping μ of a ring of sets $\mathcal{R}$ to $\mathbb{R} \cup \{+\infty\}$ is called a *finitely additive set function* if

(i) $\mu(\emptyset) = 0$,
(ii) $E, F \in \mathcal{R}$ and $E \cap F = \emptyset$ imply $\mu(E \cup F) = \mu(E) + \mu(F)$.

Furthermore, if the additivity condition (ii) is replaced by the countable additivity (ii)$'$, then μ is called a *countably additive set function* or simply a *measure*:

(ii)$'$ $E_j \in \mathcal{R}$ $(j = 1, 2, \ldots)$, $\bigcup_{j=1}^{\infty} E_j \in \mathcal{R}$ and $E_i \cap E_j = \emptyset$ $(i \neq j)$ imply

$$\mu\Big(\bigcup_{j=1}^{\infty} E_j\Big) = \sum_{j=1}^{\infty} \mu(E_j).$$

REMARK 2.1. The word measure often stands for a nonnegative, countably additive set function, but in this book the nonnegativity is not always assumed.

REMARK 2.2. Condition (ii)$'$ implies that the left-hand side is finite and the sum on the right-hand side converges to the left-hand side, or that the left-hand side is $+\infty$ and the sum diverges to $+\infty$.

THEOREM 2.1. *Let m be a set function defined by* (2.1). *Then m is a measure on $\mathcal{R}(\mathbb{R}^d)$.*

To prove Theorem 2.1 we prepare several lemmas.

LEMMA 2.1. *If $E, F \in \mathcal{R}(\mathbb{R}^d)$ and $E \subset F$, then $m(E) \leq m(F)$.*

PROOF. The set F is written as the union $F = (F - E) \cup E$ of mutually disjoint sets in $\mathcal{R}(\mathbb{R}^d)$. Thus by (2.2) we have

$$m(F) = m(F - E) + m(E) \geq m(E).$$

□

LEMMA 2.2. *If $E, E_j \in \mathcal{R}(\mathbb{R}^d)$ $(j = 1, \ldots, k)$ and $E \subset \bigcup_{j=1}^{k} E_j$, then*

$$m(E) \leq \sum_{j=1}^{k} m(E_j). \tag{2.3}$$

PROOF. Put $F_j = E \cap E_j$. We can write E as follows:

$$E = E \cap \bigcup_{j=1}^{k} E_j = \bigcup_{j=1}^{k} F_j$$
$$= F_1 \cup (F_2 - F_1) \cup (F_3 - F_1 - F_2) \cup \cdots \cup (F_k - F_1 - \cdots - F_{k-1}).$$

The right-hand side is a union of mutually disjoint sets in $\mathcal{R}(\mathbb{R}^d)$. Thus by (2.2) we have

$$m(E) = m(F_1) + m(F_2 - F_1) + m(F_3 - F_1 - F_2) + \cdots + m(F_k - F_1 - \cdots - F_{k-1}).$$

To estimate the right-hand side use the inequalities

$$m(F_1) \le m(E_1),\ m(F_2 - F_1) \le m(E_2), \ldots,\ m(F_k - F_1 - \cdots - F_{k-1}) \le m(E_k).$$

□

LEMMA 2.3. *If $E_j \in \mathcal{R}(\mathbb{R}^d)$ $(j = 1, 2, \ldots)$ and $E_1 \supset E_2 \supset \cdots \to \emptyset$, then*

$$m(E_j) \to 0 \quad \textit{as } j \to \infty.$$

PROOF. Suppose to the contrary that $m(E_j) \not\to 0$. Then there exists $\delta > 0$ such that $m(E_j) > \delta$, $j = 1, 2, \ldots$. Choose $F_j \in \mathcal{R}(\mathbb{R}^d)$ so that $E_j \supset \overline{F}_j \supset F_j$ and $m(E_j - F_j) < \delta/2^{j+1}$. By assumption,

$$\emptyset = \bigcap_{j=1}^{\infty} E_j \supset \overline{F}_1 \cap \bigcap_{j=2}^{\infty} \overline{F}_j.$$

Since every set in $\mathcal{R}(\mathbb{R}^d)$ is bounded, $\overline{F}_1$ is compact. Thus $\{\overline{F}_1 \cap \bigcap_{j=2}^{k} \overline{F}_j : k = 2, 3, \ldots\}$ does not have the finite intersection property. Therefore $\overline{F}_1 \cap \bigcap_{j=2}^{N} \overline{F}_j = \emptyset$ for some N by Theorem 1.14. Since $E_N \subset E_j$ $(j = 1, \ldots, N)$, we have

$$E_N = E_N - \bigcap_{j=1}^{N} F_j = \bigcup_{j=1}^{N} (E_N - F_j) \subset \bigcup_{j=1}^{N} (E_j - F_j).$$

Thus by Lemma 2.2 we have

$$m(E_N) \le \sum_{j=1}^{N} m(E_j - F_j).$$

Since $m(E_j - F_j) \le \delta/2^{j+1}$, the right-hand side is smaller than $\sum_{j=1}^{N} \delta/2^{j+1} < \delta/2$. On the other hand, $\delta \le m(E_N)$, which is absurd. □

PROOF OF THEOREM 2.1. Let E_j $(j = 1, 2, \ldots)$ be mutually disjoint sets in $\mathcal{R}(\mathbb{R}^d)$ and let $E = \bigcup_{j=1}^{\infty} E_j$. Assume that $E \in \mathcal{R}(\mathbb{R}^d)$. Then the sets $E - \bigcup_{j=1}^{k} E_j$ belong to $\mathcal{R}(\mathbb{R}^d)$ and decrease to $\emptyset$. Thus, by Lemma 2.3,

$$m(E - \bigcup_{j=1}^{k} E_j) \to 0 \quad \text{as } k \to \infty.$$

By the finite additivity (2.2) we have

$$m(E) = m(\bigcup_{j=1}^{k} E_j) + m(E - \bigcup_{j=1}^{k} E_j) = \sum_{j=1}^{k} m(E_j) + m(E - \bigcup_{j=1}^{k} E_j).$$

Therefore we get, by letting $k \to 0$,

$$m(E) = \sum_{j=1}^{\infty} m(E_j).$$

□

2.2. Outer measure

By definition, $\mathcal{P}(\mathbb{R}^d)$ is the family of all subsets of $\mathbb{R}^d$. For $A \in \mathcal{P}(\mathbb{R}^d)$ define

$$m^*(A) = \inf\{\sum_{j=1}^{\infty} m(E_j) : E_j \in \mathcal{R}(\mathbb{R}^d) \text{ and } A \subset \bigcup_{j=1}^{\infty} E_j\}.$$

If A belongs to a set of $\mathcal{R}(\mathbb{R}^d)$, then we have

$$m^*(A) = m(A)$$

by taking A itself as a covering of A.

DEFINITION 2.3. A set function μ^* on $\mathcal{P}(X)$, with values in $\mathbb{R} \cup \{+\infty\}$, is called a Carathéodory[1] *outer measure* or simply an outer measure if the following conditions are satisfied:

(i) $0 \leq \mu^*(A) \leq \infty$, $\mu^*(\emptyset) = 0$.
(ii) $A \subset B$ implies $\mu^*(A) \leq \mu^*(B)$.
(iii) $\mu^*(\bigcup_{j=1}^{\infty} A_j) \leq \sum_{j=1}^{\infty} \mu^*(A_j)$.

THEOREM 2.2. *m^* is an outer measure on $\mathcal{P}(\mathbb{R}^d)$.*

PROOF. The condition (i) is obvious. (ii) follows from the fact that a covering of the set B is also a covering of A.

To prove (iii) we assume that $\sum_{j=1}^{\infty} \mu^*(A_j) < \infty$. Otherwise, (iii) is always true.

Let $\varepsilon > 0$ be given. For each j choose a covering $\{E_{j,k} : k = 1, 2, \dots\}$ of the set A_j in $\mathcal{R}(\mathbb{R}^d)$ so that

$$\sum_{k=1}^{\infty} m(E_{j,k}) < m^*(A_j) + \frac{\varepsilon}{2^j}.$$

Since $\{E_{j,k} : j, k = 1, 2, \dots\}$ covers $\bigcup_{j=1}^{\infty} A_j$, we have, by the definition of m^*,

$$m^*(\bigcup_{j=1}^{\infty} A_j) \leq \sum_{j,k=1}^{\infty} m(E_{j,k}) = \sum_{j=1}^{\infty}\Big(\sum_{k=1}^{\infty} m(E_{j,k})\Big).$$

The last term is, by the above inequality, smaller than

$$\sum_{j=1}^{\infty}\Big(m^*(A_j) + \frac{\varepsilon}{2^j}\Big) = \sum_{j=1}^{\infty} m^*(A_j) + \varepsilon,$$

which implies (iii). □

[1]C. Carathéodory, 1873-1950.

EXAMPLE 2.1. The outer measure of any single-point set in $\mathbb{R}^d$ is 0. Thus the outer measure of any countable set in $\mathbb{R}^d$ is 0. In particular, the outer measure of the rational numbers $\mathbb{Q}$ is 0.

In fact, let $\{x\}$ be a single-point set. It is covered by any small interval $\prod_{j=1}^{d}[x_j - \varepsilon, x_j + \varepsilon)$. Thus $m^*(\{x\}) = 0$. Let $D = \{a_n\}$ be a countable set. Then, by(iii), $m^*(D) \leq \sum_{n=1}^{\infty} m^*(\{a_n\}) = 0$.

DEFINITION 2.4. Let μ^* be an outer measure on a set X. A subset A is said to be a *μ^*-null set* if $\mu^*(A) = 0$.

An m^*-null set is called simply a null set.

There exists an infinite null set of $\mathbb{R}$. In fact, the set $\mathbb{Q}$ is a null set by Example 2.1. Furthermore, there exists an uncountable null set.

EXAMPLE 2.2. The Cantor ternary set C (see § 1.3) is an uncountable null set.

In fact, we have shown that $\operatorname{card}(C) = \mathfrak{c}$. By definition, $C = \bigcap_{j=1}^{\infty} C_j$, where C_j consists of 2^j closed intervals of length $1/3^j$. Thus

$$m^*(C) \leq m^*(C_j) \leq 2^j \times 1/3^j$$

for every j. When $j \to \infty$, the right-hand side tends to 0. Thus we have $m^*(C) = 0$.

2.3. Measurable sets and Lebesgue measure

DEFINITION 2.5. Let E be a subset of $\mathbb{R}^d$. If

$$\text{(C)} \qquad m^*(A) = m^*(A \cap E) + m^*(A - E)$$

for all $A \in \mathcal{P}(\mathbb{R}^d)$, then E is said to be *Lebesgue*[2] *measurable* or simply *measurable*. The collection of all measurable sets is denoted by $\mathcal{M}(\mathbb{R}^d)$.

(C) is called *the Carathéodory condition.*

By a property of outer measures, we have

$$m^*(A) \leq m^*(A \cap E) + m^*(A - E)$$

for all A and E. Therefore condition (C) is equivalent to saying that

$$\text{(C)}' \qquad m^*(A) \geq m^*(A \cap E) + m^*(A - E) \quad \text{for all } A \in \mathcal{P}(\mathbb{R}^d).$$

If E is a null set, $m^*(A \cap E) = 0$. Thus condition (C)′ holds.

THEOREM 2.3. *Every null set is measurable.*

Condition (C) says that the measurability of a set E is characterized by the fact that if E divides any set into the part contained in E and the part contained in the complement of E, then the outer measure of the set is divided into the outer measures of these two parts.

The reader will find it difficult to extract individually any qualitative property of measurable sets from the Carathéodory condition. But condition (C) will be interpreted as a requirement of a quantitative relation among the measurable sets $\mathcal{M}(\mathbb{R}^d)$. Indeed, if A is the union of mutually disjoint measurable sets E_1 and E_2, then $A \cap E_1 = E_1$ and $A - E_1 = E_2$. Therefore condition (C) leads to

$$m^*(E_1 \cup E_2) = m^*(E_1) + m^*(E_2).$$

[2] H. Lebesgue, 1875-1941. Most of his work was on analysis, in particular the theory of functions of real variables.

If $\{E_1, E_2, \dots\}$ is a sequence of mutually disjoint measurable sets, then we shall obtain also

$$m^*(\bigcup_{j=1}^{\infty} E_j) = \sum_{j=1}^{\infty} m^*(E_j),$$

which will be proved later and will be the most important part of the main theorem in Chapter 2. In other words, (C) gives the condition on sets under which the outer measure m^* is countably additive on $\mathcal{M}(\mathbb{R}^d)$.

DEFINITION 2.6. A family $\mathcal{F}$ of subsets of a set X is called a *σ-ring* if the following conditions are satisfied:

(i) $E_j \in \mathcal{F}\,(j = 1, 2, \dots,)$ implies $\bigcup_{j=1}^{\infty} E_j \in \mathcal{F}$.
(ii) $E, F \in \mathcal{F}$ implies $E - F \in \mathcal{F}$.

Furthermore, if

(3) $X \in \mathcal{F}$,

then $\mathcal{F}$ is called a *σ-field* or a *σ-algebra*.

Let $\mathcal{F}$ be a σ-field. Then:

(a) $E \in \mathcal{F}$ implies $E^c = X - E \in \mathcal{F}$.
(b) If $E_j \in \mathcal{F}\,(j = 1, 2, \dots)$, then $\bigcap_{j=1}^{\infty} E_j = (\bigcup_{j=1}^{\infty} E_j^c)^c \in \mathcal{F}$.

DEFINITION 2.7.

$$\tilde{m} = m^*|_{\mathcal{M}(\mathbb{R}^d)},$$

that is, $\tilde{m}$ is the restriction of m^* to the measurable sets $\mathcal{M}(\mathbb{R}^d)$.
The set function $\tilde{m}$ is called *Lebesgue measure*.

The following is the main theorem in this chapter.

THEOREM 2.4.

(i) *$\mathcal{M}(\mathbb{R}^d) \supset \mathcal{R}(\mathbb{R}^d)$ and $\tilde{m}|_{\mathcal{R}(\mathbb{R}^d)} = m$.*
(ii) *$\mathcal{M}(\mathbb{R}^d)$ is a σ-field.*
(iii) *$\tilde{m}$ is a measure on $\mathcal{M}(\mathbb{R}^d)$.*

PROOF. First Step. *Proof of* (i).

To prove that every set E in $\mathcal{R}(\mathbb{R}^d)$ satisfies condition (C), let $A \in \mathcal{P}(\mathbb{R}^d)$ and let $\{E_j : j = 1, 2, \dots\} \subset \mathcal{R}(\mathbb{R}^d)$ be any covering of A. Then we have

$$A \cap E \subset \bigcup_{j=1}^{\infty} E_j \cap E \quad \text{and} \quad A - E \subset \bigcup_{j=1}^{\infty} (E_j - E).$$

By the finite additivity of m on $\mathcal{R}(\mathbb{R}^d)$ and the definition of the outer measure,

$$\sum_{j=1}^{\infty} m(E_j) = \sum_{j=1}^{\infty} \{m(E_j \cap E) + m(E_j - E)\}$$
$$\geq m^*(A \cap E) + m^*(A - E).$$

If we take the infimum of the left-hand side over all coverings $\{E_j\}$ of A, we get the inequality (C)$'$ for E.

As is noted in §2.2, we have $m^* = m$ on $\mathcal{R}(\mathbb{R}^d)$, and so $\tilde{m}|_{\mathcal{R}(\mathbb{R}^d)} = m^*|_{\mathcal{R}(\mathbb{R}^d)} = m$.

Second Step. *We show that*

(a) $\emptyset \in \mathcal{M}(\mathbb{R}^d)$,
(b) *if* $E \in \mathcal{M}(\mathbb{R}^d)$, *then* $E^c \in \mathcal{M}(\mathbb{R}^d)$,
(c) *if* $E, F \in \mathcal{M}(\mathbb{R}^d)$, *then* $E - F \in \mathcal{M}(\mathbb{R}^d)$.

The first statement is obvious. (b) follows directly from the Carathéodory condition. To prove (c) we must show that

$$m^*(A) = m^*(A \cap (E - F)) + m^*(A - (E - F)) \quad \text{for } A \in \mathcal{P}(\mathbb{R}^d). \tag{2.4}$$

To manage the second term on the right-hand side we divide $A - (E - F)$ into two sets:

$$[A - (E - F)] \cap F = A \cap (E \cap F^c)^c \cap F = A \cap (E^c \cup F) \cap F = A \cap F,$$

and

$$[A - (E - F)] - F = A \cap (E^c \cup F) \cap F^c = A \cap E^c \cap F^c = (A \cap F^c) - E.$$

Since F satisfies the Carathéodory condition,

$$\begin{aligned} m^*(A - (E - F)) &= m^*([A - (E - F)] \cap F) + m^*([A - (E - F)] - F) \\ &= m^*(A \cap F) + m^*((A \cap F^c) - E). \end{aligned} \tag{2.5}$$

For the first term of the right-hand side of (2.4) we have

$$A \cap (E - F) = A \cap E \cap F^c = (A \cap F^c) \cap E. \tag{2.6}$$

Substituting (2.6) and (2.5) into (2.4), the right-hand side of (2.4) is rewritten as

$$m^*((A \cap F^c) \cap E) + m^*(A \cap F) + m^*((A \cap F^c) - E). \tag{2.7}$$

We again apply the Carathéodory condition to the first and third terms of (2.7) with $E \in \mathcal{M}(\mathbb{R}^d)$. Then the last expression becomes

$$m^*(A \cap F^c) + m^*(A \cap F),$$

which is reduced to $m^*(A)$ by the assumption that $F \in \mathcal{M}(\mathbb{R}^d)$, which proves (2.4).

Third Step. *Let* $\{E_j : j = 1, 2, \dots\}$ *be a sequence of mutually disjoint sets in* $\mathcal{M}(\mathbb{R}^d)$ *Then we have*

$$m^*(A) \geq \sum_{j=1}^{\infty} m^*(A \cap E_j) + m^*(A - \bigcup_{j=1}^{\infty} E_j) \quad \textit{for all } A \in \mathcal{P}(\mathbb{R}^d). \tag{2.8}$$

To prove (2.8) we first show the inequality

$$m^*(A) \geq \sum_{j=1}^{k} m^*(A \cap E_j) + m^*(A - \bigcup_{j=1}^{k} E_j), \quad k = 1, 2, \dots, \tag{2.9}$$

for all $A \in \mathcal{P}(\mathbb{R}^d)$ by induction.

When $k = 1$, (2.9) is nothing but the definition of measurability of E_1. Suppose that (2.9) holds for all $A \in \mathcal{P}(\mathbb{R}^d)$ and an integer k. Then, by the inequality (2.9) with A replaced by $A - E_{k+1}$, we have

$$\begin{aligned} m^*(A) &= m^*(A \cap E_{k+1}) + m^*(A - E_{k+1}) \\ &\geq m^*(A \cap E_{k+1}) \\ &\quad + \left\{ \sum_{j=1}^{k} m^*((A - E_{k+1}) \cap E_j) + m^*(A - E_{k+1} - \bigcup_{j=1}^{k} E_j) \right\}. \end{aligned} \tag{2.10}$$

Since the sets E_j are mutually disjoint, we have

$$(A - E_{k+1}) \cap E_j = A \cap E_j \text{ and } (A - E_{k+1}) - \bigcup_{j=1}^{k} E_j = A - \bigcup_{j=1}^{k+1} E_j.$$

Thus we get from (2.10)

$$m^*(A) \geq m^*(A \cap E_{k+1}) + \sum_{j=1}^{k} m^*(A \cap E_j) + m^*(A - \bigcup_{j=1}^{k+1} E_j).$$

Therefore inequality (2.9) holds for $k+1$.

By (2.9), we have

$$m^*(A) \geq \sum_{j=1}^{k} m^*(A \cap E_j) + m^*(A - \bigcup_{j=1}^{\infty} E_j)$$

for all k. Letting $k \to \infty$, we get (2.8).

Fourth Step. *Let $\{E_j\}$ be a sequence of mutually disjoint sets in $\mathcal{M}(\mathbb{R}^d)$. Then $\bigcup_{j=1}^{\infty} E_j \in \mathcal{M}(\mathbb{R}^d)$.*

Indeed, since $m^*(A \cap \bigcup_{j=1}^{\infty} E_j) \leq \sum_{j=1}^{\infty} m^*(A \cap E_j)$, it follows from (2.8) that

$$m^*(A) \geq m^*(A \cap \bigcup_{j=1}^{\infty} E_j) + m^*(A - \bigcup_{j=1}^{\infty} E_j).$$

Thus $\bigcup_{j=1}^{\infty} E_j$ satisfies the Carathéodory condition.

Fifth Step. *$\mathcal{M}(\mathbb{R}^d)$ is a σ-field.*

If $E_j \in \mathcal{M}(\mathbb{R}^d)$ $(j = 1, 2, \ldots)$, then $\bigcup E_j$ is written as a union of mutually disjoint sets, that is, $\bigcup_{j=1}^{\infty} E_j = E_1 \cup (E_2 - E_1) \cup (E_3 - E_1 - E_2) \cup \cdots$. Since each term on the right-hand side belongs to $\mathcal{M}(\mathbb{R}^d)$ by the second step, the right-hand side belongs to $\mathcal{M}(\mathbb{R}^d)$ by the fourth step. Thus $\bigcup_{j=1}^{\infty} E_j \in \mathcal{M}(\mathbb{R}^d)$.

Sixth Step. *$\tilde{m}$ is a measure on $\mathcal{M}(\mathbb{R}^d)$.*

Let $\{E_j\}$ be a sequence of mutually disjoint sets in $\mathcal{M}(\mathbb{R}^d)$. Put $A = \bigcup_{j=1}^{\infty} E_j$ in the inequality (2.8). Then we have

$$\tilde{m}(\bigcup_{j=1}^{\infty} E_j) \geq \sum_{j=1}^{\infty} \tilde{m}(E_j).$$

Since the reverse inequality holds in any case by a property of outer measures, we have

$$\tilde{m}(\bigcup_{j=1}^{\infty} E_j) = \sum_{j=1}^{\infty} \tilde{m}(E_j).$$

This completes the proof. □

We call the triple $(\mathbb{R}^d, \mathcal{R}(\mathbb{R}^d), m)$ a *measure space.* Theorem 2.4 shows that the measure space $(\mathbb{R}^d, \mathcal{R}(\mathbb{R}^d), m)$ which is closed under the finite additive operations can be extended to the measure space $(\mathbb{R}^d, \mathcal{M}(\mathbb{R}^d), \tilde{m})$ closed under the countably additive operations.

In the following we write simply m for $\tilde{m}$ if there will be no confusion.

2.4. Fundamental properties of measures

In this section we shall discuss a property of measures which is equivalent to countable additivity. We don't restrict ourselves to Lebesgue measure, but treat general measures on a set X.

For a sequence of sets $\{E_j\}$ of X the limit is defined as follows:

$$\text{If } E_1 \subset E_2 \subset \cdots \subset E_j \subset \cdots, \text{ then } \lim_{j\to\infty} E_j = \bigcup_{j=1}^{\infty} E_j.$$

$$\text{If } E_1 \supset E_2 \supset \cdots \supset E_j \supset \cdots, \text{ then } \lim_{j\to\infty} E_j = \bigcap_{j=1}^{\infty} E_j.$$

For a general sequence $\{E_j\}$

$$\limsup_{j\to\infty} E_j = \bigcap_{k=1}^{\infty}\bigcup_{j=k}^{\infty} E_j \quad \text{and} \quad \liminf_{j\to\infty} E_j = \bigcup_{k=1}^{\infty}\bigcap_{j=k}^{\infty} E_j.$$

If these two limits coincide, they are simply denoted by $\lim_{j\to\infty} E_j$.

Let $\mathcal{F}$ be a σ-field of subsets of a set X, and suppose $E_j \in \mathcal{F}$. Then the sets

$$\bigcup_{j=i}^{\infty} E_j, \ \bigcap_{j=i}^{\infty} E_j, \ \limsup_{j\to\infty} E_j, \text{ and } \liminf_{j\to\infty} E_j$$

belong also to $\mathcal{F}$. Indeed, by the definition of σ-field $\bigcup_{j=k}^{\infty} E_j \in \mathcal{F}$, and $\bigcap_{j=k}^{\infty} E_j = (\bigcup_{j=k}^{\infty} E_j^c)^c \in \mathcal{F}$. Therefore it follows that $\limsup E_j, \liminf E_j \in \mathcal{F}$.

Let μ be a nonnegative measure on the σ-field $\mathcal{F}$ of subsets of X. A typical example is the Lebesgue measure m on $\mathcal{M}(\mathbb{R}^d)$.

THEOREM 2.5. *Let* $E_j \in \mathcal{F}\,(j = 1, 2, \dots)$. *Then we have:*

(i) $\mu(\bigcup_{j=1}^{\infty} E_j) \le \sum_{j=1}^{\infty} \mu(E_j)$.

(ii) *If* $E_1 \subset E_2 \subset \cdots$, *then* $\lim_{j\to\infty} \mu(E_j) = \mu(\lim_{j\to\infty} E_j)$.

(iii) *If* $E_1 \supset E_2 \supset \cdots$ *and* $\mu(E_1) < \infty$, *then* $\lim_{j\to\infty} \mu(E_j) = \mu(\lim_{j\to\infty} E_j)$.

PROOF. (i) Recall first that the set $\bigcup E_j$ can be written as a union of mutually disjoint sets of $\mathcal{F}$. For example, we have $\bigcup E_j = E_1 \cup (E_2 - E_1) \cup (E_3 - E_1 - E_2) \cup \cdots$. Thus

$$\mu(\bigcup_{j=1}^{\infty} E_j) = \mu(E_1) + \mu(E_2 - E_1) + \mu(E_3 - E_1 - E_2) + \cdots$$
$$\le \mu(E_1) + \mu(E_2) + \mu(E_3) + \cdots.$$

(ii) Suppose that $\mu(E_k) = \infty$ for some k. Since $\{E_j\}$ is assumed to be a monotone increasing sequence, $E_k \subset \lim_{j\to\infty} E_j$. Thus $\infty = \mu(E_k) \le \mu(\lim_{j\to\infty} E_j)$.

Assume $\mu(E_k) < \infty$ for every k. Expressing $\lim_{j\to\infty} E_j$ by mutually disjoint subsets, we get

$$\mu(\lim_{j\to\infty} E_j) = \mu(E_1) + \mu(E_2 - E_1) + \mu(E_3 - E_2) + \cdots.$$

Substituting $\mu(E_{j+1}) = \mu(E_{j+1} - E_j) + \mu(E_j)\ (j = 1, 2, \dots)$ in the right-hand side of the last equation, we have

$$\mu(E_1) + [\mu(E_2) - \mu(E_1)] + [\mu(E_3) - \mu(E_2)] + \cdots = \lim_{j\to\infty} (E_j).$$

(iii) Put $F_j = E_1 - E_j$. Then, since E_j is decreasing, $F_1 \subset F_2 \subset F_3 \subset \cdots$. Thus by (ii)

$$\mu(\lim_{j\to\infty} F_j) = \lim_{j\to\infty} \mu(F_j).$$

The left-hand side equals $\mu(\lim_{j\to\infty}(E_1 - E_j)) = \mu(E_1) - \mu(\lim_{j\to\infty} E_j)$ and the right-hand side equals $\lim_{j\to\infty}[\mu(E_1) - \mu(E_j)] = \mu(E_1) - \lim_{j\to\infty}\mu(E_j)$, which proves (iii). □

2.5. Measurable sets and Borel sets

Let X be a set and let $\mathcal{A}$ be a family of subsets of X. Then there exists a smallest σ-field containing $\mathcal{A}$. To prove this we first remark that $\mathcal{F} = \mathcal{P}(X)$ is a σ-field and contains $\mathcal{A}$. Let

$$\sigma(\mathcal{A}) = \bigcap \mathcal{F},$$

where the intersection runs over all σ-subfields $\mathcal{F}$ of $\mathcal{P}(X)$ which include $\mathcal{A}$. Then $\sigma(\mathcal{A})$ is the smallest σ-field including $\mathcal{A}$.

Indeed, let $\mathcal{F}$ be any σ-field such that $\mathcal{A} \subset \mathcal{F} \subset \mathcal{P}(X)$. If $\{E_j\}$ is a sequence in $\sigma(\mathcal{A})$, then $E_j \in \mathcal{F}$, $j = 1, 2, \ldots$. By the countable additivity of σ-fields we have $\bigcup E_j \in \mathcal{F}$ and thus $\bigcup E_j \in \sigma(\mathcal{A})$. In a similar way, if $E, F \in \sigma(\mathcal{A})$, then $E - F \in \sigma(\mathcal{A})$.

Therefore, for any family $\mathcal{A} \subset \mathcal{P}(X)$ there exists a unique minimal σ-field $\sigma(\mathcal{A})$ which contains $\mathcal{A}$. The proof of the uniqueness of such a σ-field is an easy exercise for the reader.

Let $\mathcal{K}(\mathbb{R}^d)$ be the family of all compact sets in $\mathbb{R}^d$. The smallest σ-field $\sigma(\mathcal{K}(\mathbb{R}^d))$ which contains $\mathcal{K}(\mathbb{R}^d)$ is called the *Borel field* and denoted by $\mathcal{B}(\mathbb{R}^d)$. A subset of the Borel field is called a *Borel set.*

Let $\mathcal{O}(\mathbb{R}^d)$ be the family of all open sets. If $O \in \mathcal{O}(\mathbb{R}^d)$, then the complement of O is a countable union of compact sets. In fact, $O^c = \bigcup_{n=1}^{\infty} O^c \cap \overline{B(0,n)}$. Thus $O \in \mathcal{B}(\mathbb{R}^d)$, which implies that $\sigma(\mathcal{O}(\mathbb{R}^d)) \subset \mathcal{B}(\mathbb{R}^d)$. On the other hand, $\mathcal{K}(\mathbb{R}^d) \subset \sigma(\mathcal{O}(\mathbb{R}^d))$, since the complement of a compact set is open. Thus $\mathcal{B}(\mathbb{R}^d) \subset \sigma(\mathcal{O}(\mathbb{R}^d))$. Combining this with the reverse inclusion, we get

$$\mathcal{B}(\mathbb{R}^d) = \sigma(\mathcal{O}(\mathbb{R}^d)).$$

Let O be a nonempty open set in $\mathbb{R}^d$. Then $O \in \mathcal{M}(\mathbb{R}^d)$, since O is a countable union of mutually disjoint right half-open intervals $\bigcup I_j$ and $I_j \in \mathcal{M}(\mathbb{R}^d)$. Thus $\mathcal{O}(\mathbb{R}^d) \subset \mathcal{M}(\mathbb{R}^d)$. Thus

$$\mathcal{B}(\mathbb{R}^d) \subset \mathcal{M}(\mathbb{R}^d).$$

Let $\mathcal{N}(\mathbb{R}^d)$ be the collection of all null sets. By Theorem 2.3,

$$\mathcal{N}(\mathbb{R}^d) \subset \mathcal{M}(\mathbb{R}^d).$$

THEOREM 2.6. *$\mathcal{M}(\mathbb{R}^d)$ is the smallest σ-field which includes $\mathcal{B}(\mathbb{R}^d)$ and $\mathcal{N}(\mathbb{R}^d)$, that is, $\mathcal{M} = \sigma(\mathcal{B} \cup \mathcal{N})$.*

The proof of Theorem 2.6 is accomplished by the following observation.

A set in a topological space is said to be a *G_δ-set*[3] if it can be written as a countable intersection of open sets, and to be an *F_σ-set* if it can be written as a countable union of closed sets.

[3]The letters G and F are often used to denote open and closed sets, respectively. δ and σ stand for intersection and union, respectively.

G_δ-sets and F_σ-sets in $\mathbb{R}^d$ are Borel sets.

THEOREM 2.7. *Let $E \in \mathcal{M}(\mathbb{R}^d)$.*

(i) *There exist a G_δ-set G and a null set N such that $E = G - N$, and $m(E) = m(G)$.*

(ii) *There exist an F_σ-set F and a null set N' such that $E = F \cup N'$, and $m(E) = m(F)$.*

PROOF OF THEOREM 2.6. Assuming Theorem 2.7 temporarily, we shall prove Theorem 2.6.

Since any measurable set can be expressed as a union of a Borel set and a null set, it is contained in the σ-field generated by Borel sets and null sets. On the other hand, $\mathcal{M}(\mathbb{R}^d)$ is a σ-field which includes $\mathcal{B}(\mathbb{R}^d)$ and $\mathcal{N}(\mathbb{R}^d)$. □

LEMMA 2.4. *For any subset A of $\mathbb{R}^d$ we have*

$$m^*(A) = \inf\{m(O) \ : A \subset O,\ O \in \mathcal{O}(\mathbb{R}^d)\}.$$

PROOF. We assume that $m^*(A) < \infty$. Otherwise the proof is trivial. By the definition of outer measure, for $n = 1, 2, \dots$ there exists a covering $\{E_j^n : j = 1, 2, \dots\}$ in $\mathcal{R}(\mathbb{R}^d)$ such that

$$A \subset \bigcup_{j=1}^{\infty} E_j^n \quad \text{and} \quad m^*(A) + \frac{1}{n} > \sum_{j=1}^{\infty} m(E_j^n).$$

Since the sets E_j^n are finite unions of right half-open intervals, there exist open sets O_j^n such that

$$E_j^n \subset O_j^n \quad \text{and} \quad m^*(E_j^n) + \frac{1}{n2^j} > m(O_j^n).$$

If $O^n = \bigcup_{j=1}^{\infty} O_j^n$, then the O^n are also open and $A \subset O^n$. Thus

$$m(A) \le m(O^n) \le \sum_{j=1}^{\infty} m(O_j^n),$$

which does not exceed, by the above two inequalities,

$$\sum_{j=1}^{\infty} \left[m(E_j^n) + \frac{1}{n2^j} \right] < m^*(A) + \frac{1}{n} + \frac{1}{n}.$$

Since n is an arbitrary integer, Lemma 2.4 is proved. □

COROLLARY 2.1. *Let E be a measurable set of finite measure. Then for every $\varepsilon > 0$ there exist mutually disjoint intervals $I_1, I_2, \dots, I_N$ such that*

$$m(E - \bigcup_{j=1}^{N} I_j) < \varepsilon \quad \textit{and} \quad m(\bigcup_{j=1}^{N} I_j - E) < \varepsilon.$$

PROOF. By Lemma 2.4 we have an open set $O \supset E$ such that $m(O - E) = m(O) - m(E) < \varepsilon$. The set O can be written as a countable union of mutually disjoint intervals $\bigcup_{j=1}^{\infty} I_j$. Thus $m(O - \bigcup_{j=1}^{N} I_j) = \sum_{j=N+1}^{\infty} m(I_j) < \varepsilon$ for sufficiently large N. Thus

$$m(E - \bigcup_{j=1}^{N} I_j) \le m(O - \bigcup_{j=1}^{N} I_j) < \varepsilon.$$

The inequality

$$m(\bigcup_{j=1}^{N} I_j - E) \leq m(O - E) < \varepsilon$$

is a direct consequence of the choice of the set O. □

PROOF OF THEOREM 2.7. For a given measurable set E put $E_k = E \cap B(0,k)$ $(k = 1, 2, \dots)$. Each set E_k is measurable and of finite measure. Thus for every $n = 1, 2, \dots$ there exists an open set O_k^n such that $E_k \subset O_k^n$ and $m(O_k^n) \leq m(E_k) + 1/n2^k$. Thus $m(O_k^n - E_k) < 1/n2^k$. Put $O^n = \bigcup_{k=1}^{\infty} O_k^n$. Then the sets O^n are open, and $E = \bigcup E_k \subset O^n$. Thus $G = \bigcap_{n=1}^{\infty} O^n$ is a G_δ-set and $E \subset G$. Note that

$$G - E \subset O^n - E = \bigcup_{k=1}^{\infty} (O_k^n - E) \subset \bigcup_{k=1}^{\infty} (O_k^n - E_k),$$

from which it follows that

$$m(G - E) \leq \sum_{k=1}^{\infty} m(O_k^n - E_k) \leq \sum_{k=1}^{\infty} \frac{1}{n2^k} = \frac{1}{n}$$

for all n. Therefore $N = G - E$ is a null set. Since $E = G - N$, we get (i).

(ii) Apply (i) to the complement E^c. □

2.6. Inner measure

In the preceding section we proved that the outer measure of $A \in \mathcal{P}(\mathbb{R}^d)$ can be defined as follows:

$$m^*(A) = \inf\{m(O) \ : A \subset O \text{ and } O \text{ is open}\}.$$

From such a viewpoint the *inner measure* of the set A is defined by

$$m_*(A) = \sup\{m(C) \ : A \supset C \text{ and } C \text{ is compact}\}.$$

The next lemma gives a relation between outer and inner measures.

LEMMA 2.5. *Suppose $m^*(A) < \infty$. Let O be any open set which is of finite measure and includes A. Then we have*

$$m_*(A) = m(O) - m^*(O - A). \tag{2.11}$$

PROOF. Let O be a given open set of finite measure which includes A. Then for any compact set C such that $C \subset A$, we have $m(C) = m(O) - m(O - C)$. Therefore we have

$$\begin{aligned} m_*(A) &= \sup\{m(C) : C \subset A \text{ and } C \text{ is compact}\} \\ &= m(O) - \inf\{m(O - C) : C \subset A \text{ and } C \text{ is compact}\}. \end{aligned} \tag{2.12}$$

Now, in the last equation we can replace the compact set C by any closed set F contained in A. In fact, for any such closed set F we have, by Theorem 2.5, (iii), $\lim_{r\to\infty} m(O - [F \cap \overline{B(O,r)}]) = m(\lim_{r\to\infty} O - [F \cap \overline{B(O,r)}]) = m(O - F)$, and $F \cap \overline{B(O,r)}$ is a compact set contained in A. Thus from (2.12) it follows that

$$m_*(A) = m(O) - \inf\{m(O - F) : F \supset A \text{ and } F \text{ is closed}\}.$$

Putting $O - F = O'$, we obtain

$$\begin{aligned} m_*(A) &= m(O) - \inf\{m(O') : O - A \subset O' \text{ and } O' \text{ is open}\} \\ &= m(O) - m^*(O - A). \end{aligned}$$

□

LEMMA 2.6. *If A and B are disjoint subsets of $\mathbb{R}^d$, then we have*

$$m_*(A \cup B) \leq m_*(A) + m^*(B) \leq m^*(A \cup B).$$

PROOF. Let O be any open subset including the set B. Then, by the definition of inner measure we have

$$\begin{aligned} m_*(A \cup B) &= \sup\{m(C) \ : C \subset A \cup B \text{ and } C \text{ is compact}\} \\ &\leq \sup\{m(C - O) + m(O) \ : C \subset A \cup B \text{ and } C \text{ is compact}\}. \end{aligned}$$

Since $C - O$ is a compact set which is contained in A, we have $m(C - O) \leq m_*(A)$. Thus

$$m_*(A \cup B) \leq m_*(A) + m(O),$$

which holds for any open set O including B. Therefore the first inequality of the lemma is proved.

To prove the second inequality it suffices to consider the case $m^*(A \cup B) < \infty$. Let O be any open set which includes $A \cup B$. Then, for any compact subset C of A we have $m(O) = m(O - C) + m(C)$. Since $O - C$ is an open set which contains B, we have $m(O - C) \geq m^*(B)$. Therefore we get

$$m(O) \geq m^*(B) + m(C).$$

Taking the supremum on compact sets C, and then the infimum on open sets O, we get the second inequality. □

THEOREM 2.8. *Let E be a subset of $\mathbb{R}^d$ with finite outer measure. Then E is measurable if and only if*

$$m^*(E) = m_*(E). \tag{2.13}$$

PROOF. Fix an open set O of finite measure which includes E.

If E is measurable, by the Carathéodory condition we have

$$m^*(O) = m(O) = m^*(O \cap E) + m^*(O - E). \tag{2.14}$$

Since $m^*(O \cap E) = m^*(E)$ and since $m^*(O - E) = m(O) - m_*(E)$ by Lemma 2.5, the right-hand side of (2.14) equals $m^*(E) + m(O) - m_*(E)$. Thus $m^*(E) = m_*(E)$.

Conversely, suppose that E satisfies (2.13). Let A be any given set. Temporarily assume that

$$m_*(O \cap E) \geq m^*(O \cap E) \tag{2.15}$$

for all open sets O containing A. Then it follows from Lemma 2.6 and (2.15) that

$$\begin{aligned} m(O) = m^*(O) &\geq m_*(O \cap E) + m^*(O - E) \\ &\geq m^*(O \cap E) + m^*(O - E). \end{aligned}$$

Since the last term is not less that $m^*(A \cap E) + m^*(A - E)$, we have $m^*(A) \geq m^*(A \cap E) + m^*(A - E)$ for all $A \in \mathcal{P}(\mathbb{R}^d)$, which is nothing but the Carathéodory condition for E. Thus E is measurable.

To prove (2.15), notice that open sets are measurable and thus satisfy the Carathéodory condition

$$m^*(E) = m^*(O \cap E) + m^*(E - O).$$

On the other hand, by Lemma 2.6

$$m_*(E) \leq m_*(O \cap E) + m^*(E - O).$$

Since $m^*(E) = m_*(E)$ by our assumption, (2.15) follows from the last two inequalities. □

2.7. The existence of nonmeasurable sets

In this section we discuss the fundamental problem of measurability of sets: "Is every subset of $\mathbb{R}^d$ measurable?" The answer is negative. In fact, we shall prove that there exists a nonmeasurable set in $\mathbb{R}$. Zorn's lemma is essentially applied in the proof.

A subset B of $\mathbb{R}$ is called a *Hamel basis* if it satisfies the following:

(i) B is linearly independent over the rational number field $\mathbb{Q}$,
(ii) $\text{span}_{\mathbb{Q}} B = \mathbb{R}$.

(i) means by definition that for every finite subset $\{x_1, \dots, x_N\}$ of distinct elements of B and for every subset $\{r_1, \dots, r_N\}$ of $\mathbb{Q}$, the equality $r_1x_1 + \dots + r_Nx_N = 0$ implies $r_1 = \dots = r_N = 0$.

(ii) implies that for every $x \in \mathbb{R}$ there exist a finite subset $\{x_1, \dots, x_N\}$ of B and a subset $\{r_1, \dots, r_N\}$ of $\mathbb{Q}$ such that

$$x = r_1x_1 + \dots + r_Nx_N. \tag{2.16}$$

LEMMA 2.7. *A Hamel basis exists.*

PROOF. Let $\mathcal{I}$ be the collection of all subsets of $\mathbb{R}$ linearly independent over $\mathbb{Q}$. For example, the single-point set $\{\sqrt{2}\}$ belongs to $\mathcal{I}$. The set $\mathcal{I}$ is a partially ordered set with respect to the inclusion relation " $\subset$ ".

If $\{I_\alpha : \alpha \in A\}$ is a linearly ordered subset of $\mathcal{I}$, then it has an upper bound. In fact, $\bigcup_{\alpha \in A} I_\alpha$ includes every I_α and belongs to $\mathcal{I}$. For, let $x_{\beta_1}, x_{\beta_2}, \dots, x_{\beta_N}$ be any given finite elements of $\bigcup_{\alpha \in A} I_\alpha$. Then $x_{\beta_i} \in I_{\alpha_i}$ for some $\alpha_i \in A\,(i = 1, 2, \dots, N)$. Let I_{α_j} be the largest element of $I_{\alpha_1}, I_{\alpha_2}, \dots, I_{\alpha_N}$. Since $\{x_{\beta_1}, x_{\beta_2}, \dots, x_{\beta_N}\} \subset I_{\alpha_j}$, $\{x_{\beta_1}, x_{\beta_2}, \dots, x_{\beta_N}\}$ is linearly independent. Thus $\bigcup_{\alpha \in A} I_\alpha$ is linearly independent over $\mathbb{Q}$. Therefore any linearly ordered subset of $\mathcal{I}$ has an upper bound.

By Zorn's lemma (see Appendix A.1) there exists a maximal element in $\mathcal{I}$.

We shall show that any maximal element B of $\mathcal{I}$ is a Hamel basis. Condition (i) is obviously satisfied. If B does not satisfy (ii), then there exists $x \in \mathbb{R}$ such that equation (2.16) does not hold for any $x_1, \dots, x_N \in B$ and $r_1, \dots, r_N \in \mathbb{Q}$. Thus $B \cup \{x\}$ is a linearly independent set over $\mathbb{Q}$ and properly contains B, which contradicts the maximality of B. Therefore B satisfies all conditions of a Hamel basis. □

THEOREM 2.9. *There exists a nonmeasurable set in* $\mathbb{R}$.

PROOF. Let B be a Hamel basis and fix an element $a \in B$. Put

$$A = \text{span}_{\mathbb{Q}}(B - \{a\}),$$

that is, A consists of the real numbers whose expressions as the linear combinations of elements of B do not contain a. Indeed, A is a nonmeasurable set, as we shall prove by contradiction. Assume that A is a measurable set.

First Step. *We have* $m(A) > 0$.

For $r \in \mathbb{Q}$ let A_r be the translation of A by ra, that is, $ra+A = \{ra+x : x \in A\}$. Then A_r is measurable, since we assumed that A is measurable, and $m(A_r) = m(A)$.

By the definition of A we have $\mathbb{R} = \bigcup_{r\in\mathbb{Q}} A_r$. Thus $\sum_{r\in\mathbb{Q}} m(A_r) \geq m(\mathbb{R}) = \infty$, which implies that $m(A) \neq 0$.

Second Step. *There exist distinct rational numbers* r *and* s *such that* $A_r \cap A_s \neq \emptyset$.

To prove the second step, put

$$A_{r,n} = ra + A \cap [n-1, n), \quad n \in \mathbb{Z}.$$

Then we have $m(A_{r,n}) = m(A \cap [n-1, n))$.

Suppose that the sets $A_r (r \in \mathbb{Q})$ are mutually disjoint. Then we have

$$m(\bigcup\{A_{r,n} : 0 < r < 1, r \in \mathbb{Q}\}) = \sum_{\{0<r<1:r\in\mathbb{Q}\}} m(A_{r,n}).$$

Since $\bigcup_{\{0<r<1:r\in\mathbb{Q}\}} A_{r,n}$ is contained in the finite interval $[n-1-|a|, n+|a|]$, it follows that $m(A_{r,n}) = 0$ for all $0 < r < 1$, $r \in \mathbb{Q}$, and $n \in \mathbb{Z}$. Thus $m(A_r) = m(A \cap [n-1.n)) = 0$ for all n. Therefore $m(A) = 0$, which contradicts the first step.

Third Step. Suppose that $A_r \cap A_s \neq \emptyset$ for some distinct rational numbers r and s. Then there exist $x, y \in A$ such that $x + ra = y + sr$. Thus $x - y = (s-r)a$. Since linear combinations of elements of A belong to A, it follows that $x - y \in A$, and $(s-r)a \in A$, which contradicts that the linear combinations of elements in A do not contain a.

The contradiction is caused by the assumption that the set A is measurable. □

REMARK 2.3. Both sets A and A^c are dense in $\mathbb{R}$.

Indeed, since the set $\mathbb{Q}a = \{ra : r \in \mathbb{Q}\}$ is dense in $\mathbb{R}$ and $\mathbb{Q}a \subset A^c$, the set A^c is dense in $\mathbb{R}$. On the other hand, if b is an element of B which is different from a, then $\mathbb{Q}b \subset A$.

To conclude this section we construct a particular function as an application of the Hamel basis. This function is an example of nonmeasurable functions, whose definition is given later.

Fix again an element $a \in B$. For a real number x such that $x = r_a a + r_1 x_1 + \cdots + r_N x_N$, with $r_a, r_1, \ldots, r_N \in \mathbb{Q}$ and $x_1, \ldots, x_N \in B$, define

$$f(x) = r_a.$$

Clearly we have

$$f(x+y) = f(x) + f(y) \quad (x, y \in \mathbb{R}). \tag{2.17}$$

The function f is everywhere discontinuous, and the range of f is $\mathbb{Q}$. In fact, we have $f(a) = 1$ and $f(x) = 0\,(x \in A)$. Since A is dense in $\mathbb{R}$, f is not continuous at a.

To prove that f is discontinuous at a given point x_0, observe that

$$f(x_0 + x) = f(a + x) + f(x_0 - a).$$

Since $f(a+x)$ is discontinuous at 0, f is discontinuous at x_0.

If f satisfies equation (2.17) and is continuous at a point, then f is continuous everywhere, and furthermore

$$f(x) = f(1)x$$

(see Problem 2.7).

2.8. Problems

2.1. Let $\{E_j\}$ be a sequence of sets. Prove that

$$\limsup_{j\to\infty} \chi_{E_j}(x) = \chi_{\limsup E_j}(x) \quad \text{and} \quad \liminf_{j\to\infty} \chi_{E_j}(x) = \chi_{\liminf E_j}(x).$$

2.2. Let $\{E_j\}$ be a sequence of measurable sets such that $\sum_{j=0}^{\infty} m(E_j) < \infty$. Prove that $\limsup_{j\to\infty} E_j$ is a null set.

2.3. Let $E \in \mathcal{M}(\mathbb{R}^d)$. For $a > 0$ and $y \in \mathbb{R}^d$ put

$$aE = \{ax : x \in E\} \quad \text{and} \quad E + y = \{x + y : x \in E\}.$$

Prove that

(i) $m(aE) = a^d m(E)$

(ii) $m(E + y) = m(E)$.

2.4. Let E and F be measurable subsets of $[0,1]$. If $m(F) = 1$, then

$$m(E \cap F) = m(E).$$

2.5. Prove that if E is a measurable set, then

$$m(E) = \sup\{m(C) : C \subset E \text{ and } C \text{ is compact}\}.$$

2.6. Let N_1 and N_2 be subsets in $\mathbb{R}$. The set $N_1 + N_2 = \{x+y : x \in N_1 \text{ and } y \in N_2\}$ is not always null, even if N_1 and N_2 are null. (Hint: We know that the Cantor ternary set C is null. Prove that $C + C = [0,2]$, whose measure is 2.)

2.7. Let f be a function in $\mathbb{R}$ which satisfies

$$f(x) + f(y) = f(x+y) \quad (x, y \in \mathbb{R})$$

and which is continuous at a point. Then prove that f has the expression

$$f(x) = f(1)x.$$

CHAPTER 3

The Lebesgue Integral on Euclidean Spaces

Let us consider the trigonometric series

$$\frac{a_0}{2} + \sum_{n=1}^{\infty}(a_n \cos 2\pi nx + b_n \sin 2\pi nx).$$

The series does not always converge, and even if it converges everywhere, the limit function is not always continuous. Furthermore, it is known that for a given null set there exists a trigonometric series which represents a continuous function in a sense but diverges on the null set.

In other words, even if we deal basically with continuous functions, it is inevitable, in order to manage countably infinite arithmetic operations, to introduce new families of sets, new spaces of functions, and a concept of integration which fits the countably infinite calculations.

The Lebesgue integration theory is the one which can be applied successfully to the problems which we meet in applying countable arithmetic operations. The convergence theorems in §§3 and 6 show well the characteristics of this theory.

The argument in this chapter makes it possible to extend the concept of integration on Euclidean spaces to more general spaces, which will be described in the following chapters.

3.1. Measurable functions

DEFINITION 3.1. Let f be an extended real valued function in $\mathbb{R}^d$. f is said to be *measurable* if the set

$$\{x : f(x) > a\}$$

is measurable for all $a \in \mathbb{R}$.

A complex valued function f is said to be measurable if both its real and imaginary parts are measurable.

We shall restrict the object of the integration to measurable functions and show that certain spaces of measurable functions, called the Lebesgue spaces, are closed under countably infinite arithmetic operations.

EXAMPLE 3.1. If f is a real valued continuous function, then the set $\{x : f(x) > a\}$ is measurable because it is open. Thus any continuous function is measurable.

EXAMPLE 3.2. Let $E_j, j = 1, 2, \ldots, k$, be mutually disjoint sets and a_j be distinct real numbers. If the function

$$f(x) = \sum_{j=1}^{k} a_j \chi_{E_j}(x)$$

is measurable, the sets E_j are all measurable. Conversely, if every E_j is measurable, then the set $\{x : f(x) > a\} = \bigcup\{E_j : a_j > a\}$ is measurable. Thus the function f is measurable.

The assertion is clearly valid if the real numbers a_j are replaced by complex numbers.

DEFINITION 3.2. Let f be an extended real valued function defined in a topological space X. f is said to be *lower semicontinuous*, if

(a) $f(x) \neq -\infty$ in X,
(b) $\{x \in X : f(x) > a\}$ is open for all $a \in \mathbb{R}$.

By the same way, f is said to be *upper semicontinuous* if $f(x) \neq \infty$ in X and $\{x \in X : f(x) < a\}$ is open for all $a \in \mathbb{R}$.

The lower semicontinuous funcions in $\mathbb{R}^d$ are measurable, which is an immediate consequence of the definition, and the upper semicontinuous functions are also measurable by Lemma 3.1 (iii) below.

In the rest of this chapter, the functions are always assumed to be defined on $\mathbb{R}^d$ and have values in the extended real numbers, unless otherwise specified.

LEMMA 3.1. *For a function f the following conditions are equivalent.*

(i) *f is measurable, that is,* $\{x : f(x) > a\} \in \mathcal{M}(\mathbb{R}^d) \quad (a \in \mathbb{R})$.
(ii) $\{x : f(x) \geq a\} \in \mathcal{M}(\mathbb{R}^d) \quad (a \in \mathbb{R})$.
(iii) $\{x : f(x) < a\} \in \mathcal{M}(\mathbb{R}^d) \quad (a \in \mathbb{R})$.
(iv) $\{x : f(x) \leq a\} \in \mathcal{M}(\mathbb{R}^d) \quad (a \in \mathbb{R})$.

PROOF. (i) $\Rightarrow$ (ii). Observe that $\{x : f(x) \geq a\} = \bigcap_{n=1}^{\infty} \{x : f(x) > a - \frac{1}{n}\}$. The right-hand side belongs to $\mathcal{M}(\mathbb{R}^d)$.

In the same way the rest follows from the following equations.

(ii) $\Rightarrow$ (iii). $\{x : f(x) < a\} = \{x : f(x) \geq a\}^c$.
(iii) $\Rightarrow$ (iv). $\{x : f(x) \leq a\} = \bigcap_{n=1}^{\infty} \{x : f(x) < a + \frac{1}{n}\}$.
(iv) $\Rightarrow$ (i). $\{x : f(x) > a\} = \{x : f(x) \leq a\}^c$. □

If f is measurable, $\{x : f(x) = a\}$ is measurable, which is a consequence of Lemma 3.1 (i) and (iv).

THEOREM 3.1. *A function f is measurable, if and only if*

$$f^{-1}(O) \in \mathcal{M}(\mathbb{R}^d) \quad \text{for all } O \in \mathcal{O}(\mathbb{R}). \tag{3.1}$$

PROOF. If O is an infinite interval (a, ∞), then (3.1) is nothing but the definition of measurability of f.

Suppose that f is measurable. If $O = (a, b)$, then by Lemma 3.1 (iii) we have

$$f^{-1}((a,b)) = \{x : f(x) < b\} \cap \{x : f(x) > a\} \in \mathcal{M}(\mathbb{R}^d).$$

In general, any open set O is written as a countable union of mutually disjoint open intervals $\bigcup_{j=1}^{\infty}(a_j, b_j)$. Therefore

$$f^{-1}(O) = f^{-1}\Big(\bigcup_{j=1}^{\infty}(a_j, b_j)\Big) = \bigcup_{j=1}^{\infty} f^{-1}((a_j, b_j)),$$

and the last union belongs to $\mathcal{M}(\mathbb{R}^d)$. □

DEFINITION 3.3. A property $\mathbf{P}(x)$ on a set E is said to hold *almost everywhere* in E, or, in abbreviated form, $\mathbf{P}(x)$ a.e.[1] in E, if it holds for all x in E except for those x in some null subset of E.

LEMMA 3.2. *If a function f is measurable and if $f = g$ a.e., then g is also measurable.*

PROOF. By the assumption the set $N = \{x : f(x) \neq g(x)\}$ is null, and we have

$$\{x : g(x) > a\} = [\{x : f(x) > a\} - N] \cup [\{x : g(x) > a\} \cap N].$$

Since the null sets are measurable, each term on the right-hand side is measurable. Thus g is measurable. □

THEOREM 3.2. *Let $\varphi(s,t)$ be a continuous function in $\mathbb{R}^2$. If functions f and g are measurable and almost everywhere finite, then $\varphi(f(x), g(x))$ is measurable.*

PROOF. By the assumption the set $\{x : f(x) = \pm\infty\} \cup \{x : g(x) = \pm\infty\}$ is null. Thus we can assume that f and g are finite everywhere by Lemma 3.2. Since $\{(s,t) : \varphi(s,t) > a\}$ is an open set in $\mathbb{R}^2$, it is a countable union of mutually disjoint right half-open intervals $[a_j, b_j) \times [c_j, d_j)$. Thus,

$$\begin{aligned}\{x \in \mathbb{R}^d : \varphi(f(x), g(x)) > a\} &= \bigcup_{j=1}^{\infty} \{x : (f(x), g(x)) \in [a_j, b_j) \times [c_j, d_j)\} \\ &= \bigcup_{j=1}^{\infty} \{x : a_j \leq f(x) < b_j\} \cap \{x : c_j \leq g(x) < d_j\}.\end{aligned}$$

The last sum is measurable since each summand is measurable. □

COROLLARY 3.1. (i) *The set of all complex valued measurable functions is a vector space; that is, if f and g are measurable functions and $a, b \in \mathbb{C}$, then $af + bg$ is measurable.*

(ii) *If f and g are measurable, $f \cdot g$ is measurable. Furthermore, if $g \neq 0$ a.e., then f/g is measurable.*

(iii) *If f is measurable and $p > 0$, then $|f|^p$ is measurable.*

(iv) *If f is a real valued measurable function, then the positive part $f^+ = \max(f, 0)$ is measurable.*

PROOF. We assume always that the functions are real valued, since the complex valued function case is reduced to the real valued function case by decomposing the functions into their real and imaginary parts (see Problem 3.1).

(i) Apply Theorem 3.2 to the measurable functions f and g with $\varphi(s,t) = as + bt$. The other properties of vector spaces are immediate.

(ii) The first part follows again from Theorem 3.2 with $\varphi(s,t) = st$. To prove the second part, observe that

$$\begin{aligned}\{x : f(x)/g(x) > a\} \cap \{x : g(x) \neq 0\} &= [\{x : f(x) - ag(x) > 0\} \cap \{x : g(x) > 0\}] \\ &\quad \cup [\{x : f(x) - ag(x) < 0\} \cap \{x : g(x) < 0\}].\end{aligned}$$

Since the right-hand side is measurable and $\{x : g(x) = 0\}$ is null, the set $\{x : f(x)/g(x) > a\}$ is measurable.

[1] A.e. is used also for "almost every". In probability theory the phrase "almost surely" is usually used for almost everywhere and is abbreviated as a.s.

(iii) follows from Theorem 3.2 with $\varphi(s,t) = |s|^p$. In particular, $|f|$ is measurable. Thus by (i) $f^+ = (|f| + f)/2$ is measurable. □

THEOREM 3.3. *If f_j ($j = 1, 2, \dots$) are extended real valued measurable functions, then the following limits are all measurable:*

$$\sup_j f_j(x), \quad \inf_j f_j(x), \quad \limsup_{j\to\infty} f_j(x), \quad \textit{and} \quad \liminf_{j\to\infty} f_j(x).$$

In particular, if $\lim f_j$ exists almost everywhere, then $\lim f_j$ is measurable, and if the sum $u(x) = \sum_{j=1}^{\infty} u_j(x)$ of measurable functions u_j converges almost everywhere, then $u(x)$ is measurable.

PROOF. Observe that

$$\{x : \sup_j f_j(x) > a\} = \bigcup_{j=1}^{\infty} \{x : f_j(x) > a\},$$

and

$$\{x : \inf_j f_j(x) \geq a\} = \bigcap_{j=1}^{\infty} \{x : f_j(x) \geq a\}.$$

Since the sets on the right-hand side are measurable, $\sup_j f_j$ and $\inf_j f_j$ are measurable.

Remark that $\limsup_{j\to\infty} f_j(x) = \inf_{k\geq 1} \sup_{j\geq k} f_j(x)$. The last term is measurable by the above argument. The measurability of $\liminf_{j\to\infty} f_j(x)$ follows in the same way. □

3.2. Simple functions and measurable functions

DEFINITION 3.4. An extended real valued function is said to be *simple* if it has finitely many values. More generaly, a function f is said to be simple if $f = g + ih$, where g and h are extended real valued simple functions.

Any extended real valued simple function is expressed as $f(x) = \sum_{j=1}^{k} a_j \chi_{E_j}(x)$, where the sets E_j are mutually disjoint and a_j are distinct extended real numbers. Furthermore, f is measurable if and only if every set E_j is measurable.

THEOREM 3.4. *Let f be a nonnegative measurable function. Then there exists a sequence of measurable simple functions $\{s_j\}$ such that*

(i) $0 \leq s_1(x) \leq s_2(x) \leq \cdots \to f(x)$ *for every* x,
(ii) *each simple function s_j is bounded and $m(\{x : s_j(x) \neq 0\}) < \infty$.*

PROOF. For $j = 1, 2, \dots$ define $s_j(x)$ as follows:

$$s_j(x) = \begin{cases} (k-1)2^{-j} & (|x| \leq j \text{ and } (k-1)2^{-j} \leq f(x) < k \cdot 2^{-j}), \\ & \qquad k = 1, 2, \dots, j \cdot 2^j, \\ j & (|x| \leq j \text{ and } f(x) \geq j), \\ 0 & (|x| > j). \end{cases}$$

(ii) holds obviously. For every point x and $j = 1, 2, \dots$, we have

$$0 \leq s_j(x) \leq s_{j+1}(x) \leq f(x).$$

Furthermore, if $|x| \leq j$ and $f(x) < j$, then

$$0 \leq f(x) - s_j(x) \leq 2^{-j}.$$

Thus $s_j(x) \to f(x)$ when $f(x) < \infty$. In the case $f(x) = \infty$, we have $s_j(x) = j \to \infty$. Thus (i) is proved. □

3.3. Definition of the Lebesgue integral

In the following, every subset of $\mathbb{R}^d$ and every function in $\mathbb{R}^d$ are assumed to be measurable unless otherwise mentioned.

Let $s(x)$ be a nonnegative simple function. Let $s(x) = \sum_{j=1}^{k} a_j \chi_{E_j}(x)$, where the E_j are mutually disjoint and the a_j are distinct positive numbers or ∞. Then the integral of s is defined by

$$\int_{\mathbb{R}^d} s(x)dm(x) = \sum_{j=1}^{k} a_j m(E_j). \tag{3.2}$$

The value $\int_{\mathbb{R}^d} s(x)dm(x)$ might be ∞.

For a nonnegative function $f(x)$ choose a sequence of simple functions $\{s_p\}$ such that

$$0 \le s_1(x) \le s_2(x) \le \cdots \to f(x) \tag{3.3}$$

for every x, and define

$$\int_{\mathbb{R}^d} f(x)dm(x) = \lim_{p\to\infty} \int_{\mathbb{R}^d} s_p(x)dm(x). \tag{3.4}$$

There exists at least one sequence of simple functions $\{s_p\}$ satisfying (3.3) by Theorem 3.4, and the limit of (3.4) exists (it may be infinite).

Sometimes the variable x and the domain of the integral $\mathbb{R}^d$ are omitted. The notations dm or dx are used for $dm(x)$.

If $f = u + iv$, where u and v are extended real valued functions, we decompose u and v into nonnegative functions as follows:

$$u = u_+ - u_-, \quad v = v_+ - v_-,$$

where $u_+(x) = \max[u(x), 0]$ and $u_-(x) = u_+(x) - u(x)$. $v_+(x)$ and $v_-(x)$ are defined similarly.

If the four integrals

$$\int u_+ dm, \quad \int u_- dm, \quad \int v_+ dm, \quad \text{and} \quad \int v_- dm$$

are finite, f is said to be *integrable in the sense of Lebesgue, Lebesgue integrable* or, simply, *integrable*, and the *Lebesgue integral* is defined by

$$\int f dm = \int u_+ dm - \int u_- dm + i\Big(\int v_+ dm - \int v_- dm\Big).$$

Let $E \in \mathcal{M}(\mathbb{R}^d)$. The integral of f on the set E is defined by

$$\int_E f dm = \int_{\mathbb{R}^d} f \cdot \chi_E dm.$$

In order to ascertain that the above definition is legitimate, we need to show the following:

(I) The right-hand side of (3.2) does not depend on the expression of the simple function $s(x)$.

(II) The right-hand side of (3.4) does not depend on the choice of the sequence $\{s_p\}$ of simple functions.

PROOF OF (I). Let $s(x) = \sum_{j=1}^k a_j\chi_{E_j}(x) = \sum_{p=1}^q b_p\chi_{F_p}(x)$ be two expressions of the nonnegative simple function $s(x)$, where $a_j, b_p > 0$. Observe that $\bigcup E_j = \bigcup F_p$. Let Δ be the collection of the sets of the form

$$A_1 \cap \cdots \cap A_k \cap B_1 \cap \cdots \cap B_q,$$

where $A_j = E_j$ or E_j^c, and $B_p = F_p$ or F_p^c.

First we show that the sets in Δ are mutually disjoint and that $E_j = \bigcup\{D \in \Delta : D \subset E_j\}$ for any E_j.

Indeed, let $D, D' \in \Delta$. If $D = A_1 \cap \cdots \cap A_k \cap B_1 \cap \cdots \cap B_q$ and $D' = A_1' \cap \cdots \cap A_k' \cap B_1' \cap \cdots \cap B_q'$, and if $D \neq D'$, then $A_j \neq A_j'$ for some j or $B_p \neq B_p'$ for some p. Thus $A_j \cap A_j' = \emptyset$ or $B_p \cap B_p' = \emptyset$. Therefore $D \cap D' = \emptyset$. Thus the sets in Δ are mutually disjoint. Next, suppose $x \in E_j$. Put $A_l = E_l$ if $x \in E_l$ and $A_l = E_l^c$ if $x \notin E_l$ $(1 \leq l \leq k)$, and $B_p = F_p$ if $x \in F_p$ and $B_p = F_p^c$ if $x \notin F_p$ $(1 \leq p \leq q)$. Then $x \in D = A_1 \cap \cdots \cap A_k \cap B_1 \cap \cdots \cap B_q$. Thus $E_j \subset \bigcup\{D \in \Delta : D \subset E_j\}$. Since the reverse inclusion is obvious, we get $E_j = \bigcup\{D \in \Delta : D \subset E_j\}$. Therefore we have

$$m(E_j) = \sum_{\{D \in \Delta : D \subset E_j\}} m(D).$$

Now, interchanging the sums, we get

$$\sum_{j=1}^k a_j m(E_j) = \sum_{j=1}^k a_j \Bigg(\sum_{\{D \in \Delta : D \subset E_j\}} m(D) \Bigg) = \sum_{D \in \Delta} \Bigg(\sum_{\{j : D \subset E_j\}} a_j \Bigg) m(D).$$

If $x \in D$, then $s(x) = \sum_{\{j:D\subset E_j\}} a_j$ and also $s(x) = \sum_{\{p:D\subset F_p\}} b_p$. Thus the right-hand side equals

$$\sum_{D \in \Delta} \Bigg(\sum_{\{p : D \subset F_p\}} b_p \Bigg) m(D).$$

In the same way as above this is equal to

$$\sum_{p=1}^q b_p \Bigg(\sum_{\{D \in \Delta : D \subset F_p\}} m(D) \Bigg) = \sum_{p=1}^q b_p m(F_p).$$

Therefore we get (I). □

In order to prove (II) we need two lemmas.

LEMMA 3.3. *Let $s(x)$ and $t(x)$ be nonnegative simple functions. Then we have*

$$\text{(i)} \qquad \int (s+t)dm = \int sdm + \int tdm.$$

Furthermore if $s(x) \leq t(x)$, we have

$$\text{(ii)} \qquad \int sdm \leq \int tdm.$$

PROOF. (i) Let $s(x) = \sum_i a_i\chi_{E_i}(x)$ and $t(x) = \sum_j b_j\chi_{F_j}(x)$, where $a_i, b_j \geq 0$, and $\{E_i\}$ and $\{F_j\}$ are families of mutually disjoint sets, respectively, such that $\bigcup E_i = \bigcup F_j = \mathbb{R}^d$. Then $E_i \cap F_j$ are mutually disjoint, and $E_i = \bigcup_j E_i \cap F_j$ and

$F_j = \bigcup_i E_i \cap F_j$. Thus $s(x)+t(x) = \sum_{i,j}(a_i+b_j)\chi_{E_i\cap F_j}(x)$. Thus by the definition of integral we have

$$\int (s+t)dm = \sum_{i,j}(a_i+b_j)m(E_i \cap F_j)$$
$$= \sum_{i,j} a_i m(E_i \cap F_j) + \sum_{i,j} b_j m(E_i \cap F_j).$$

Since $\sum_j m(E_i \cap F_j) = m(E_i)$ and $\sum_i m(E_i \cap F_j) = m(F_j)$, the right-hand side equals

$$\sum_i a_i m(E_i) + \sum_j b_j m(F_j) = \int sdm + \int tdm.$$

(ii) Put $t(x) = s(x)+u(x)$. Then $u(x) = t(x)-s(x)$ is nonnegative and simple. Therefore by (i) we have

$$\int tdm = \int sdm + \int udm \geq \int sdm.$$

□

LEMMA 3.4. *Let $\{s_p\}$ and $\{t_q\}$ be nondecreasing sequences of nonnegative simple functions. If*

$$\lim_{q\to\infty} t_q(x) \leq \lim_{p\to\infty} s_p(x)$$

for $x \in \mathbb{R}^d$, then we have

$$\lim_{q\to\infty} \int t_q dm \leq \lim_{p\to\infty} \int s_p dm.$$

PROOF. Fix a positive integer q and denote $t_q(x) = \sum_{j=1}^k b_j \chi_{F_j}(x)$, where $b_j > 0$ and $\{F_j\}$ is a disjoint sequence.

First assume that $\int t_q dm < \infty$. Then $m(\bigcup F_j) < \infty$. Fix an $\varepsilon > 0$. Let $F = \bigcup F_j$, and put

$$E_p = \{x \in F : s_p(x) + \varepsilon > t_q(x)\}$$

for $p = 1, 2, \ldots$.

Since $t_q = t_q\chi_{F-E_p} + t_q\chi_{E_p}$, we have, by Lemma 3.3,

$$\int t_q dm = \int t_q \chi_{F-E_p} dm + \int t_q \chi_{E_p} dm.$$

By the definition of the set E_p the right-hand side is bounded by

$$\sup_x t_q(x) \cdot m(F - E_p) + \int [s_p(x) + \varepsilon]\chi_{E_p}(x)dm$$
$$\leq \sup_x t_q(x) m(F - E_p) + \int s_p dm + \varepsilon \cdot m(E_p).$$

Since $\{E_p\}$ is a nondecreasing sequence which converges to F as $p \to \infty$, we have $m(F - E_p) \to 0\ (p \to \infty)$ by Theorem 2.5,(iii). Therefore we get

$$\int t_q dm \leq \varepsilon m(F) + \lim_{p\to\infty} \int s_p dm.$$

Since $\varepsilon > 0$ is arbitrary, let $\varepsilon \to 0$ and then $q \to \infty$. Then we obtain the desired inequality.

If $\int t_q dm = \infty$ for some q, then for any large number N we can choose $a > 0$ and a set F so that

$$a\chi_F(x) < t_q(x) \ (x \in \mathbb{R}^d) \quad \text{and} \quad N < \int a\chi_F dm < \infty. \tag{3.5}$$

If we assume (3.5) temporarily, by the last argument

$$N < \int a\chi_F dm \leq \lim_{p\to\infty} \int s_p dm.$$

Since N is arbitrary, $\lim_{p\to\infty} \int s_p dm = \infty$. Therefore we get the lemma.

Proof of (3.5). Let $t_q = \sum b_j \chi_{F_j}$, where the F_j are mutually disjoint and $b_j > 0$. By our assumption $b_j m(F_j) = \infty$ for some j. Put $F_j^n = \{|x| < n\} \cap F_j$. Then, since $F_j^n \nearrow F_j$ as $n \to \infty$, we have $m(F_j^n) \to m(F_j)$ by Theorem 2.5. Thus if $a = \min\{b_j, n\}$ and $F = F_j^n$, then $am(F) > N$ for a large n. □

PROOF OF (II). We shall show that the right-hand side of (3.4) does not depend on the choice of s_j. Let $\{s_p\}$ and $\{t_q\}$ be nondecreasing sequences of nonnegative simple functions such that $\lim s_p = \lim t_q = f$. By Lemma 3.4,

$$\lim_{p\to\infty} \int s_p dm = \lim_{q\to\infty} \int t_q dm.$$

□

Therefore the integral (3.4) is well-defined.

3.4. Fundamental properties of the integral

For a subset E of $\mathbb{R}^d$ let $\mathcal{L}(E, dm)$ be the set of all Lebesgue integrable functions on E. The notation $\mathcal{L}(E, dm)$ is often abbreviated as $\mathcal{L}(E)$ or simply $\mathcal{L}$.

LEMMA 3.5. *Let f and g be measurable functions such that $0 \leq f \leq g$. If $g \in \mathcal{L}$, then $f \in \mathcal{L}$ and*

$$0 \leq \int f dm \leq \int g dm.$$

PROOF. The lemma follows immediately from Lemma 3.4. □

THEOREM 3.5. *Let f be a measurable function. Then $f \in \mathcal{L}$ if and only if $|f| \in \mathcal{L}$, that is,*

$$\int |f| dm < \infty.$$

In this case

$$\left| \int f dm \right| \leq \int |f| dm. \tag{3.6}$$

PROOF. Let u and v be the real and imaginary parts of f, respectively, and let $u_+ = \max(u, 0)$ and $u_- = \max(-u, 0)$. Then we have $f = u + iv = (u_+ - u_-) + i(v_+ - v_-)$ and $0 \leq u_\pm, v_\pm \leq |f|$. If $|f| \in \mathcal{L}$, then $u_\pm, v_\pm \in \mathcal{L}$ by Lemma 3.5. Thus $f \in \mathcal{L}$ by the definition of integrability.

Conversely, suppose that f is integrable. Since $0 \leq |f| \leq u_+ + u_- + v_+ + v_- \in \mathcal{L}$, we again apply Lemma 3.5 to get $|f| \in \mathcal{L}$.

To prove (3.6) let $\{s_p^{\pm}\}$ and $\{t_p^{\pm}\}$ be the nondecreasing sequences of nonnegative simple functions which are used to define the integrals of $u_{\pm}$ and $v_{\pm}$, respectively. For each p we can express $s_p^{\pm}$ and $t_p^{\pm}$ as follows:

$$s_p^{\pm} = \sum a_m^{\pm} \chi_{Em} \quad \text{and} \quad t_p^{\pm} = \sum b_m^{\pm} \chi_{Em}$$

with common sets $\{E_m\}$, and $a_m^{\pm}$, $b_m^{\pm} \geq 0$. Indeed, if $s_p^{\pm} = \sum a_j^{\pm} \chi_{E_j^{\pm}}$ and $t_p^{\pm} = \sum b_k^{\pm} \chi_{F_k^{\pm}}$, denote the family $\{E_i^+ \cap E_j^- \cap E_k^+ \cap F_l^-\}$ by $\{E_m\}$. Then we have

$$\begin{aligned} \left| \int \{(s_p^+ - s_p^-) + i(t_p^+ - t_p^-)\} dm \right| &= \left| \sum \{(a_m^+ - a_m^-) + i(b_m^+ - b_m^-)\} m(E_m) \right| \\ &\leq \sum |\{(a_m^+ - a_m^-) + i(b_m^+ - b_m^-)\}| m(E_m). \end{aligned}$$

Letting $p \to \infty$, we get the inequality (3.6). □

LEMMA 3.6.
(i) *If $f \in \mathcal{L}(\mathbb{R}^d)$, then $f \in \mathcal{L}(E)$ for every measurable subset E.*
(ii) *If $E \cap F = \emptyset$, then*

$$\int_{E \cup F} f dm = \int_E f dm + \int_F f dm.$$

PROOF. (i) is obvious. (ii) follows directly from the definition of the integral. □

THEOREM 3.6.
(i) *$\mathcal{L}$ is a vector space over $\mathbb{C}$.*
(ii) *The integration is a linear functional on $\mathcal{L}$, that is, the integration is a mapping of $\mathcal{L}$ to $\mathbb{C}$ such that*

$$\int (af + bg) dm = a \int f dm + b \int g dm \quad (f, g \in \mathcal{L}, \ a, b \in \mathbb{C}). \tag{3.7}$$

PROOF. (i) follows from (ii) and Theorem 3.5. To prove (ii) we divide the assertion into two equations,

$$\int (af) dm = a \int f dm \quad (f \in \mathcal{L}, \ a \in \mathbb{C}), \tag{3.8}$$

and

$$\int (f + g) dm = \int f dm + \int g dm \quad (f, g \in \mathcal{L}). \tag{3.9}$$

First Step. *Equation (3.8) holds for any real valued function f and real number a.*

In fact, (3.8) is a direct consequence of the Lebesgue integral if $f \geq 0$ and $a > 0$. If f is a real valued function, decompose f to $f_+ - f_-$, and then apply the nonnegative function case.

Second Step. *Equality (3.9) holds for real valued functions f and g*
Put

$$E_1 = \{x : f \geq 0 \text{ and } g \geq 0\},$$

$$E_2 = \{x : f < 0 \text{ and } g < 0\},$$

$$E_3 = \{x : f \geq 0, \ g < 0 \text{ and } f + g \geq 0\},$$

$$E_4 = \{x : f \geq 0,\ g < 0 \text{ and } f+g < 0\},$$

$$E_5 = \{x : f < 0,\ g \geq 0 \text{ and } f+g \geq 0\},$$

and

$$E_6 = \{x : f < 0,\ g \geq 0 \text{ and } f+g < 0\}.$$

The sets $E_1, \ldots, E_6$ are mutually disjoint and $E_1 \cup \cdots \cup E_6 = \mathbb{R}^d$. Thus by Lemma 3.6 we have

$$\int (f+g)dm = \sum_{j=1}^{6} \int_{E_j} (f+g)dm,$$

which reduces the proof to showing that

$$\int_{E_j} (f+g)dm = \int_{E_j} fdm + \int_{E_j} gdm \quad (j = 1, 2, \ldots, 6). \tag{3.10}$$

When $j = 1$, (3.10) is nothing but Lemma 3.3. When $j = 2$, apply the case $j = 1$ to $-(f+g)$, and then use the first step. When $j = 3$, note that $f + g \geq 0$ and $-g \geq 0$ on E_3. Thus

$$\begin{aligned}\int_{E_3} fdm &= \int_{E_3} [(f+g) + (-g)]dm \\ &= \int_{E_3} (f+g)dm + \int_{E_3} (-g)dm = \int_{E_3} (f+g)dm - \int_{E_3} gdm,\end{aligned}$$

from which (3.9) follows.

The cases $j = 4, 5$ and 6 are similarly proved, and are left to the reader.

Third Step. *Equalities* (3.8) *and* (3.9) *hold for any complex valued functions* f, g *and any complex number* a.

In order to prove (3.8) decompose f and a to real and imaginary parts respectively and apply (3.8) and (3.9) for real functions and real numbers. To prove (3.9) for complex valued functions, decompose f and g into real and imaginary parts. □

THEOREM 3.7. *Let* $f \in \mathcal{L}(\mathbb{R}^d)$. *Then for each* $\varepsilon > 0$ *there exist complex numbers* $a_1, \ldots, a_k$ *and intervals* $I_1, \ldots, I_k$ *such that*

$$\int |f - \sum_{j=1}^{k} a_j \chi_{I_j}| dm < \varepsilon.$$

PROOF. We shall prove the case where f is a nonnegative real valued function. The other cases reduce to this case. By the definition of the integral, for a given $\varepsilon > 0$ there exists a nonnegative simple function $s = \sum_{j=1}^{k} a_j \chi_{E_j}$ such that

$$\int |f - s| dm < \varepsilon/2,$$

where the E_j are mutually disjoint measurable sets.

Choose an $\varepsilon' > 0$ so that $(a_1 + \cdots + a_k)\varepsilon' < \varepsilon/2$. By Corollary 2.1, for each $j = 1, \ldots, k$ there exists a finite sequence of disjoint intervals $I_1^j, I_2^j, \ldots, I_{N_j}^j$ such that

$$m(E_j - \bigcup_{n=1}^{N_j} I_n^j) + m(\bigcup_{n=1}^{N_j} I_n^j - E_j) < \varepsilon'. \tag{3.11}$$

The step function $S = \sum_{j=1}^{k}\sum_{n=1}^{N_j} a_j \chi_{E_{I_n^j}}$ satisfies the desired condition of the theorem. In fact, we have

$$\int |f - S|dm \le \int |f - s|dm + \int |s - S|dm.$$

Since $s - S = \sum_{j=1}^{k} a_j(\chi_{E_j} - \sum_{n=1}^{N_j}\chi_{E_{I_n^j}})$, we have

$$\int |s - S|dm \le \sum_{j=1}^{k} a_j \int \left|\chi_{E_j} - \sum_{n=1}^{N_j}\chi_{E_{I_n^j}}\right| dm.$$

Observe that the last integral is equal to the left-hand side of (3.11). Therefore $\int |f - S|dm < \sum_{j=1}^{k} a_j\varepsilon' < \varepsilon/2$, which proves the theorem. □

THEOREM 3.8. *If $f \in \mathcal{L}$, then we have, for any $\lambda > 0$,*

$$m(\{x : |f(x)| > \lambda\}) \le \frac{1}{\lambda}\int |f|dm. \qquad (\textit{Chebyshev's inequality})$$

PROOF. Put $E_\lambda = \{x : |f(x)| > \lambda\}$. Observe that $|f(x)| > \lambda$ on E_λ. Then we have

$$\int |f|dm \ge \int_{E_\lambda} |f|dm \ge \int_{E_\lambda} \lambda dm = \lambda m(E_\lambda).$$

Dividing both sides by λ, we get (3.8). □

THEOREM 3.9. *If $f \in \mathcal{L}$, then $|f(x)| < \infty$ a.e.*

PROOF. Put $N = \{x : |f(x)| = \infty\}$. Then $N \subset \{x : |f(x)| > \lambda\}$ for all $\lambda > 0$, and thus by Chebyshev's inequality

$$m(N) \le m(\{x : |f(x)| > \lambda\}) \le \frac{1}{\lambda}\int |f|dm \quad \text{for all } \lambda > 0,$$

which implies that $m(N) = 0$. □

THEOREM 3.10. *If $f \in \mathcal{L}$ and if $f = g$ a.e., then $g \in \mathcal{L}$, and*

$$\int f dm = \int g dm.$$

PROOF. The measurability of g follows immediately from Lemma 3.2. Let $N = \{x\,; f(x) \ne g(x)\}$ and let $u(x) = 0\,(x \notin N)$, $= \infty\,(x \in N)$. Then $\int u dm = 0$, since the set N is null. We have

$$\int |g|dm \le \int (|f| + u)dm = \int |f|dm < \infty$$

and

$$\left|\int f dm - \int g dm\right| \le \int |f - g|dm \le \int u dm = 0.$$

Thus g is integrable and $\int f dm = \int g dm$. □

For meaurable functions f and g define the relation "$\sim$" by

$$f \sim g \iff f(x) = g(x) \text{ a.e.}$$

"$\sim$" yields an equivalence relation in the set of measurable functions. In fact, the following relations hold obviously:

$$f \sim f,$$

$$f \sim g \Rightarrow g \sim f,$$

and

$$f \sim g,\ g \sim h \Rightarrow f \sim h.$$

For $f \in \mathcal{L}$ let $\dot{f} = \{g : f \sim g\}$ be the equivalence class which contains f. Denote the set of all equivalence classes of functions in $\mathcal{L}$ by $L^1(\mathbb{R}^d, \mathcal{M}, dm)$; in other words,

$$L^1(\mathbb{R}^d, \mathcal{M}, dm) = \mathcal{L}/\sim .$$

Remark that if $f' \sim f''$ and $g' \sim g''$, then $f' + g' \sim f'' + g''$ and $af' \sim af''$ for $a \in \mathbb{C}$. Therefore we define $\dot{f} + \dot{g}$ by $(\dot{f+g})$ and $(\dot{af})$ by $a\dot{f}$. Therefore $L^1(\mathbb{R}^d, \mathcal{M}, dm)$ is a vector space. When no confusion occurs, $\mathbb{R}^d$, $\mathcal{M}$ or dm are omitted and $L^1(\mathbb{R}^d, \mathcal{M}, dm)$ is denoted by $L^1(\mathbb{R}^d)$, $L^1(dm)$, L^1 and so on.

Let $\dot{f} \in L^1$. If $f', f'' \in \dot{f}$, then

$$\int f' dm = \int f'' dm,$$

by Theorem 3.10. Therefore we can define the integral of $\dot{f}$ by

$$\int \dot{f} dm = \int f' dm,$$

where f' is any function of $\dot{f}$.

For the rest of this book, the equivalence class which contains a measurable function f is denoted simply by f. When there is no additional note, f may denote a function itself in one case and the equivalence class in another.

3.5. Convergence of sequences of functions

One purpose of the Lebesgue integration theory is to construct an integration theory which is closed under countably infinite arithmetic operations on functions and sets.

In this section, we define various kinds of convergence of sequences of functions. In the next section we shall show how the Lebesgue integration theory fits the problems involved in these limitting operations.

Let $\{f_n(x)\}$ be a sequence of functions on a set E.

$\{f_n\}$ is said to *converge pointwise* to f on E if there exists a function $f(x)$ such that

$$f_n(x) \to f(x) \quad (n \to \infty) \quad \text{for each } x \in E.$$

$\{f_n\}$ is said to *converge almost everywhere* on E to f if there exist a null set N and a function $f(x)$ on E such that

$$f_n(x) \to f(x) \quad (n \to \infty) \quad \text{for } x \in E - N,$$

which is denoted simply as $f_n \to f$ a.e.

$\{f_n\}$ is said to *converge uniformly* on E to f if for every $\varepsilon > 0$ there exists an integer $m > 0$ such that

$$|f_n(x) - f(x)| < \varepsilon \quad \text{for all} \quad x \in E \text{ and } n \geq m.$$

$\{f_n\}$ is said to *converge in measure* on E to f if for every $\varepsilon > 0$

$$m(\{x \in E : |f_n(x) - f(x)| > \varepsilon\}) \to 0 \quad \text{as } n \to \infty.$$

This is denoted simply by $f_n \to f$ in measure on E.

$\{f_n\}$ is said to *converge in* $L^1(E)$ to f if f_n and f are integrable and

$$\int_E |f_n - f| dm \to 0 \ (n \to \infty).$$

It is often said to *converge strongly* or to *converge in L^1-norm* instead of "converge in L^1", and expressed as $f_n \to f$ in L^1, $\lim f_n = f$ in L^1, or $\|f_n - f\|_1 \to 0\,(n \to \infty)$.

3.5.1. Strong convergence and a.e. convergence.

THEOREM 3.11. *If $\{f_n\}$ converges in $L^1(E)$ to f, then it converges in measure on E to f.*

PROOF. Fix any $\varepsilon > 0$. By Chebyshev's inequality

$$m(\{x \in E : |f_n(x) - f(x)| > \varepsilon\}) \leq \frac{1}{\varepsilon} \int_E |f_n(x) - f(x)| dm.$$

By assumption the right-hand side converges to 0 as $n \to \infty$. □

THEOREM 3.12. *Assume $m(E) < \infty$. Then, $\{f_n\}$ converges in measure on E to f if and only if*

$$\int_E \frac{|f_n(x) - f(x)|}{|f_n(x) - f(x)| + 1} dm \to 0 \ (n \to \infty). \tag{3.12}$$

PROOF. Suppose f_n converges in measure to f. Let $\varepsilon > 0$ and put $A_\varepsilon = \{x \in E : |f_n(x) - f(x)| > \varepsilon\}$. Then $E - A_\varepsilon = \{x \in E : |f_n(x) - f(x)| \leq \varepsilon\}$.

Since $t/(t+1)$ is monotonously increasing in $t \geq 0$ and bounded by 1, we have $|f_n(x) - f(x)|/(|f_n(x) - f(x)| + 1) \leq \varepsilon/(\varepsilon + 1) < \varepsilon$ if $x \in E - A_\varepsilon$ and ≤ 1 if $x \in A_\varepsilon$. Therefore dividing the integral domain E into $E - A_\varepsilon$ and A_ε, and applying these two inequalities, we get

$$\int_E \frac{|f_n - f|}{|f_n - f| + 1} dm \leq \int_{E - A_\varepsilon} \varepsilon dm + \int_{A_\varepsilon} 1 dm.$$

Thus we get (3.12) since $m(A_\varepsilon) \to 0$ as $n \to \infty$ by assumption, and since ε is an arbitrarily given positive number.

Again by the monotonousness of $t/(t+1)$ $(t > 0)$ we have

$$\int_E \frac{|f_n - f|}{|f_n - f| + 1} dm \geq \int_{A_\varepsilon} \frac{\varepsilon}{1 + \varepsilon} dm = \frac{\varepsilon}{1 + \varepsilon} m(A_\varepsilon).$$

Therefore if the left-hand side converges to 0, then $m(A_\varepsilon) \to 0$ for any fixed $\varepsilon > 0$. □

3.5.2. Uniform convergence. Let $C(E)$ be the set of all continous functions in E. For $f \in C(E)$ let

$$\|f\|_\infty = \sup_{x \in E} |f(x)|,$$

and call it the *uniform norm* of f. In terms of the uniform norm a sequence $\{f_j\}_{j=1}^\infty$ in $C(E)$ converges uniformly to f if and only if

$$\forall \varepsilon > 0 \quad \exists N \quad : \quad \|f_j - f\|_\infty < \varepsilon \quad (j \geq N). \tag{3.13}$$

THEOREM 3.13. *If a sequence $\{f_j\}$ of $C(E)$ converges uniformly to f, then $f \in C(E)$.*

PROOF. For a given $\varepsilon > 0$ take N so that (3.13) holds. Since f_N is continuous on E, for a given point x in E there exists $\delta > 0$ such that

$$|f_N(x) - f_N(y)| < \varepsilon \quad (y \in E, \ |x - y| < \delta). \tag{3.14}$$

On the other hand, we have

$$|f(x) - f(y)| \leq |f(x) - f_N(x)| + |f_N(x) - f_N(y)| + |f_N(y) - f(y)|.$$

If $y \in E$ and $|x - y| < \delta$, the first and the third terms are less than ε by (3.13) and the second by (3.14). Thus $|f(x) - f(y)| < 3\varepsilon$, which implies that f is continuous at x. □

THEOREM 3.14 (Dini's theorem). *Let K be a compact subset of $\mathbb{R}^d$. If*

(a) $f_j, f \in C(K)$,
(b) $f_1(x) \leq f_2(x) \leq \cdots \to f(x)$,

then $\{f_j\}$ converges uniformly to f.

PROOF. Put $g_j(x) = f(x) - f_j(x)$. We need to prove that $g_j(x) \to 0$ uniformly on K.

Let $\varepsilon > 0$. Since $\{g_j(x)\}$ converges monotonously to 0, for each $x \in K$ there exists j_x such that

$$0 \leq g_{j_x}(x) < \varepsilon/2,$$

where j_x is an integer which may depend on x. For a fixed x, $g_{j_x}(y)$ is a continuous function of y. Therefore there exists $\delta_x > 0$ such that

$$|g_{j_x}(y) - g_{j_x}(x)| < \varepsilon/2 \quad \text{for } y \in K \cap B(x, \delta_x).$$

Thus, when $y \in B(x, \delta_x) \cap K$,

$$0 \leq g_{j_x}(y) \leq |g_{j_x}(y) - g_{j_x}(x)| + g_{j_x}(x) < \varepsilon. \tag{3.15}$$

The family $\{B(x, \delta_x) : x \in K\}$ is an open covering of the compact set K. Thus we can choose a finite subcovering $B(x_n, \delta_{x_n})$, $n = 1, 2, \ldots, N$, of K.

Put $J = \max(j_{x_1}, \ldots, j_{x_N})$. For any $y \in K$ pick $B(x_n, \delta_{x_n})$ which contains y. Since $\{g_j(x)\}$ is nonincreasing in j, we have

$$0 \leq g_J(y) \leq g_{j_{x_n}}(y) < \varepsilon$$

by (3.15). Thus $0 \leq g_J(y) < \varepsilon$ for all $y \in K$. Thus $0 \leq g_j(y) < \varepsilon$ for all $j \geq J$ and $y \in K$, which implies that $g_j \to 0$ uniformly on K. □

3.5.3. Egorov's theorem and Lusin's theorem.

THEOREM 3.15 (Egorov's theorem). *Let E be a measurable set of finite measure. Suppose that f_n and f are almost everywhere finite measurable functions on E, and the f_n converge almost everywhere on E to f. Then for every $\varepsilon > 0$ there exists a measurable set E_0 such that*

(i) *$E_0 \subset E$ and $m(E - E_0) < \varepsilon$,*
(ii) *the f_n converge uniformly to f on E_0.*

PROOF. By excluding a null set we can suppose that f_n and f are everywhere finite and $f_n(x) \to f(x)$ for $x \in E$.

For p, $q = 1, 2, \ldots$ put

$$E_p(q) = \{x \in E : |f(x) - f_j(x)| < 1/q \text{ for all } j \geq p\}.$$

Then we have, for each q,

$$m(E - E_p(q)) \to 0 \quad \text{as } p \to \infty. \tag{3.16}$$

To prove (3.16), observe that

$$E_p(q) = \bigcap_{j=p}^{\infty} \{x \in E : |f(x) - f_j(x)| < 1/q\}.$$

Thus $E_p(q) \in \mathcal{M}(\mathbb{R}^d)$. Furthermore, $E_1(q) \subset E_2(q) \subset \cdots$ and $\lim_{p\to\infty} E_p(q) = E$ by our asumption. Therefore we get (3.16) by Theorem 2.5(ii).

Let $\varepsilon > 0$. For each q choose a positive integer p_q so large that

$$m(E - E_{p_q}(q)) < \varepsilon/2^q,$$

which is possible by (3.16). Then the set $E_0 = \bigcap_{q=1}^{\infty} E_{p_q}(q)$ satisfies the conditions (i) and (ii). In fact,

$$m(E - E_0) = m(\bigcup_{q=1}^{\infty} [E - E_{p_q}(q)]) \leq \sum_{q=1}^{\infty} m(E - E_{p_q}(q)) < \sum_{q=1}^{\infty} \frac{\varepsilon}{2^q} = \varepsilon.$$

By the definition of E_0, we have $E_0 \subset E_{p_q}(q)$ for every $q > 1$. Thus for $x \in E_0$ we have

$$|f(x) - f_j(x)| \leq 1/q \quad (j \geq p_q).$$

□

REMARK 3.1. The set E_0 can be replaced by a compact set. Indeed, there exists a compact subset K of E_0 such that $m(E_0 - K)$ is smaller than a given positive number.

THEOREM 3.16 (Lusin's theorem). *Let E be a measurable set of finite measure and f be an almost everywhere finite measurable function in E. Then for $\varepsilon > 0$ there exists a compact set K_0 such that*

(i) *$K_0 \subset E$, $m(E - K_0) < \varepsilon$,*
(ii) *f is continuous on K_0.*

PROOF. We can suppose that f is almost everywhere real valued and, furthermore, almost everywhere nonnegative.

First consider the case where f is a simple function: $f(x) = \sum_{j=1}^{k} a_j \chi_{E_j}(x)$, where the E_j are mutually disjoint. Let C_j be compact sets such that $C_j \subset$

E_j and $m(E_j - C_j) < \varepsilon/k\,(j = 1, 2, \ldots, k)$. Then f is continuous in each C_j, since f is constant in C_j. Thus the set $C = \bigcup_{j=1}^{k} C_j$ has the required properties.

For the general case take a nondecreasing sequence $\{s_n\}$ of nonnegative simple functions which converges almost everywhere to f. For each $n = 1, 2, \ldots$ there exista a compact set $K_n \subset E$ such that

$$m(E - K_n) < \varepsilon/2^{n+1}, \text{ and } s_n \text{ is continuous on } K_n.$$

Now put $K = \bigcap_{n=1}^{\infty} K_n$. Then K is a compact subset of K_n and $m(E - K) = m(\bigcup_{n=1}^{\infty}(E - K_n)) \leq \sum_{n=1}^{\infty} m(E - K_n) < \varepsilon/2$.

By Egorov's theorem there exists a compact set K_0 such that

$$K_0 \subset K, \quad m(K - K_0) < \varepsilon/2, \text{ and } s_n \to f \text{ uniformly on } K_0$$

Since s_n is continuous in K_0, $f = \lim s_n$ is continuous in K_0 by Theorem 3.13.

We have $m(E - K_0) = m(E - K) + m(K - K_0) < \varepsilon$. Thus the proof is complete. □

3.5.4. Convergence in measure.

THEOREM 3.17. *Let E be a measurable set of finite measure. If $\{f_j\}$ converges almost everywhere on E to f, then it converges in measure. The converse is not always true.*

PROOF. By Egorov's theorem, for $\varepsilon > 0$ there exist $E_0 \subset E$ and an integer N such that

$$m(E - E_0) < \varepsilon \quad \text{and} \quad |f_j(x) - f(x)| < \varepsilon \ (x \in E_0, \ j > N).$$

Thus

$$\{x \in E : |f_j(x) - f(x)| > \varepsilon\} \subset E - E_0 \quad (j > N).$$

Thus $m(\{x \in E : |f_j(x) - f(x)| > \varepsilon\}) \leq m(E - E_0) < \varepsilon\,(j > N)$, and therefore $\{f_j\}$ converges in measure to f.

Now we mention an example for which the converse of the theorem does not hold. For $j = 1, 2, \ldots$ and $k = 1, 2, \ldots, 2^j$, let $\psi_j^k(x) = 1\,(x \in [(k-1)/2^j, k/2^j))$, and $= 0$ otherwise (see Figure 3.1).

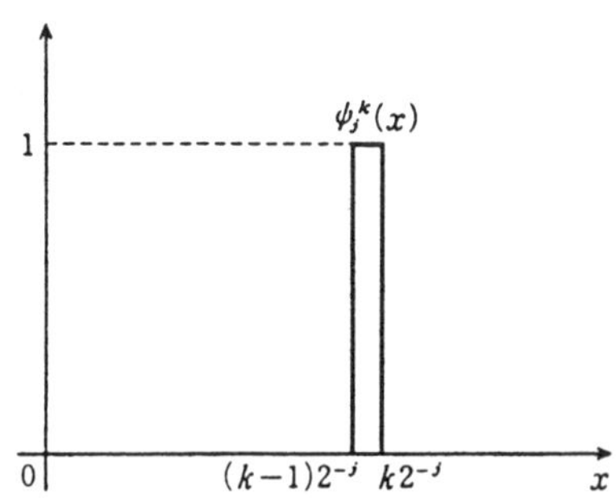

FIGURE 3.1. $y = \psi_j^k(x)$

Denote the sequence

$$\psi_1^1, \psi_2^1, \psi_2^2, \psi_3^1, \psi_3^2, \ldots, \psi_3^{2^3}, \ldots, \psi_j^1, \psi_j^2, \ldots, \psi_j^{2^j}, \ldots$$

by $f_1, f_2, \ldots$.

For $x \in [1,0)$ we have $f_j(x) = 1$ for infinitely many j. Thus $\limsup f_j(x) = 1$. On the other hand, $\liminf f_j(x) = 0$. Thus the sequence $\{f_j(x)\}$ does not converge everywhere in $[1,0)$, but $\{f_j\}$ converges in measure to 0. In fact, we have $\{x : \psi_j^k(x) \neq 0\} = [(k-1)/2^j, k/2^j)$. Thus for any $0 < \varepsilon < 1$ we have

$$m(\{x \in [0,1] : |\psi_j^k(x)| > \varepsilon\}) = 2^{-j} \to 0 \text{ as } j \to \infty.$$

□

The converse of Theorem 3.17 can be made to hold by taking a subsequence.

THEOREM 3.18. *If a sequence $\{f_j\}$ converges in measure on a set E to f, then there exists a subsequence $\{f_{n_k}\}$ of $\{f_j\}$ which converges almost everywhere to f.*

PROOF. Since $f_j \to f$ in measure, for every k there exists an n_k such that

$$m(\{x \in E : |f_j(x) - f(x)| > 2^{-k}\}) < 2^{-k} \quad (j \geq n_k).$$

We can assume that $n_1 < n_2 < \cdots$. Put $N_k = \{x \in E : |f_{n_k}(x) - f(x)| > 2^{-k}\}$, and $N = \limsup_{k\to\infty} N_k$. Then the set N is null (cf. Problem 2.2) and for $x \notin N$ we have $f_{n_k} \to f(x)$ as $k \to \infty$.

In fact, if $x \notin N$, that is, $x \in \bigcup_{l=1}^{\infty} \bigcap_{k=l}^{\infty} N_k^c$, then $x \in \bigcap_{k=l}^{\infty} N_k^c$ for some l. Thus $x \notin N_k$ for all $k \geq l$. Therefore we have $|f_{n_k}(x) - f(x)| \leq 2^{-k}$ for $k \geq l$. □

3.6. Convergence theorems

Let us reconsider the definition of Lebesgue integration. First, we introduced the family of measurable sets, which is closed under the countable union operations. This is equivalent to saying, under the difference condition (see Definition 2.6,(ii)), that if $\{E_j\}$ is an increasing sequence of measurable sets, then $\lim_{j\to\infty} E_j$ is also measurable. Lebesgue measure, which is defined for the measurable sets $\mathcal{M}(\mathbb{R}^d)$, preserves also the limits for increasing sequences $\{E_j\}$; that is, we have $\lim_{j\to\infty} m(E_j) = m(\lim_{j\to\infty} E_j)$ (Theorem 2.5,(ii)).

The Lebesgue integral inherits this nature, too. B. Levi's theorem and the Lebesgue convergence theorem are the typical examples. In fact, the theory of Lebesgue integration adapts well to countable arithmetic operations. Another remarkable example is Fubini's theorem, which is also deeply connected with the property mensioned above. It enables us to exchange the order of integrations, and is shown in Chapter 5.

THEOREM 3.19 (B. Levi's theorem). *Let u_j, $j = 1, 2, 3, \ldots$, be nonnegative functions. Then we have*

$$\int \Big(\sum_{j=1}^{\infty} u_j\Big) dm = \sum_{j=1}^{\infty} \int u_j dm. \tag{3.17}$$

When one side of (3.17) is infinite, the other side is infinite, too.

PROOF. Put $f = \sum_{j=1}^{\infty} u_j$. Since $\sum_{j=1}^{n} u_j \leq f$, we have

$$\sum_{j=1}^{n} \int u_j dm = \int \Big(\sum_{j=1}^{n} u_j\Big) dm \leq \int f dm,$$

from which it follows that

$$\sum_{j=1}^{\infty}\int u_j dm \leq \int f dm. \tag{3.18}$$

To prove the reverse inequality, choose sequences $\{s_j^p(x)\}_{p=1}^{\infty}$ of simple functions such that for each j and x we have $0 \leq s_j^1(x) \leq s_j^2(x) \leq \cdots \to u_j(x)$. We have shown the existence of such simple functions. Put

$$S^p(x) = \sum_{j=1}^{p} s_j^p(x).$$

Then $S^p(x)$ are simple, nonnegative and increase monotonously to $f(x)$. Indeed, since $0 \leq s_j^p(x) \leq s_j^{p+1}(x)$, we have

$$S^p(x) \leq \sum_{j=1}^{p} s_j^{p+1}(x) \leq \sum_{j=1}^{p+1} s_j^{p+1}(x) = S^{p+1}(x) \leq f(x).$$

Furthermore, for a fixed p we have

$$\sum_{j=1}^{p} u_j(x) = \lim_{q\to\infty}\sum_{j=1}^{p} s_j^q(x) \leq \lim_{q\to\infty}\sum_{j=1}^{q} s_j^q(x) = \lim_{q\to\infty} S^q(x).$$

Letting $p \to \infty$, we get $f(x) \leq \lim_{q\to\infty} S^q(x)$. Thus $f(x) = \lim_{q\to\infty} S^q(x)$.

By the definition of the Lebesgue integral,

$$\int f dm = \lim_{q\to\infty}\int S^q(x)dm. \tag{3.19}$$

Since $S^p(x) \leq \sum_{j=1}^{p} u_j(x)$, we have

$$\lim_{p\to\infty}\int S^p(x)dm \leq \lim_{p\to\infty}\sum_{j=1}^{p}\int u_j(x)dm = \sum_{j=1}^{\infty}\int u_j(x)dm. \tag{3.20}$$

By (3.18), (3.19) and (3.20) we get (3.17). □

THEOREM 3.20. *If* $0 \leq f_1(x) \leq f_2(x) \leq \cdots$, *then*

$$\int(\lim_{j\to\infty} f_j)dm = \lim_{j\to\infty}\int f_j dm.$$

PROOF. Put $u_1 = f_1$ and $u_j = f_{j+1} - f_j$ $(j \geq 2)$. Then $u_j \geq 0$ and $\sum_{j=1}^{\infty} u_j = \lim_{k\to\infty} f_k$. Thus the theorem follows immediately from Theorem 3.19. □

THEOREM 3.21 (Fatou's lemma). *If* $\{f_j\}$ *is a sequence of nonnegative functions, then*

$$\int(\liminf_{j\to\infty} f_j)dm \leq \liminf_{j\to\infty}\int f_j dm.$$

PROOF. Put $g_k = \inf_{j\geq k} f_j$. Then

$$0 \leq g_1(x) \leq g_2(x) \leq \cdots \to \lim_{k\to\infty} g_k = \liminf_{j\to\infty} f_j.$$

Applying Theorem 3.20, we get

$$\int(\liminf_{j\to\infty} f_j)dm = \int(\lim_{k\to\infty} g_k)dm = \lim_{k\to\infty}\int g_k dm. \tag{3.21}$$

Since $g_k \leq f_j \ (j \geq k)$, we have $\int g_k dm \leq \int f_j dm \ (j \geq k)$, and thus $\int g_k dm \leq \inf_{j \geq k} \int f_j dm$. Therefore

$$\lim_{k\to\infty} \int g_k dm \leq \liminf_{j\to\infty} \int f_j dm. \tag{3.22}$$

Thus the lemma follows from (3.21) and (3.22). □

REMARK 3.2. In Fatou's lemma, the inequality sign cannot be replaced by equality. For example, let $f_j(x) = j$ if $0 \leq x \leq 1/j$, and $= 0$ otherwise. Then, $\liminf f_j(x) = 0 \ (x \neq 0)$. Thus we have $\int \liminf f_j dm = 0$, but $\int f_j dm = 1 \ (j = 1, 2, \ldots)$.

THEOREM 3.22 (Lebesgue's dominated convergence theorem). *Let $\{f_j\}$ be a sequence of functions which satisfies the following conditions:*

(a) *$\{f_j\}$ converges almost everywhere.*
(b) *There exists g in L^1 such that*

$$|f_j(x)| \leq g(x) \ \ a.e., \quad j = 1, 2, \ldots .$$

Then

$$\lim_{j\to\infty} \int f_j(x) dm = \int \lim_{j\to\infty} f_j(x) dm. \tag{3.23}$$

REMARK 3.3. In the dominated convergence theorem the existence of the integrable function g is important. In fact, the sequence $\{f_j\}$ mentioned in Remark 3.2 converges to 0 for $x \neq 0$ but (3.23) does not hold.

PROOF. It is enough to prove the theorem for real valued functions.

By assumption (b), $g(x) + f_j(x) \geq 0$ and $g(x) - f_j(x) \geq 0$. Applying Fatou's lemma, we get

$$\int \liminf(g + f_j) dm \leq \liminf \int (g + f_j) dm.$$

Since $\{f_j\}$ is convergent almost everywhere, it follows that

$$\int \lim f_j dm \leq \liminf \int f_j dm. \tag{3.24}$$

In the same way we have

$$\int \liminf(g - f_j) dm \leq \liminf \int (g - f_j) dm,$$

and thus

$$\limsup \int f_j dm \leq \int \lim f_j dm. \tag{3.25}$$

The theorem follows from (3.24) and (3.25). □

COROLLARY 3.2 (Lebesgue's bounded convergence theorem). *Let E be a measurable set of finite measure, and let $\{f_j\}$ be a sequence of functions on E satisfying the following conditions:*

(a) *f_j converges almost everywhere on E,*
(b) *$|f_j(x)| \leq M$ a.e. on E $(j = 1, 2, \ldots)$, where M is a constant.*

Then

$$\lim_{j\to\infty} \int f_j(x)dm = \int \lim_{j\to\infty} f_j(x)dm.$$

PROOF. Put $g(x) = M\chi_E(x)$. Then $g \in L^1$, since $m(E) < \infty$. Now the corollary follows from Lebesgue's dominated convergence theorem. □

In Corollary 3.2 the condition $m(E) < \infty$ cannot be removed. In fact, if $f_j(x) = 1$ for $x \in [j, j+1)$, and $= 0$ otherwise, then $\lim f_j(x) = 0$ for all x, but $\int f_j dm = 1\,(j = 1, 2, \ldots)$.

THEOREM 3.23. *Let $E = \bigcup_{j=1}^{\infty} E_j$ be a disjoint union. Let $f \in L^1(E)$. Then*

$$\int_E f dm = \sum_{j=1}^{\infty} \int_{E_j} f dm. \tag{3.26}$$

PROOF. Put $g_k(x) = \sum_{j=1}^{k} f(x)\chi_{E_j}(x)$. Observe that $|g_k(x)| \le |f(x)|$, and $\lim_{k\to\infty} g_k(x) = f(x)$. Then the theorem follows from Lebesgue's dominated convergence theorem. □

For $f \in L^1(\mathbb{R}^d)$ put

$$\mu_f(E) = \int_E f dm.$$

Then μ_f is countably additive by Theorem 3.26, that is,

$$\mu_f(\bigcup_{j=0}^{\infty} E_j) = \mu_f(E_1) + \mu_f(E_2) + \cdots,$$

for every sequence $\{E_j\}$ of disjoint sets. Therefore μ_f is a measure on $\mathcal{M}(\mathbb{R}^d)$.

We remark that if f is nonnegative but not necessarily integrable, then μ_f is a nonnegative measure. This follows by applying B. Levi's theorem in place of Lebesgue's convergence theorem.

EXAMPLE 3.3. We give a condition which ensures the exchange of integration and differentiation operations as an application of Lebesgue's convergence theorem.

Suppose that a function $f(x, y)$ in $\mathbb{R}^2$ satisfies the conditions

(a) $f(x, y) \in L^1(\mathbb{R}, dy)$ for each x,

(b) $\partial f(x, y)/\partial x$ exists almost everywhere, and there exists $g \in L^1(\mathbb{R})$ such that

$$\left|\frac{\partial f(x, y)}{\partial x}\right| \le g(y) \quad (x, y) \in \mathbb{R}^2.$$

Then

$$\frac{d}{dx}\int_{-\infty}^{\infty} f(x, y)dm(y) = \int_{-\infty}^{\infty} \frac{\partial f(x, y)}{\partial x} dm(y).$$

In fact, by the mean value theorem

$$\left|\frac{f(x+h, y) - f(x, y)}{h}\right| = \left|\frac{\partial f(x+\theta h, y)}{\partial x}\right| \le g(y),$$

where $0 < \theta < 1$. Therefore by Lebesgue's dominated convergence theorem we have

$$\lim_{h\to 0}\int_{-\infty}^{\infty} \frac{f(x+h, y) - f(x, y)}{h} dm(y) = \int_{-\infty}^{\infty} \lim_{h\to 0} \frac{f(x+h, y) - f(x, y)}{h} dm(y).$$

REMARK 3.4. The sequences treated in Lebesgue's dominated convergence theorem are parametrized by the integers. However, the application for the continuous parameter h as above is justified by taking any sequence h_j tending to 0.

3.7. The Riemann integral and the Lebesgue integral

Under what condition will a function be Riemann integrable? Is a function Lebesgue integrable if it is Riemann integrable, and do both integrals coincide? These problems are considered in this section.

To make the argument simple, we assume that the domain under consideration is a finite interval and the functions are real valued and bounded but not necessarily measurable.

Let $\Delta = \{I_j : j = 1, \dots, k\}$ be a partition of the interval $[a, b]$. The Darboux upper and lower integrals of f are defined by

$$\overline{\int} f dx = \inf_{\Delta} \overline{s}(f, \Delta) \quad \text{and} \quad \underline{\int} f dx = \sup_{\Delta} \underline{s}(f, \Delta),$$

respectively, where

$$\overline{s}(f, \Delta) = \sum_{j=1}^{k} \sup_{x \in I_j} f(x) m(I_j) \quad \text{and} \quad \underline{s}(f, \Delta) = \sum_{j=1}^{k} \inf_{x \in I_j} f(x) m(I_j).$$

Put

$$\overline{f}(x) = \inf_{\delta > 0} \sup_{|x-y|<\delta,\, y \in [a,b]} f(y) \quad \text{and} \quad \underline{f}(x) = \sup_{\delta > 0} \inf_{|x-y|<\delta,\, y \in [a,b]} f(y).$$

LEMMA 3.7. *$\overline{f}$ is upper semicontinuous and $\underline{f}$ is lower semicontinuous. Thus $\overline{f}$ and $\underline{f}$ are measurable.*

PROOF. Let us prove that $\underline{f}$ is lower semicontinuous. The case for $\overline{f}$ will be proved similarly.

Let $\underline{f}(a) > \lambda$. By definition we have $\inf\{f(y) : |a - y| < 2\delta, y \in [a, b]\} > \lambda$ for some $\delta > 0$. Suppose $|a - x| < \delta$. If $|x - y| < \delta$ and $y \in [a, b]$, then $|a - y| < 2\delta$. Thus $\inf\{f(y) : |a - y| < 2\delta, y \in [a, b]\} > \lambda$. Thus $\underline{f}(x) > \lambda$, which implies that a is an inner point of $\{t : \underline{f}(t) > \lambda\}$. Therefore the set $\{t : \underline{f}(t) > \lambda\}$ is open and thus $\underline{f}$ is lower semicontinuous. □

For now we denote the Riemann and Lebesgue integrals by

$$\mathcal{R}\int \quad \text{and} \quad \mathcal{L}\int,$$

respectively, to distinguish them clearly.

LEMMA 3.8. *Let f be a bounded function in the finite interval $[a, b]$. Then*

$$\overline{\int} f dx = \mathcal{L}\int \overline{f} dm \quad \textit{and} \quad \underline{\int} f dx = \mathcal{L}\int \underline{f} dm.$$

PROOF. By the definition of the Darboux integrals there exists a sequence of partitions $\Delta^n : a = a_0^n < a_1^n < \cdots < a_{k_n}^n = b$ of $[a, b]$ such that

$$\overline{\int} f dx = \lim_{n\to\infty} \overline{s}(f, \Delta^n). \tag{3.27}$$

If

$$f^n(x) = \sup_{y \in [a^n_{j-1}, a^n_j)} f(y), \quad \text{for } x \in [a^n_{j-1}, a^n_j),$$

$j = 1, 2, \dots, k_n$, then the functions f^n are simple and measurable. Furthermore,

$$\overline{s}(f, \Delta^n) = \mathcal{L} \int f^n dm. \tag{3.28}$$

We have, therefore, by (3.27) and (3.28),

$$\overline{\int} f dx = \lim_{n\to\infty} \mathcal{L} \int f^n dm. \tag{3.29}$$

Let $M = \sup_y |f(y)|$ and put $N = \{a^n_j : j = 1, 2, \dots, k_n, n = 1, 2, \dots\}$. The set N is null, since it is countable. Therefore we have

(a) $|f^n(x)| \le M < \infty, \ x \in [a, b]$,
(b) if $x \notin N$, then $f^n(x) \to \overline{f}(x)$.

Applying the bounded convergence theorem to (3.29), we get

$$\overline{\int} f dx = \mathcal{L} \int \overline{f} dm.$$

In the same way we have

$$\underline{\int} f dx = \mathcal{L} \int \underline{f} dm.$$

□

THEOREM 3.24. *Let f be a bounded function in a finite interval $[a, b]$. Then f is Riemann integrable if and only if f is almost everywhere continuous.*

In this case we have

$$\mathcal{R} \int_a^b f dx = \mathcal{L} \int_a^b f dx. \tag{3.30}$$

PROOF. A necessary and sufficient condition for f to be Riemann integrable is that

$$\overline{\int} f dx = \underline{\int} f dx. \tag{3.31}$$

By Lemma 3.8, (3.31) is equivalent to the equation $\mathcal{L} \int (\overline{f} - \underline{f}) dm = 0$. Since $\overline{f} \ge \underline{f}$, the last equation is equivalent to the fact that $\overline{f}(x) = \underline{f}(x)$ a.e., that is, f is continuous almost everywhere in $[a, b]$.

Conversely, if $\overline{f} = \underline{f}$ a.e., then

$$\mathcal{R} \int f dx = \mathcal{L} \int f dx$$

by Lemma 3.8. □

For example, the characteristic function $\chi_{\mathbb{Q}}(x)$ of $\mathbb{Q}$ is everywhere discontinuous, so it is not Riemann integrable on $[0, 1]$. On the other hand the characteristic function of the set $\bigcup_{j=1}^{\infty} [1/(2j+1), 1/2j]$ is Riemann integrable, because the set of points of discontinuity $\{0\} \cup \{\dots, 1/(k+1), 1/k, \dots, 1/2\}$ is null.

Theorem 3.24 implies, furthermore, that any bounded function which is continuous except on a countable set is Riemann integrable over any finite interval. In particular, a monotone function is Riemann integrable on any finite interval.

3.8. The Lebesgue integral and the Riemann-Stieltjes integral

3.8.1. Functions of bounded variation. Let f be a function on the real axis. For a partition Δ of $[a, x]$ with the points of division $a = a_0 < a_1 < \cdots < a_k = x$ put $V_f(\Delta) = \sum_{j=1}^k |f(a_j) - f(a_{j-1})|$. Define

$$V_f([a, x]) = \sup_{\Delta} V_f(\Delta),$$

where Δ runs over all partitions of $[a, x]$. When $V_f([a, x]) < \infty$, f is said to be *of bounded variation* on $[a, x]$ and $V_f([a, x])$ is called the *total variation* of f or simply the *variation* of f. When $a = -\infty$, we define $V_f((-\infty, x]) = \lim_{a \to -\infty} V_f([a, x])$. The case where $x = \infty$ is similar.

THEOREM 3.25. *If f is a function of bounded variation in an interval $[a, b]$ and if $a < c < b$, then*

$$V_f([a, b]) = V_f([a, c]) + V_f([c, b]).$$

PROOF. For a given $\varepsilon > 0$ let Δ_1, Δ_2 and Δ_0 be partitions of $[a, c]$, $[c, b]$ and $[a, b]$, respectively, such that

$$V_f([a, c]) - \frac{\varepsilon}{2} < V_f(\Delta_1) \le V_f([a, c]), \text{ and } V_f([c, b]) - \frac{\varepsilon}{2} < V_f(\Delta_2) \le V_f([c, b]),$$

$$V_f([a, b]) - \varepsilon < V_f(\Delta_0) \le V_f([a, b]). \tag{3.32}$$

Note that if the partition Δ is finer than Δ', that is, any points of division of Δ' are contained in the points of division of Δ, then $V_f(\Delta') \le V_f(\Delta)$. Thus we can suppose that $\Delta_0 = \Delta_1 \cup \Delta_2$ by refining the partitions. Then we have $V_f(\Delta_0) = V_f(\Delta_1) + V_f(\Delta_2)$. Thus

$$V_f([a, c]) + V_f([c, b]) - \varepsilon < V_f(\Delta_0) \le V_f([a, c]) + V_f([c, b]). \tag{3.33}$$

From (3.32) and (3.33) it follows that

$$V_f([a, c]) + V_f([c, b]) - \varepsilon < V_f([a, b]) \le V_f([a, c]) + V_f([c, b]) + \varepsilon,$$

which completes the proof. □

THEOREM 3.26. *Every function of bounded variation is expressed as a difference of two bounded nondecreasing functions.*

PROOF. Let f be a function of bounded variation in an interval $[a, b]$, where a and/or b may be infinite. Put

$$2P_f([a, x]) = V_f([a, x]) + f(x) - f(a),$$

and

$$2N_f([a, x]) = V_f([a, x]) - f(x) + f(a).$$

Then we have

$$f(x) = f(a) + P_f([a, x]) - N_f([a, x]).$$

$P_f([a,x])$ is a nondecreasing function of x. Indeed, for $h > 0$

$$\begin{aligned}&P_f([a,x+h]) - P_f([a,x])\\&= \{V_f([a,x+h]) - V_f([a,x])\} + \{f(x+h) - f(x)\}.\end{aligned}$$

The first term on the right-hand side is equal to $V_f([x,x+h])$ by Theorem 3.25, which is not less than $|f(x+h) - f(x)|$. Thus the left-hand side is nonnegative.

In the same way, $N_f([a,x])$ is nondecreasing. □

$P_f([a,b])$ and $N_f([a,b])$ are called the *positive and negative variations* respectively.

A function f in $[a, a+\delta)$ is said to be *continuous on the right* at a if $f(a) = f(a+)$. Continuity on the left is defined in the same way.

REMARK 3.5. If a function f is of bounded variation in an interval $[a-\delta, a+\delta]$ and continuous at a, then $P_f([a-\delta, x])$ and $N_f([a-\delta, x])$ are continuous at a. More precisely, if f is continuous on the right (left) at a point a, then $P_f([a-\delta, x])$ and $N_f([a-\delta, x])$ are continuous on the right (left) at a.

In fact, suppose that f is continuous on the right at a point a. To show that $P_f([a-\delta, x])$ and $N_f([a-\delta, x])$ are continuous on the right at a, it suffices to prove that $V_f([a, a+h]) \to 0\,(h \to +0)$ when $f(a+0) = 0$, by virtue of Theorem 3.25. The continuity on the left will be proved in the same way.

Suppose that our assertion is false. Then for any $h > 0$ there exist $\delta_0 > 0$ and a partition of $[a, a+h]$ with the points of division $a = a_0 < a_1 < \cdots < a_k = a+h$ such that

$$\sum_{j=1}^{k} |f(a_j) - f(a_{j-1})| > \delta_0.$$

Since f is continuous on the right, there exists $h' > 0$ such that $a < a+h' < a_1$ and

$$\sum_{j=2}^{k} |f(a_j) - f(a_{j-1})| + |f(a_1) - f(a+h')| > \delta_0.$$

Thus $V_f([a+h', a+h]) > \delta_0$.

The next step is to apply the above argument to the interval $[a, a+h']$, getting $0 < h'' < h'$ such that $V_f([a+h'', a+h']) > \delta_0$. In the same way we have $h > h' > h'' > h''' > \cdots > 0$ such that $V_f([a+h^{(j+1)}, a+h^{(j)}]) > \delta_0$. Therefore we have, by Theorem 3.25,

$$\begin{aligned}V_f([a, a+h]) &= V_f([a+h', a+h]) + V_f([a+h'', a+h'])\\&\quad + \cdots + V_f([a+h^{(k)}, a+h^{(k-1)}]) + V_f([a, a+h^{(k+1)}]),\end{aligned}$$

which is larger than $k\delta_0$ for any k. This contradicts that f is of bounded variation.

Let f be a function defined in a neighborhood of x. If f is of bounded variation, then both limits $f(x \pm 0)$ exist, because f is expressed as a linear combination of monotone functions, and for a monotone function f the limits $f(x \pm 0)$ always exist.

If a function f is of bounded variation on the real axis, then the points of discontinuity are at most countable. In fact, the points of discontinuity of any monotone function are at most countable (see Remark 1.2).

EXAMPLE 3.4. Let f be a real valued function in $L^1([a,b])$. If

$$\mu_f(x) = \int_a^x f(t)dt,$$

then

$$V_{\mu_f}([a,b]) = \int_a^b |f(t)|dt, \tag{3.34}$$

and therefore

$$P_{\mu_f}([a,x]) = \frac{1}{2}\left(\int_a^x |f(t)|dt - \int_a^x f(t)dt\right) = \int_a^x f_+(t)dt.$$

In the same way,

$$N_{\mu_f}([a,x]) = \int_a^x f_-(t)dt.$$

In fact, for any partition $a = a_0 < a_1 < \cdots < a_N = b$ of $[a,b]$ we have

$$\sum_{j=1}^N |\mu_f(a_j) - \mu_f(a_{j-1})| = \sum_{j=1}^N \left|\int_{a_{j-1}}^{a_j} f(t)dt\right|. \tag{3.35}$$

Since the right-hand side is at most $\int_a^b |f|dt$, we have

$$V_{\mu_f}([a,b]) \leq \int_a^b |f(t)|dt.$$

To prove the reverse inequality let $\varepsilon > 0$, and let $s(x) = \sum_j c_j \chi_{I_j}(x)$ be a simple function such that

$$\int |f - s| < \varepsilon,$$

where $\{I_j\}$ is a finite, mutually disjoint sequence of intervals. Then the right-hand side of (3.35) is not less than

$$\sum_{j=1}^N \left|\int_{a_{j-1}}^{a_j} s(t)dt\right| - \sum_{j=1}^N \left|\int_{a_{j-1}}^{a_j} (f-s)dt\right|.$$

The first sum equals $\int_a^b |s|dt$ when the points of division $\{a_j\}$ include all end points of the intervals I_j. On the other hand, the second sum is at most $\int_a^b |f - s|dt$. Therefore we get

$$V_{\mu_f}([a,b]) \geq \sum_{j=1}^N |\mu_f(a_j) - \mu_f(a_{j-1})| \geq \int_a^b |s|dt - \int_a^b |s - f|dt.$$

Since $\int_a^b |f|dm - \int_a^b |s|dm \leq \int_a^b |f-s|dm < \varepsilon$, we obtain $V_{\mu_f}([a,b]) \geq \int_a^b |f|dm - 2\varepsilon$. Therefore equation (3.34) is proved.

3.8.2. The Riemann-Stieltjes integral. For convenience we assume that any functions of bounded variatin are continuous on the right in the rest of this section. Let $I = [a, b]$ be a finite interval and μ be a nondecreasing function in I. For a partition Δ with the points of division $a = a_0 < a_1 < \cdots < a_k = b$ put

$$\Delta_j\mu = \mu(a_j) - \mu(a_{j-1}), \quad j = 1, 2, \ldots, k.$$

For a real valued function f in I let

$$\overline{s}_\mu(f, \Delta) = \sum_{j=1}^{k} \overline{f}_j \Delta_j\mu \quad \text{and} \quad \underline{s}_\mu(f, \Delta) = \sum_{j=1}^{k} \underline{f}_j \Delta_j\mu,$$

where $\overline{f}_j = \sup_{x \in [a_{j-1}, a_j]} f(x)$ and $\underline{f}_j = \inf_{x \in [a_{j-1}, a_j]} f(x)$, and define

$$\overline{\int} f d\mu = \inf_{\Delta} \overline{s}_\mu(f, \Delta) \quad \text{and} \quad \underline{\int} f d\mu = \sup_{\Delta} \underline{s}_\mu(f, \Delta).$$

As in the Riemann integral, they are called the Darboux upper and lower integrals of f with respect to μ, respectively. If these two integrals coincide, then f is said to be Riemann-Stieltjes[2] integrable, and the common value is called the *Riemann-Stieltjes integral* of f with respect to μ and denoted by

$$\int_I f d\mu.$$

In particular, if $\mu(x) = x$, the Riemann-Stieltjes integral is identical with the Riemann integral.

If μ is a function of bounded variation, write $\mu = (\nu_1 - \nu_2) + i(\lambda_1 - \lambda_2)$ with nondecreasing functions ν_1, ν_2, λ_1, and λ_2, and define

$$\int f d\mu = \left(\int f d\nu_1 - \int f d\nu_2\right) + i\left(\int f d\lambda_1 - \int f d\lambda_2\right).$$

It is easily shown that the integral does not depend on the decomposition of μ.

THEOREM 3.27. *Let μ be a nondecreasing function in a bounded interval I.*

(i) *If f is a continuous function in I, then f is Riemann-Stieltjes integrable with respect to μ.*

(ii) *If f is of bounded variation and, furthermore, μ is continuous, then f is Riemann-Stieltjes integrable with respect to μ.*

PROOF. (i) Let Δ be a partition $\{I_1, I_2, \ldots, I_k\}$ of I. Let $\varepsilon > 0$. Since f is uniformly continuous on I, we have

$$|f(x) - f(y)| < \varepsilon \quad (x, y \in I_j),$$

$j = 1, 2, \ldots, k$, if Δ is fine enough. Thus $\overline{f}_j - \underline{f}_j \leq \varepsilon$. Therefore we have

$$\begin{aligned} \overline{s}_\mu(f, \Delta) - \underline{s}_\mu(f, \Delta) &= \sum_{j=1}^{k} [\overline{f}_j - \underline{f}_j] \Delta_j\mu \leq \sum_{j=1}^{k} \varepsilon \Delta_j\mu \\ &\leq \varepsilon[\mu(b) - \mu(a)], \end{aligned}$$

which implies that f is Riemann-Stieltjes integrable.

[2]T. J. Stieltjes, 1856–1894.

(ii) We can assume that f is nondecreasing. For $\varepsilon > 0$ choose a partition so that $\Delta_j \mu < \varepsilon$. Then we have

$$\begin{aligned} \overline{s}_\mu(f,\Delta) - \underline{s}_\mu(f,\Delta) &= \sum_{j=1}^{k}[\overline{f}_j - \underline{f}_j]\Delta_j\mu \\ &\le \sum_{j=1}^{k}[f(a_j+0) - f(a_{j-1})]\varepsilon = [f(b)-f(a)]\varepsilon, \end{aligned}$$

which proves (ii). □

THEOREM 3.28. *Let $w \in L^1([a,b])$. If μ is a measure defined by $\mu(x) = \int_a^x w(t)dt$, then we have*

$$\int_I f(t)d\mu(t) = \int_I f(t)w(t)dt \tag{3.36}$$

for any bounded Riemann integrable function f.

The right-hand side of (3.36) is called a *weighted integral* with the weight w.

PROOF. We may assume that $w \ge 0$. In the notation of the preceeding proof,

$$\overline{s}_\mu(f,\Delta) = \sum_{j=1}^{k}\overline{f}_j\Delta_j\mu = \sum_{j=1}^{k}\overline{f}_j\int_{a_{j-1}}^{a_j} w(t)dt.$$

Let $\overline{f}^\Delta(t) = \overline{f}_j$ for $t \in [a_{j-1}, a_j)$, $j = 1, 2, \ldots, k$. Then the right-hand side is written as $\int_a^b \overline{f}^\Delta(t)w(t)dt$ with the simple function $\overline{f}^\Delta$. By the assumption, $\overline{f}^\Delta(t) \to f(t)$ a.e. when $\max(a_j - a_{j-1}) \to 0$. Thus by the dominated convergence theorem $\int_a^b \overline{f}^\Delta(t)w(t)dt \to \int_a^b f(t)w(t)dt$. □

THEOREM 3.29 (Integration by parts). *If f and μ are functions of bounded variation in $[a,b]$, then*

$$\int_a^b f d\mu = f(b)\mu(b) - f(a)\mu(a) - \int_a^b \mu\, df.$$

PROOF. We assume that f and μ are nondecreasing. The other cases will be deduced from this case. Thus $\sup_{a_{j-1}\le t\le a_j} f(t) = f(a_j)$.

For the partition Δ with the points of division $a = a_0 < a_1 < \cdots < a_k = b$, we have

$$\begin{aligned} \overline{s}_\mu(f,\Delta) &= \sum_{j=1}^{k} f(a_j)[\mu(a_j) - \mu(a_{j-1})] \\ &= f(b)\mu(b) - f(a_1)\mu(a) - \sum_{j=1}^{k}\mu(a_{j-1})[f(a_j) - f(a_{j-1})]. \end{aligned}$$

The left-hand side converges to $\int_a^b f d\mu$ and the sum on the right-hand side converges to $\int_a^b \mu\, df$ when $\max(a_j - a_{j-1}) \to 0$. □

3.8.3. A representation of the Lebesgue integral. For a real valued function f let

$$\lambda(a) = \lambda_{f,E}(a) = m(\{x \in E : f(x) > a\}) \quad (a \in \mathbb{R}).$$

$\lambda(a)$ is called the *distribution function* of f.

LEMMA 3.9. *Let f be a real valued function in a set E.*

(i) *The distribution function λ is continuous on the right and nonincreasing.*
(ii) *If $m(E) < \infty$, then λ is bounded.*

PROOF. (i) The set $\{x \in E : f(x) > b\}$ increases and converges to $\{x \in E : f(x) > a\}$ as b decreases to a. Therefore, by Theorem 2.5(ii), $\lambda(b) \to \lambda(a)$ as b decreases to a.

(ii) It is obvious that $\lambda(a)$ is bounded by $m(E)$. □

LEMMA 3.10. *If f is a measurable function in a set E and $a \le f(x) < b$ a.e., then*

$$\int_E f dm = -\int_a^b s\, d\lambda(s).$$

If one side is infinite, so is the other side.

PROOF. For a partition $a = a_0 < a_1, \cdots < a_k = b$ of $[a, b]$ put $E_j = \{x \in E : a_{j-1} < f(x) \le a_j\}$, $j = 1, 2, \ldots, k$, so that E is the disjoint union of $\{E_j\}$. Therefore we have

$$\int_E f dm = \sum_{j=1}^k \int_{E_j} f dm.$$

On the other hand,

$$\sum_{j=1}^k a_{j-1} m(E_j) \le \sum_{j=1}^k \int_{E_j} f dm \le \sum_{j=1}^k a_j m(E_j).$$

Observe that $m(E_j) = \lambda(a_{j-1}) - \lambda(a_j) = -\Delta_j \lambda$ and that the sums on the left- and right-hand sides converge to the Riemann-Stieltjes integral $-\int_a^b s\, d\lambda(s)$, when $\max(a_j - a_{j-1}) \to 0$. □

THEOREM 3.30. *Let f be a real valued measurable function in a set E. Then $f \in L^1(E)$ if and only if the limit*

$$\lim_{a \to -\infty,\, b \to \infty} \int_a^b s\, d\lambda_{f,E}(s)$$

exists and is finite. In this case

$$\int_E f dm = -\int_{-\infty}^{\infty} s\, d\lambda_{f,E}(s).$$

PROOF. Put $E^b = \{x \in E : 0 \le f(x) < b\}$ for $b > 0$. By Lemma 3.10 we have

$$\int_{E^b} f dm = -\int_0^b s\, d\lambda_{f,E^b}(s) = -\int_0^b s\, d\lambda_{f,E}(s).$$

Let $b \to \infty$. By B. Levi's theorem

$$\int_E f_+ dm = -\int_0^\infty s d\lambda_{f,E}(s). \tag{3.37}$$

In the same way we get

$$\int_E f_- dm = -\int_{-\infty}^0 s\, d\lambda_{f,E}(s). \tag{3.38}$$

By definition, $f \in L^1(E)$ if and only if $f_+, f_- \in L^1(E)$. Thus the theorem follows from (3.37) and (3.38). □

Theorem 3.30 is generalized as follows. The proof is similar to the preceding case, so it is left to the reader.

THEOREM 3.31. *Let φ be a nondecreasing function on the real axis. If f is a real valued function and $\varphi(f) \in L^1(E)$, then*

$$\int_E \varphi(f) dm = -\lim_{a \to -\infty,\, b \to \infty} \int_a^b \varphi(s) d\lambda_{f,E}(s).$$

In particular, when $\varphi(s) = |s|^p$, $p > 0$, and $|f|^p \in L^1(E)$,

$$\int_E |f|^p dm = -\int_0^\infty s^p d\lambda_{f,E}(s) = p\int_0^\infty s^{p-1}\lambda_{f,E}(s) ds. \tag{3.39}$$

The last equality follows from integration by parts. In fact,

$$-\int_0^\infty s^p d\lambda_{f,E}(s) = \lim_{a \to +0, b \to \infty} \left\{ \left[-s^p \lambda_{f,E}(s) \right]_a^b + p\int_a^b s^{p-1}\lambda_{f,E}(s) ds \right\}.$$

By Chebyshev's inequality

$$b^p \lambda_{f,E}(b) \le \int_{\{x : |f(x)| > b\}} |f(x)|^p dm \to 0 \quad (b \to \infty).$$

On the other hand, put $f_a(x) = 2a$ if $|f(x)| > 2a$, $= f(x)$ otherwise. Then, $|f_a(x)| \le |f(x)|$ and $f_a(x) \to 0$ as $a \to 0$. Therefore, by Chebyshev's inequality and Lebesgue's dominated convergence theorem we get

$$a^p \lambda_{f,E}(a) = a^p \lambda_{f_a,E}(a) \le \int_E |f_a(x)|^p dm \to 0 \ (a \to +0).$$

Thus $[-s^p \lambda_{f,E}(s)]_{+0}^{+\infty} = 0$.

3.9. Problems

3.1. Prove Corollary 3.1 (i),(ii) and (iii) for complex valued measurable functions.

3.2. Give an example of a sequence $\{f_j\}$ in $C([0,1])$ which converges everywhere to a continuous function but does not converge uniformly.

3.3. Even if $|f|$ is measurable, f is not always measurable. Show it by example.

3.4. Let $f_\iota(x)$, $\iota \in I$, be a family of lower semicontinuous functions. Then $\sup_{\iota \in I} f_\iota(x)$ is also lower semicontinuous.

3.5. Let f be a nonnegative function. Prove that if $\int f dm = 0$, then $f = 0$ a.e.

3.6. Let f_j and f be real valued integrable functions. Suppose that $|f_j(x)| \leq |f(x)|$ a.e. $(j = 1, 2, \ldots)$. Then prove that

$$\int \limsup_{j\to\infty} f_j dm \geq \limsup_{j\to\infty} \int f_j dm \geq \liminf_{j\to\infty} \int f_j dm \geq \int \liminf_{j\to\infty} f_j dm.$$

3.7. Is Fatou's lemma valid without the assumption $f_j \geq 0$? If not, give a counterexample.

3.8. Let $\{E_j\}$ be a sequence of measurable sets

(i) Prove that

$$\liminf_{j\to\infty} m(E_j) \geq m(\liminf_{j\to\infty} E_j).$$

(ii) If there exists a set E of finite measure such that $E_j \subset E$, $j = 1, 2, \ldots$, prove that

$$\limsup_{j\to\infty} m(E_j) \leq m(\limsup_{j\to\infty} E_j).$$

(iii) In general, (ii) is not valid. Give a counterexample.

3.9. Let f be a measurable function. For $k \in \mathbb{Z}$ put $E_k = \{x \in E : 2^k < |f(x)| \leq 2^{k+1}\}$. Show that $f \in L^1(E)$ if and only if

$$\sum_{k=-\infty}^{\infty} 2^k m(E_k) < \infty.$$

3.10. Let $f \in L^1(\mathbb{R})$. Prove the following.

(i) If E_j are measurable and $m(E_j) \to 0$, then

$$\lim_{j\to\infty} \int_{E_j} f dm = 0.$$

(ii) Let $F(t) = \int_{-\infty}^{t} f dm$. Then F is continuous in t and $F(t) \to 0 \, (t \to -\infty)$.

3.11. If $f \in L^1(\mathbb{R})$, prove that

$$\lim_{y\to\infty} \int_{\mathbb{R}} |f(x+y) - f(x)| dm(x) = 2 \int_{\mathbb{R}} |f(x)| dm(x).$$

3.12. Let $f \in L^1(0,1)$. Then prove that $x^n f(x)$ is integrable for each $n = 1, 2, \ldots$ and that

$$\lim_{n\to\infty} \int_0^1 x^n f(x) dm = 0.$$

3.13. Suppose that $f \in L^1(\mathbb{R})$ and φ is a bounded continuous function. Put

$$F(x) = \int_{-\infty}^{\infty} f(y)\varphi(x-y) dm(y).$$

Prove that $F(x)$ is bounded and continuous.

3.14. In the last problem assume, furthermore, that φ' is bounded and continuous. Then prove that $F(x)$ is differentiable, and that

$$\frac{dF(x)}{dx} = \int_{-\infty}^{\infty} f(y)\varphi'(x-y) dm(y).$$

3.15. Suppose that $f \in L^1(\mathbb{R})$. If $xf(x) \in L^1(\mathbb{R})$, then prove that the integral on the left-hand side of the following equation is differentiable and

$$\frac{d}{d\xi}\int_{-\infty}^{\infty} f(x)e^{-2\pi i\xi x}dm(x) = -2\pi i\int_{-\infty}^{\infty} xf(x)e^{-2\pi i\xi x}dm(x).$$

3.16. Let $0 < p$. Show that

$$\int_0^1 \frac{x^p}{1-x}\log\frac{1}{x}dx = \sum_{n=1}^{\infty}\frac{1}{(p+n)^2}.$$

3.17. Let $u_n(x) = e^{-nx} - 2e^{-2nx}$. Then prove that

$$\sum_{n=1}^{\infty}\int_0^{\infty} u_n(x)dm \neq \int_0^{\infty}\sum_{n=1}^{\infty} u_n(x)dm.$$

CHAPTER 4

Differentiation

Differentiation and integration are inverse operations to each other on the space of smooth functions; that is, if f is a continuous function on the real axis, we have

$$\frac{d}{dx}\int_a^x f(y)dy = f(x).$$

Conversely, if the derivative is continuous, integration of the derivative coincides with the original function:

$$\int_a^x f'(y)dy = f(x) - f(a).$$

In the theory of the Lebesgue integral the first relaton holds in the sense of "almost everywhere". The second relation contains a delicate and interesting problem. In Chapter 5 the functions for which the second relation holds are characterized. In fact, this problem is treated in a more general setting, and the Radon-Nikodým derivative is introduced.

4.1. Definition of differentiation

Let f be a real valued function defined in a neighborhood of $x \in \mathbb{R}$. Define

$$D^+f(x) = \limsup_{h\to+0}\frac{f(x+h)-f(x)}{h},$$

$$D_+f(x) = \liminf_{h\to+0}\frac{f(x+h)-f(x)}{h},$$

$$D^-f(x) = \limsup_{h\to+0}\frac{f(x)-f(x-h)}{h},$$

and

$$D_-f(x) = \liminf_{h\to+0}\frac{f(x)-f(x-h)}{h}.$$

These four limits are called the *Dini derivates* of f at x; more precisely, these four extended real numbers are called the upper right derivate, the upper left derivate, the lower right derivate and the lower left derivate, respectively. If

$$D^+f(x) = D_+f(x),$$

f is said to have a right derivate at x. The left derivate is defined analogously. If the four derivates are finite and equal, f is said to be *differentiable* at x, and the common value, which is called the *derivate* at x, is denoted by

$$Df(x), \quad \frac{d}{dx}f(x), \quad \text{or} \quad f'(x).$$

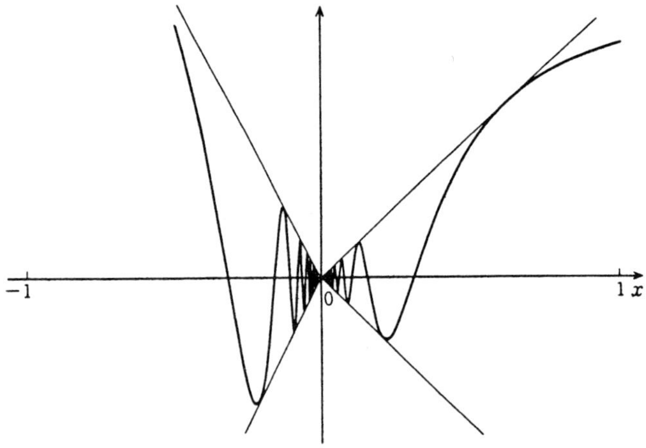

FIGURE 4.1. The four derivates of f at 0 are different.

Even if f is continuous, it can happen that the four derivates are all different. For example, let

$$f(x) = \begin{cases} x \sin \frac{1}{x} & (x > 0), \\ 0 & (x = 0), \\ 2x \sin \frac{1}{x} & (x < 0). \end{cases}$$

Then it is not difficult to see that f is continuous and

$$D^+ f(0) = 1, \quad D_+ f(0) = -1, \quad D^- f(0) = 2, \quad \text{and} \quad D_- f(0) = -2$$

(see Figure 4.1).

More strongly, there exists a continuous function which is not differentiable everywhere. Let $b \geq 2$ be any integer, and let $0 < a < 1$. The function f defined by the series

$$f(x) = \sum_{n=1}^{\infty} a^{-n} \cos 2\pi b^n x$$

is called the *Weierstrass*[1] *function*. Since the series converges uniformly, f is continuous and periodic with period 1, but it is known that f is not differentiable everywhere if $ab \geq 1$ (see [Ka] for the proof).

We shall give a similar example, called the Takagi function, whose nondifferentiability is proved by a direct computation.

Let

$$f(x) = \sum_{n=0}^{\infty} \frac{\operatorname{dist}(2^n x, \mathbb{Z})}{2^n},$$

where $\operatorname{dist}(a, \mathbb{Z})$ denotes for the distance from the point a to the integers $\mathbb{Z}$. The function $\Delta(x) = \operatorname{dist}(x, \mathbb{Z})$ is continuous and periodic with period 1. Thus $f(x)$ is again continuous and periodic with period 1. We observe that $0 \leq \Delta(x) \leq 1/2$ and $|D^{\pm}_{\pm} \Delta(x)| = 1$. The graph of f is obtained by piling the triangles in Figure 4.1 one after another.

[1] K. Weierstrass, 1815–1897. One of the progenitors of function theory.

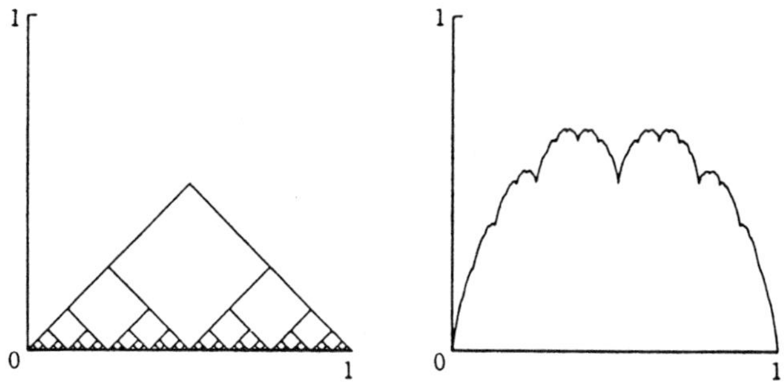

FIGURE 4.2. The Takagi function

Let us consider the derivates at $0 \le x < 1$. First, assume that x has a finite dyadic expansion, that is, $x = \sum_{j=1}^{\infty} x_j 2^{-j} = 0.x_1 x_2 \ldots$, where $x_j = 0$ or 1, and $x_j = 0\,(j \ge N)$ for some N.

Let $k \ge N$ and write

$$\frac{f(x+2^{-k}) - f(x)}{2^{-k}}$$

$$= \left(\sum_{n=1}^{N-1} + \sum_{n=N}^{k-1} + \sum_{n=k}^{\infty} \right) \frac{\Delta(2^n(x+2^{-k})) - \Delta(2^n x)}{2^{n-k}}. \tag{4.1}$$

The third sum on the right-hand side of (4.1) vanishes. Indeed,

$$\operatorname{dist}(2^n(x+2^{-k}), \mathbb{Z}) = \operatorname{dist}(2^n x, \mathbb{Z})$$

for $n \ge k$, and $2^n x$ is an integer. When $k \to \infty$ the first sum is bounded by

$$\limsup_{k\to\infty} \left| \sum_{n=1}^{N-1} \frac{\Delta(2^n x + 2^{n-k}) - \Delta(2^n x)}{2^{n-k}} \right| \le \sum_{n=1}^{N-1} |D_\pm^\pm \Delta(2^n x)| \le N.$$

Observe that if $N \le n < k$, we have $\operatorname{dist}(2^n x + 2^{n-k}, \mathbb{Z}) = \operatorname{dist}(2^{n-k}, \mathbb{Z}) = 2^{n-k}$, and $\operatorname{dist}(2^n x, \mathbb{Z}) = 0$. Then the second sum of (4.1) equals

$$\sum_{n=N}^{k-1} 1 = k - N \to \infty\,(k \to \infty).$$

Therefore we have $D_+ f(x) = \infty$. In the same way we can prove that $D_- f(x) = -\infty$. Thus f is not differentiable at x.

Next, suppose that $x = 0.x_1 x_2 \ldots$ has no finite dyadic expansion. By the above computation

$$\frac{f(x+2^{-k}) - f(x)}{2^{-k}} = \sum_{n=1}^{k-1} \frac{\Delta(2^n(x+2^{-k})) - \Delta(2^n x)}{2^{n-k}}.$$

Choose an integer k such that $x_k = 0$. By the assumption on x, infinitely many such k's exist. Let $n < k$. If $x_{n+1} = 0$, then the part of the dyadic fraction of $2^n(x+2^{-k})$ equals $0.x_{n+1} \ldots x_{k-1}\, 1\, x_{k+1} \ldots \le 1/2$. Thus

$$\frac{\Delta(2^n(x+2^{-k})) - \Delta(2^n \cdot x)}{2^n} = \frac{2^{n-k}}{2^n} = 2^{-k}.$$

If $x_{n+1} = 1$, since $0.x_{n+1} \dots x_{k-1}\, 1\, x_{k+1} \dots \geq 1/2$, we have

$$\frac{\Delta(2^n(x+2^{-k})) - \Delta(2^n \cdot x)}{2^n} = \frac{-2^{n-k}}{2^n} = -2^{-k}.$$

Thus

$$\frac{f(x+2^{-k}) - f(x)}{2^{-k}} = \sum_{n=1}^{k-1} \pm 1$$

for some sequence of $\pm$. In any case the right-hand side does not converge. Thus $D_{\pm} f(x)$ do not exist.

Therefore we conclude that f is not differentiable everywhere.

4.2. Differentiation theorem

4.2.1. The Hardy-Littlewood maximal theorem. For a function f on $\mathbb{R}^d$ define

$$Mf(x) = \sup_{r>0} \frac{1}{m(B(x,r))} \int_{B(x,r)} |f(y)|dy,$$

where $B(x,r) = \{y : |x-y| < r\}$. The function $Mf(x)$ is called the *Hardy-Littlewood*[2] *maximal function.*

For a fixed positive number r the function $I_r(x) = \int \chi_{B(x,r)}(y)|f(y)|dy$ is continuous in x. In fact, by Lebesgue's convergence theorem $I_r(x) \to I_r(x_0)$ as $x \to x_0$. Thus $Mf(x) = \sup I_r(x)/m(B(x,r))$ is lower semicontinuous (see Problem 3.4), and thus measurable.

THEOREM 4.1 (Hardy-Littlewood maximal theorem).

(i) *Let $f \in L^1(\mathbb{R}^d)$. For every $a > 0$ we have*

$$m(\{x : Mf(x) > a\}) \leq \frac{C}{a} \int_{\mathbb{R}^d} |f(y)|dy.$$

(ii) *Let $1 < p < \infty$ and $f \in L^p(\mathbb{R}^d)$. Then we have*

$$\int_{\mathbb{R}^d} (Mf)^p dx \leq \frac{2Cp}{p-1} \int_{\mathbb{R}^d} |f|^p dx.$$

Here C denotes a constant not depending on $f, a,$ and p.

For $r > 0$ choose an integer k so that $2^{-k-1} \leq r < 2^{-k}$. Put

$$M^r f(x) = \sup_Q \frac{1}{m(B(x,r))} \int_Q |f(y)|dy, \quad x \in \mathbb{R}^d,$$

where the supremum runs over all dyadic cubes Q with side length 2^{-k} such that $Q \cap B(x,r) \neq \emptyset$. Let $\sum_Q$ be the sum over such Q. Since the number of such cubes is at most 3^d, we have

$$\frac{1}{m(B(x,r))} \int_{B(x,r)} |f|dy \leq \sum_Q \frac{1}{m(B(x,r))} \int_Q |f|dy \leq 3^d M^r f(x).$$

Define

$$M^* f(x) = \sup_{r>0} M^r f(x).$$

---

[2] G. H. Hardy, 1877–1947, J. E. Littlewood, 1885–1977.

Then, we obtain

$$Mf(x) \le 3^d M^* f(x). \tag{4.2}$$

Therefore to prove the theorem it suffices to show the inequalities with M^*f in place of Mf.

PROOF OF THE THEOREM. (i) Put $E = \{x : M^*f(x) > a\}$. For every $x \in E$ there exist $r > 0$ and $Q \in \mathcal{Q}_k$ such that

$$\frac{1}{m(B(x,r))}\int_Q |f| dy > a \quad \text{and} \quad Q \cap B(x,r) \neq \emptyset, \tag{4.3}$$

where k is chosen so that $2^{-k-1} \le r < 2^{-k}$.

Let $\mathcal{A}$ be the collection of all such cubes Q, where x runs over the set E, and let $\mathcal{A}_0 = \{Q_j : j = 1, 2, \dots\}$ be the set of all maximal elements in $\mathcal{A}$. Denote by $3Q$ the cube concentric with Q and having three times the side length.

(a) The Q_j are nonoverlapping and $E \subset \bigcup_j 3Q_j$.

Indeed, for any $x \in E$ there exists $Q \in \mathcal{A}$ such that $x \in 3Q$. Furthermore, there exists $Q_j \in \mathcal{A}_0$ such that $Q \subset Q_j$. Therefore we get

$$m(E) \le \sum_{j=1}^{\infty} m(3Q_j) = 3^d \sum_{j=1}^{\infty} m(Q_j).$$

(b) By (4.3) we have

$$\Omega_d m(Q) a \le \Omega_d 2^d r^d a < 2^d \int_Q |f| dy \quad \text{for } Q \in \mathcal{A},$$

where Ω_d denotes the volume of the unit ball.

From (a) and (b) it follows that

$$m(E) \le \frac{3^d \cdot 2^d}{\Omega_d a} \sum_{j=1}^{\infty} \int_{Q_j} |f| dy \le \frac{6^d}{\Omega_d a} \int_{\mathbb{R}^d} |f| dy.$$

Therefore we get (i) with the constant $C = 6^d/\Omega_d$.

(ii) For $a > 0$ define $f^a(x) = f(x)$ if $|f(x)| > a/2$, and $= 0$ otherwise. Put $f_a = f - f^a$. Since $|f_a| \le a/2$, we have $Mf(x) \le Mf^a(x) + Mf_a(x) \le Mf^a(x) + a/2$. Thus

$$\{x : Mf(x) > a\} \subset \{x : Mf^a(x) > a/2\}.$$

Applying (i) to f^a, we get

$$m(\{x : Mf(x) > a\}) \le m(\{x : Mf^a(x) > a/2\}) \le \frac{C}{a/2} \int |f^a(y)| dy.$$

Therefore by the formula (3.39) and the last inequality we have

$$\begin{aligned}\int (Mf)^p dx &= p \int_0^{\infty} a^{p-1} m(\{x : Mf(x) > a\}) da \\ &\le p \int_0^{\infty} a^{p-1} \frac{2C}{a} \left(\int |f^a(y)| dy \right) da.\end{aligned}$$

Observe that the inner integral equals $\int_{\{y : |f(y)| > a/2\}} |f(y)| dy$. Changing the order of integration, we obtain

$$\int (Mf)^p dx \le 2Cp \int dy \int_0^{2|f(y)|} a^{p-2} da = \frac{2Cp}{p-1} \int |f(y)|^p dy.$$

This completes the proof. □

4.2.2. The Lebesgue differentiation theorem. We deduce the following differentiation theorem from the Hardy-Littlewood maximal theorem.

THEOREM 4.2 (Lebesgue's differentiation theorem). *If f is an integrable function in $\mathbb{R}^d$, then we have*

$$\lim_{r\to 0}\frac{1}{m(B(x,r))}\int_{B(x,r)} f(y)dy = f(x) \quad a.e.$$

PROOF. Let $f \in L^1(\mathbb{R}^d)$. For any $\varepsilon > 0$ put

$$E_\varepsilon = \left\{x : \limsup_{r\to 0}\left|\frac{1}{m(B(x,r))}\int_{B(x,r)} f(y)dy - f(x)\right| > \varepsilon\right\}.$$

Then we shall show that if $\varepsilon > 0$ is sufficiently small,

$$m(E_\varepsilon) < \varepsilon. \tag{4.4}$$

We assume (4.4) for a moment and prove the theorem. Put

$$N = \bigcap_{k=1}^{\infty}\bigcup_{j=k}^{\infty} E_{2^{-j}}.$$

Then N is a null set, since $\sum m(E_{2^{-j}}) < \infty$ (see Problem 2.2). If $x \notin N$, there exists k such that $x \notin E_{2^{-j}}$ $(j \geq k)$, that is,

$$\limsup_{r\to 0}\left|\frac{1}{m(B(x,r))}\int_{B(x,r)} f(y)dy - f(x)\right| \leq 2^{-j} \quad (\text{for all } j \geq k),$$

which implies the theorem.

To prove (4.4), choose a simple function $s = \sum_{j=1}^n a_j\chi_{I_j}$ so that $\int |f-s|dx < \varepsilon^3$, and the I_j's are intervals, which is possible by Theorem 3.7. Since the theorem is obviously valid for the characteristic functions of intervals, we have

$$\lim_{r\to 0}\frac{1}{m(B(x,r))}\int_{B(x,r)} s(y)dy = s(x) \quad \text{a.e.}$$

Therefore we have

$$\begin{aligned}
&\limsup_{r\to 0}\left|\frac{1}{m(B(x,r))}\int_{B(x,r)} f(y)dy - f(x)\right| \\
&= \limsup_{r\to 0}\left|\frac{1}{m(B(x,r))}\int_{B(x,r)} [f(y)-s(y)]dy - [f(x)-s(x)]\right| \\
&\leq M(f-s)(x) + |f(x)-s(x)|.
\end{aligned}$$

Thus

$$m(E_\varepsilon) \leq m(\{x : M(f-s)(x) > \varepsilon/2\}) + m(\{x : |f(x)-s(x)| > \varepsilon/2\}).$$

By the Hardy-Littlewood maximal theorem we have

$$m(\{x : M(f-s)(x) > \varepsilon/2\}) \leq 2C\varepsilon^{-1}\int |f-s|dx < 2C\varepsilon^2,$$

and by Chebyshev's inequality

$$m(\{x : |f(x)-s(x)| > \varepsilon/2\}) \leq 2\varepsilon^{-1}\int |f-s|dx < 2\varepsilon^2.$$

Therefore

$$m(E_\varepsilon) \le 2C\varepsilon^2 + 2\varepsilon^2.$$

The last term is less than ε if ε is small. Thus (4.4) is proved. □

Lebesgue's differentiation theorem holds if f is *locally integrable* in $\mathbb{R}^d$, that is, if f is integrable on all compact sets.

For example, if a set E is measurable, the characteristic function χ_E is locally integrable. Therefore by Lebesgue's differentiation theorem we have

$$\lim_{r\to 0} \frac{m(E \cap B(x,r))}{m(B(x,r))} = \chi_E(x) = 1 \text{ a.e.}$$

A point x for which the limit exists and equals 1 is called a *point of density* of E.

Thus almost every point of a measurable set E is a point of density.

A point x is called a *Lebesgue point* of f if

$$\lim_{r\to 0} \frac{1}{m(B(x,r))} \int_{B(x,r)} |f(y) - f(x)|dy = 0.$$

THEOREM 4.3. *If f is a locally integrable function in $\mathbb{R}^d$, then almost every point is a Lebesgue point of f.*

PROOF. Suppose that f is a real valued locally integrable function. For a real number s let N_s be the set of x's for which the following relation fails:

$$\lim_{r\to 0} \frac{1}{m(B(x,r))} \int_{B(x,r)} |f(y) - s|dy = |f(x) - s|. \tag{4.5}$$

By Theorem 4.2 the set N_s is null. Therefore the countable union $N = \bigcup\{N_s : s \in \mathbb{Q}\}$ is also null, and if $x \notin N$, then (4.5) is valid for every $s \in \mathbb{Q}$.

Let $x \notin N$. Since $|f(y) - f(x)| \le |f(y) - s| + |s - f(x)|$, we have

$$\begin{aligned} &\frac{1}{m(B(x,r))} \int_{B(x,r)} |f(y) - f(x)|dy \\ &\le \frac{1}{m(B(x,r))} \int_{B(x,r)} |f(y) - s|dy + |f(x) - s|, \end{aligned}$$

from which we have

$$\limsup_{r\to 0} \frac{1}{m(B(x,r))} \int_{B(x,r)} |f(y) - f(x)|dy \le 2|f(x) - s|$$

for all $s \in \mathbb{Q}$. Therefore the left-hand side must be zero. □

In particular, for an integrable function f of a single variable we have

$$\lim_{h\to 0} \frac{1}{h} \int_x^{x+h} f(y)dy = f(x) \text{ a.e.}$$

Therefore, we have

$$\frac{d}{dx} \int_a^x f(y)dy = f(x) \text{ a.e.}$$

4.3. The Vitali covering theorem and differentiation

Any characteristic function of an interval in the real line is almost everywhere differentiable and the derivative is 0 almost everywhere, although it is not represented by indefinite integral. In this section we shall discuss the differentiability of functions more precisely.

DEFINITION 4.1. Let $\mathcal{V}$ be a family of subsets of $\mathbb{R}^d$. $\mathcal{V}$ is said to cover a set E in the Vitali sense, if for every $x \in E$ and $\varepsilon > 0$ there exists $V \in \mathcal{V}$ such that

$$x \in V \text{ and } \operatorname{diam}(V) < \varepsilon,$$

where $\operatorname{diam}(V) = \sup_{x,y \in V} |x - y|$ is the diameter of V. In this case, the family $\mathcal{V}$ is said to be a *Vitali covering* of E.

THEOREM 4.4 (The Vitali covering theorem). *Let E be a subset of $\mathbb{R}^d$ with finite outer measure, and let $\mathcal{V}$ be a Vitali covering of E consisting of closed balls. Then there exists a countable disjoint subfamily $\{V_n\}_{n=1}^{\infty}$ of $\mathcal{V}$ such that*

$$m^*(E - \bigcup_{n=1}^{\infty} V_n) = 0.$$

PROOF. Fix any open set O of finite measure which includes E, and put

$$\mathcal{V}_0 = \{V \in \mathcal{V} : V \subset O\}.$$

$\mathcal{V}_0$ is clearly a Vitali covering of E, too. We shall choose the sets $\{V_n\}$ inductively from $\mathcal{V}_0$.

First Step. Pick any $V_1 \in \mathcal{V}_0$ and fix it. If $E \subset V_1$, then the proof is complete. Otherwise, put

$$O_1 = O - V_1 \quad \text{and} \quad \mathcal{V}_1 = \{V \in \mathcal{V}_0 : V \subset O_1\}.$$

The set O_1 is open and contains a point of E. Thus $\mathcal{V}_1 \neq \emptyset$ and any ball of $\mathcal{V}_1$ does not intersect V_1. Put $\delta_1 = \sup\{\operatorname{diam}(V) : V \in \mathcal{V}_1\}$ and choose $V_2 \in \mathcal{V}_1$ so that

$$\operatorname{diam}(V_2) > \frac{1}{2}\delta_1.$$

Second Step. Suppose that we have chosen $V_1, \ldots, V_n \in \mathcal{V}_0$, which are mutually disjoint. If $E \subset \bigcup_{j=1}^{n} V_j$, the argument stops. Otherwise, put

$$O_n = O - (V_1 \cup \cdots \cup V_n) \quad \text{and} \quad \mathcal{V}_n = \{V \in \mathcal{V}_0 : V \subset O_n\}.$$

The set O_n is open and contains a point of E. Thus $\mathcal{V}_n \neq \emptyset$. Put $\delta_n = \sup\{\operatorname{diam}(V) : V \in \mathcal{V}_n\}$ and choose $V_{n+1} \in \mathcal{V}_n$ so that

$$\operatorname{diam}(V_{n+1}) > \frac{1}{2}\delta_n.$$

The family $\{V_j\}$ which we have chosen possesses the following properties:

(a) $\{V_j\}$ is a finite or infinite sequence of mutually disjoint closed balls.

(b) Since $V_n \subset O$, $m(\bigcup V_j) = \sum m(V_j) \leq m(O) < \infty$. Thus $\delta_n \to 0$.

Third Step. We have also the property:

(c) Let $5V$ be the closed ball concentric with V and with 5 times the radius. Then

$$E - \bigcup_{j=1}^{\infty} V_j \subset \bigcup_{j=N}^{\infty} 5V_j \quad (N = 1, 2, \ldots).$$

In fact, let $x \in E - \bigcup_{j=1}^{\infty} V_j$. Then $x \in E - \bigcup_{j=1}^{N} V_j$ for any N. Thus $x \in O_N$. Choose a set $V \in \mathcal{V}_N$ which contains x. By property (b) we have $\delta_p < \operatorname{diam}(V)$ for a large p. Therefore by definition of δ_p, $V \notin \mathcal{V}_p$. Let p be the smallest integer such that $V \notin \mathcal{V}_p$. Then $p > N$, and we have

$$V \notin \mathcal{V}_p \quad \text{and} \quad V \in \mathcal{V}_{p-1},$$

which is equivalent to

$$V \cap \bigcup_{j=1}^{p} V_j \neq \emptyset \quad \text{and} \quad V \cap \bigcup_{j=1}^{p-1} V_j = \emptyset.$$

Thus

$$V \cap V_p \neq \emptyset.$$

Since $V \in \mathcal{V}_{p-1}$, we have $\operatorname{diam}(V) \leq \delta_{p-1}$ and $\delta_{p-1}/2 < \operatorname{diam}(V_p)$ by the definitions of δ_{p-1} and V_p, from which it follows that $V \subset 5V_p$. Since $p > N$ we have $V \subset \bigcup_{j=N}^{\infty} 5V_j$, and thus $x \in \bigcup_{j=N}^{\infty} 5V_j$, which implies (c).

Fourth Step. From (c) and (b) it follows that

$$m^*(E - \bigcup_{j=1}^{\infty} V_j) \leq \sum_{j=N}^{\infty} m(5V_j) = \sum_{j=N}^{\infty} 5^d m(V_j) \to 0 \ (N \to \infty),$$

which completes the proof of Theorem 4.4. □

REMARK 4.1. Theorem 4.4 holds when the family $\mathcal{V}$ of closed balls is replaced by a family of open balls. Indeed, the interior of V_j satisfies the conclusion of the theorem. Furthermore, the theorem is true when $\mathcal{V}$ consists of cubes, which is proved by a minor modification of the proof.

REMARK 4.2. Under the condition of Theorem 4.4, for any $\varepsilon > 0$ there exists a finite number N such that

$$m^*(E - \bigcup_{j=1}^{N} V_j) < \varepsilon.$$

A measure μ defined in the Borel field $\mathcal{B}(\mathbb{R}^d)$ is called a *Borel measure* if $\mu(K)$ is finite for every compact set K.

For example, the Lebesgue measure is a Borel measure. If f is a nonnegative and locally integrable function in $\mathbb{R}^d$, then the measure defined by

$$\mu_f(E) = \int_E f dm$$

is a Borel measure. Clearly $\mu_f(\emptyset) = 0$, and the countable additivity follows from B. Levi's theorem.

The differentiation theorem described in §4.2 is generalized as follows.

THEOREM 4.5 (Lebesgue's differentiation theorem). *Let μ be a nonnegative Borel measure. Then the limit*

$$\lim_{r \to 0} \frac{\mu(B(x,r))}{m(B(x,r))}$$

exists almost everywhere.

PROOF. First Step. Put

$$\overline{D}\mu(x) = \limsup_{B\to x} \frac{\mu(B)}{m(B)} \quad \text{and} \quad \underline{D}\,\mu(x) = \liminf_{B\to x} \frac{\mu(B)}{m(B)},$$

where B denotes any open ball containing x. Let

$$N = \{x \in \mathbb{R}^d : \overline{D}\mu(x) > \underline{D}\,\mu(x)\}.$$

Then the assertion follows from $m^*(N) = 0$.

For $R > 0$ and $a > b$ put

$$E(R, a, b) = \{x \in \mathbb{R}^d : \overline{D}\mu(x) > a > b > \underline{D}\mu(x),\ |x| < R\}.$$

Then $N = \bigcup_{a,b,R} E(R, a, b)$, where $a > b$ and $R > 0$ run over rational numbers. Now, we suppose contrarily that $m^*(N) > 0$. Then at least one $E(R, a, b)$ has positive outer measure, which we denote by E.

Second Step. Fix a bounded open set O which includes E. Let $\mathcal{V}$ be the family of all open balls B such that $B \subset O$ and $bm(B) > \mu(B)$. Then $\mathcal{V}$ is a Vitali covering of E. Therefore we can choose a sequence of disjoint open balls $B_1, B_2, \ldots$ from $\mathcal{V}$ which satisfies

$$m^*(E - \bigcup_{j=1}^{\infty} B_j) = 0 \quad \text{and} \quad \bigcup_{j=1}^{\infty} B_j \subset O. \tag{4.6}$$

Thus

$$\mu(\bigcup_j B_j) = \sum_j \mu(B_j) \le b \sum_j m(B_j) = bm(\bigcup_j B_j) \le bm(O).$$

Since O is an arbitrary open set containing E, we get

$$\mu(\bigcup B_j) \le bm^*(E). \tag{4.7}$$

Set $E' = E \cap \bigcup B_j$. Then $m^*(E') = m^*(E)$, since $E - E' = E - \bigcup B_j$ which is null by (4.6). Fix an open set O' containing E', and let $\mathcal{V}'$ be the family of open balls B such that $B \subset O'$ and $\mu(B) > am(B)$. Then $\mathcal{V}'$ is a Vitali covering of E'. Thus we get again a sequence of disjoint open balls $\{B'_k\}$ in $\mathcal{V}'$ such that $m^*(E' - \bigcup_k B'_k) = 0$, from which it follows that

$$m(\bigcup_k B'_k) \ge m^*(E') = m^*(E).$$

On the other hand,

$$\mu(O') \ge \mu(\bigcup_k B'_k) = \sum_k \mu(B'_k) \ge a \sum_k m(B'_k) = am(\bigcup_k B'_k).$$

Therefore, combining the above two inequalities, we obtain

$$\mu(O') \ge am^*(E). \tag{4.8}$$

In particular, if $O' = \bigcup_j B_j$, then by (4.7) and (4.8) we get

$$\mu(\bigcup B_j) \ge am^*(E) > bm^*(E) \ge \mu(\bigcup B_j),$$

which is absurd. □

Let $f(x)$ be a nondecreasing bounded function in $\mathbb{R}$. We assume that f is continuous on the right. For the interval $I = [a, b)$ define

$$\mu(I) = f(b) - f(a).$$

In the same way as in the case of the Lebesgue measure m, the function μ is extended to a finitely additive set function on $\mathcal{R}(\mathbb{R})$. Furthermore we shall show, in Example 5.3, that μ is extended to a measure on the Borel field $\mathcal{B}(\mathbb{R})$. Temporarily assuming this fact, we have

COROLLARY 4.1. *Every nondecreasing function is almost everywhere differentiable.*

In the Lebesgue theory of integration one must introduce a new notion of measures to establish the formula $f(x) = \int_{-\infty}^{x} f'(x)dt$. As we shall see in the following example, this formula is not valid even if f ia continuous, and f' exists almost everywhere and is integrable.

EXAMPLE 4.1 (The Cantor singular function). Let us define a set G in the square $[0,1]\times[0,1]$ as follows. Let G_{01} be the "middle-third" line segment $\{(x, 1/2) : 1/3 < x < 2/3\}$. Next, consider two rectangles $[0, 1/3] \times [0, 1/2]$ and $[2/3, 1] \times [1/2, 1]$. Let $G_{1,1}$ and $G_{1,2}$ be the middle-third segments $\{(x, 1/2^2) : 1/3^2 < x < 2/3^2\}$ and $\{(x, 1/2 + 1/2^2) : 2/3 + 1/3^2 < x < 2/3 + 2/3^2\}$ contained in these two rectangles, respectively (see Figure 4.3).

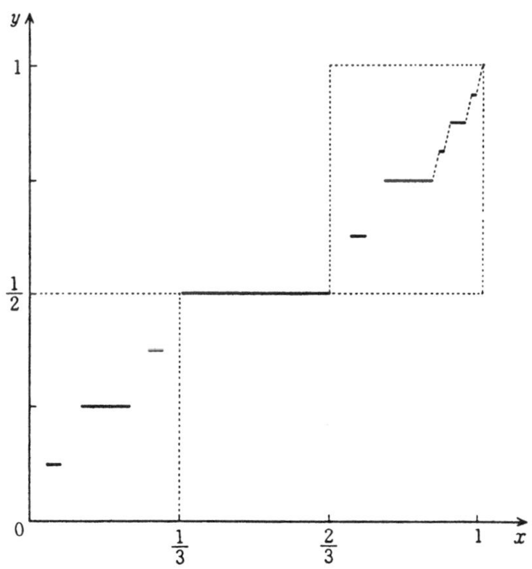

FIGURE 4.3. The Cantor singular function

In the same way we define the line segments $G_{j,k}$ $(k = 1, 2, \ldots, 2^j,\ j = 1, 2, \ldots)$ in the 2^j rectangles of size $3^{-j} \times 2^{-j}$.

Let G be the closure of $\bigcup_{j,k} G_{j,k}$. The Cantor singular function f is defined by the graph G. The function $f(x)$ is single valued and nondecreasing on the set $[0,1] - C$, which is a union of the middle-third open intervals. If x is in the Cantor set C and if $\{x_n\}$ is a sequence in $[0,1] - C$ which converges to x, then

$f(x) = \lim_{n\to\infty} f(x_n)$. Thus f is single valued and continuous in $[0,1]$, and constant on each middle-third interval.

If $x \notin C$, that is, if x is contained in a middle-third inteval, then $f'(x) = 0$. Since $m(C) = 0$, we have $f'(x) = 0$ a.e.

Therefore, f is continuous, and f' exists almost everywhere and is integrable, but

$$1 = f(1) - f(0) \neq \int_0^1 f' dm = 0.$$

4.4. Problems

4.1. Give an example of real valued functions f and φ in $\mathbb{R}$ such that f is nonmeasurable and φ is continuous but $\varphi(f(x))$ is measurable.

4.2. Let f be the Cantor singular function. Show that $f^{-1}(A)$ is measurable for every (not necessarily measurable) subset A of $[0,1]$.

Thus the continuous image of a measurable set is not necessarily measurable.

4.3. Put $\varphi(x) = 3x\,(0 \leq x \leq 1/3)$, $= 0\,(1/3 < x < 2/3)$, and $= 3x - 2$ $(2/3 \leq x \leq 1)$. Let $\varphi^0(x) = x$, and $\varphi^n(x) = \varphi(\varphi^{n-1}(x))\,(n \geq 1)$. Then prove that the Cantor singular function f is expressed as

$$f(x) = \sum_{n=0}^{\infty} 2^{-n-1}\chi(\varphi^n(x)),$$

where χ is the characteristic function of the interval $(1/3, 1]$.

4.4. Let f be a real valued continuous function on the real axis. Prove that the Dini derivative $D^+f(x)$ is measurable.

4.5. A function f in $[a,b]$ is said to satisfy a *Lipschitz condition* on $[a,b]$ if there exists a constant $C < \infty$ such that

$$|f(x) - f(y)| \leq C|x - y| \quad (x, y \in [a,b]).$$

Prove that any Lipschitz function is almost everywhere differentiable (Rademacher's theorem).

CHAPTER 5

Measures in Abstract Spaces

In the previous chapters we have introduced the σ-field of measurable sets of d-dimensional Euclidean space, Lebesgue measure and the Lebesgue integration. In the present chapter, we give our attention to the fundamental properties of these notions and try to extend them to abstract sets.

In the Euclidean space case, first we introduced the triple $(\mathbb{R}^d, \mathcal{R}(\mathbb{R}^d), m)$ and then extended it to the outer measure space $(\mathbb{R}^d, \mathcal{P}(\mathbb{R}^d), m^*)$. Finally it was reuduced to the Lebesgue measure space $(\mathbb{R}^d, \mathcal{M}(\mathbb{R}^d), \tilde{m})$ via the Carathéodory condition. This method works in the general setting $(X, \mathcal{R}(X), \mu)$, where X is an abstract set, $\mathcal{R}(X)$ is a ring of subsets of X and μ is a countably additive set function, without any essential change in the arguments.

For example, let X be the set of all infinite sequence $x = (x_1, x_2, \dots)$, where $x_j = 0$ or 1. Each x corresponds to making infinite tosses of a coin. For a given set of finite numbers $e_{k_1}, \dots, e_{k_n}$ ($e_{k_j} = 0$ or 1), let $E(e_{k_1}, \dots, e_{k_n})$ be the set of $x = (x_1, x_2, \dots) \in X$ such that $x_{k_j} = e_{k_j}$ for $j = 1, \dots, n$ but the x_i are arbitrary for $i \neq k_1, \dots, k_n$, where 0 stands for heads and 1 for tails. If $\mathcal{R}(X)$ denotes the family which consists of all such sets $E(e_{k_1}, \dots, e_{k_n})$, it is a ring of sets. The set function μ defined by $\mu(E(e_{k_1}, \dots e_{k_n})) = 2^{-n}$ is a finitely additive set function in $\mathcal{R}(X)$. The measure $\mu(E(e_{k_1}, \dots e_{k_n}))$ is interpreted as the probability of the event that we get the previously-specified head or tail of a coin at the $k_1, \dots, k_{n-1}$ and k_nth steps. Actually the triple $(X, \mathcal{R}(X), \mu)$ is extended to a measure space $(X, \mathcal{M}(X), \tilde{\mu})$.

Extentions of Euclidean space to abstract sets and of Lebesgue measure to general countably additive set functions suggests a measure theoretic treatment of probability theory and many important applications to functional analysis. In fact, Kolmogorov[1] laid the foundation of probability theory based on the abstract measure theory.

5.1. Measures in an abstract space

In order to make the foundations clear we repeat the definitions of σ-field and measure which are given in § 3.1.

Let X be a set. $\mathcal{P}(X)$ denotes the family of all subsets of X. A subfamily $\mathcal{R}$ of $\mathcal{P}(X)$ is said to be a ring of sets if for $E, F \in \mathcal{R}$ we have $E - F \in \mathcal{R}$ and $E \cup F \in \mathcal{R}$.

A set function μ with values in $\mathbb{R} \cup \{\infty\}$ on a ring $\mathcal{R}$ is said to be finitely additive if

(i) $\mu(\emptyset) = 0$,

(ii) $E, F \in \mathcal{R}$ and $E \cap F = \emptyset$ imply $\mu(E \cup F) = \mu(E) + \mu(F)$.

[1] A. N. Kolmogorov, 1903–1987. The author of Grundbegriffe der Wahrscheinlichkeitsrechnung (English translation: Foundation of the Theory of Probability).

Furthermore, if μ satisfies condition (iii), μ is called a countably additive set function or a measure:

(iii) For any sequence E_j $(j = 1, 2, \dots)$ of mutually disjoint sets in $\mathcal{R}$,

$$\mu(\bigcup_{j=1}^{\infty} E_j) = \sum_{j=1}^{\infty} \mu(E_j).$$

A finitely additive set function is not necessarily countably additive.

A finitely additive set function μ in a ring $\mathcal{R}$ is said to be *continuous from below* at $E \in \mathcal{R}$, if for every nondecreasing sequence $\{E_j\}$ in $\mathcal{R}$ such that $\lim_{j\to\infty} E_j = E$ we have $\lim_{j\to\infty}\mu(E_j) = \mu(E)$. Similarly, μ is said to be *continuous from above* at E, if for every nonincreasing sequence $\{E_j\}$ such that $\lim_{j\to\infty} E_j = E$ and $\mu(E_k) < \infty$ for some k, we have $\lim_{j\to\infty}\mu(E_j) = \mu(E)$.

As we have shown in § 2.4, the measures on a ring of sets are continous both from below and from above.

THEOREM 5.1. *Let μ be a finitely additive set function on the ring $\mathcal{R}$. If μ is either continuous from below, or finite-valued and continuous from above at $\emptyset$, then μ is countably additive on $\mathcal{R}$.*

PROOF. To prove the countable additivity let $\{E_j\}$ be a disjoint sequence in $\mathcal{R}$ whose union $E = \bigcup_{j=1}^{\infty} E_j$ also belongs to $\mathcal{R}$.

Suppose that μ is continuous from below. Put $F_k = \bigcup_{j=1}^{k} E_j$. Since $F_1 \subset F_2 \subset \cdots \subset F_k \to E$, we have $\lim \mu(F_k) = \mu(E)$ by assumption. On the other hand, by finite additivity we have $\mu(F_k) = \mu(E_1) + \cdots + \mu(E_k)$. Thus $\lim_{k\to\infty} \sum_{j=1}^{k} \mu(E_j) = \mu(E)$, which implies that μ is countably additive.

Next, suppose that μ is finite-valued and continuous from above at $\emptyset$. Since $E - F_1 \supset E - F_2 \supset \cdots \supset E - F_k \to \emptyset$,

$$\lim \mu(E - F_k) = \mu(\emptyset) = 0.$$

Thus $\lim[\mu(E) - \mu(F_k)] = 0$ or, equivalently, $\mu(E) = \lim \mu(F_k)$, since $m(E)$ is finite. □

EXAMPLE 5.1. If $\mathcal{R}$ is the collection of all countable subsets of a set X, $\mathcal{R}$ is a ring of sets. The set function μ defined by $\mu(E) = \text{card}(E)$ is a measure on $\mathcal{R}$.

EXAMPLE 5.2 (Lebesgue measures on Euclidean spaces). Let $\mathcal{R}(\mathbb{R}^d)$ be the collection of all finite union of right half-open intervals in $\mathbb{R}^d$. Then $\mathcal{R}(\mathbb{R}^d)$ is a ring of sets. Let us define the set function m as follows. For the right half-open interval $I = [a_1, b_1) \times \cdots \times [a_d, b_d)$ put $m(I) = \prod_{j=1}^{d}(b_j - a_j)$. If E is a finite union of disjoint right half-open intervals $I_1, \dots, I_k$, then define $m(E) = m(I_1) + \cdots + m(I_k)$. The set function m is finitely additive on $\mathcal{R}(\mathbb{R}^d)$ (see § 2.1). Furthermore, it is continuous from above at $\emptyset$ (see Lemma 2.3). Thus m is a measure on $\mathcal{R}(\mathbb{R}^d)$.

EXAMPLE 5.3. Let μ be a bounded nondecreasing function on $\mathbb{R}$ which is continuous on the right. Put

$$\mu([a, b)) = \mu(b) - \mu(a).$$

If E is a set which is a union of mutually disjoint right half-open intervals $I_1, \dots, I_k$, define $\mu(E) = \mu(I_1) + \cdots + \mu(I_k)$. The value $\mu(E)$ is independent of the expressions of E, as is proved in the same way as in § 2.1 with μ in place of m. Since each set of $\mathcal{R}(\mathbb{R})$ has such an expression, μ is a finitely additive set function on $\mathcal{R}(\mathbb{R})$.

Furthermore, μ is continuous from above at $\emptyset$. In fact, Lemma 2.3 is valid when m is replaced by μ. Thus μ is a measure.

5.2. Extension of measures

Let $\mathcal{F}$ be a ring of sets in $\mathcal{P}(X)$. If $\mathcal{F}$ is closed under countable union, that is, if $E_j \in \mathcal{F}$ $(j = 1, 2, \dots)$ imply $\bigcup_{j=1}^{\infty} E_j \in \mathcal{F}$, then $\mathcal{F}$ is called a σ-ring. Furthermore, if $X \in \mathcal{F}$, it is called a σ-algebra or σ-field.

In Chapter 2 we extended the measure m on $\mathcal{R}(\mathbb{R}^d)$ to a measure on the σ-field $\mathcal{M}(\mathbb{R}^d)$ via an outer measure and the Carathéodory condition. In the present section, we shall show that if μ is a nonnegative measure on a ring of sets $\mathcal{R}(X)$ in $\mathcal{P}(X)$, then it can be extended to a measure on a σ-field $\mathcal{M}_\mu(X)$. The proof is accomplished by tracing through §§ 2.2 and 2.3 word by word.

Let $\mathcal{R}(X)$ be a ring of sets in $\mathcal{P}(X)$ and μ be a nonnegative measure on $\mathcal{R}(X)$. For $A \in \mathcal{P}(X)$ define

$$\mu^*(A) = \inf\{\sum_{j=1}^{\infty} \mu(E_j) : A \subset \bigcup_{j=1}^{\infty} E_j \text{ and } E_j \in \mathcal{R}(X)\}.$$

If A has no countable covering of sets in $\mathcal{R}(X)$, then $\mu^*(A) = \infty$ by definition.

It is clear that if $E \in \mathcal{R}(X)$, then $\mu^*(E) = \mu(E)$.

THEOREM 5.2. *μ^* is an outer measure on $\mathcal{P}(X)$.*

PROOF. The reader will find the definition of outer measure in § 2.2. A proof follows from the proof of Theorem 2.2 if m^* is replaced by μ^*. □

Let $\mathcal{M}_\mu(X)$ be the collection of the subsets E of X which satisfy the Carathéodory condition, that is,

$$\text{(C)} \qquad \mu^*(A) = \mu^*(A \cap E) + \mu^*(A - E)$$

for all $A \in \mathcal{P}(X)$.

DEFINITION 5.1. A set in $\mathcal{M}_\mu(X)$ is called a *μ-measurable set.*

Define the set function $\tilde{\mu}$ by

$$\tilde{\mu} = \mu^*|_{\mathcal{M}_\mu(X)}.$$

THEOREM 5.3. *Let $\mathcal{R}(X)$ be a ring of sets in $\mathcal{P}(X)$ and μ be a nonnegative measure on $\mathcal{R}(X)$. Then:*

(i) *$\mathcal{M}_\mu(X) \supset \mathcal{R}(X)$.*
(ii) *$\mathcal{M}_\mu(X)$ is a σ-field.*
(iii) *$\tilde{\mu}$ is a measure on $\mathcal{M}_\mu(X)$.*

PROOF. The proof is done by following word by word the proof of Theorem 2.4, where we started from the triple $(\mathbb{R}^d, \mathcal{R}(\mathbb{R}^d), m)$ to construct the collection $\mathcal{M}(\mathbb{R}^d)$ of measurable sets. Replace the triple $(\mathbb{R}^d, \mathcal{R}(\mathbb{R}^d), m)$ by $(X, \mathcal{R}(X), \mu)$, $(\mathbb{R}^d, \mathcal{P}(\mathbb{R}^d), m^*)$ by $(X, \mathcal{P}(X), \mu^*)$, and $(\mathbb{R}^d, \mathcal{M}(\mathbb{R}^d), \tilde{m})$ by $(X, \mathcal{M}(X), \tilde{\mu})$. Then every argument of the proof of Theorem 2.4 is carried forward without any change. □

DEFINITION 5.2. A subset N of X with $\mu^*(N) = 0$ is called a *μ-null set* or, simply, a null set.

Every μ-null set is a μ-measurable set.

A property $\mathbf{P}(x)$ on a set E is said to hold *μ-almost everywhere* in E or, simply, $\mathbf{P}(x)$ μ-a.e., if it holds for all x in E except in some null subset of E.

μ is called a *finite measure* if $\mu(X) < \infty$. If there exists a countable family of measurable sets $\{X_n\}$ of finite measure which covers X, then μ is called a *σ-finite measure* .

Note that if μ is a σ-finite measure in $\mathcal{R}(X)$, then $\tilde{\mu}$ is σ-finite in $\mathcal{M}_\mu(X)$, too.

THEOREM 5.4. *Let μ be a nonnegative σ-finite measure on the ring $\mathcal{R}(X)$. Then $\mathcal{M}_\mu(X)$ is the smallest σ-field which contains $\mathcal{R}(X)$ and all μ-null sets.*

We shall generalize the notion of G_δ-set as follows. We call a subset A of X a G_δ-set if A is expressed as $A = \bigcap_{j=1}^\infty G_j$, where the G_j are countable unions of sets in $\mathcal{R}(X)$.

Let $\sigma(\mathcal{R})(X)$ be the smallest σ-field which includes $\mathcal{R}(X)$. Evidently every G_δ-set is contained in $\sigma(\mathcal{R})(X)$.

PROOF. The proof is similar to that of Theorem 2.6.

We have already shown that $\mathcal{M}_\mu(X)$ is a σ-field which includes $\mathcal{R}(X)$, and contains all μ-null sets. Therefore to complete the proof it suffices to show that if $E \in \mathcal{M}_\mu(X)$, there exist G in $\sigma(\mathcal{R})(X)$ and a μ-null set N such that $E = G - N$.

Let $E \in \mathcal{M}_\mu(X)$. First, assume that $\tilde{\mu}(E) < \infty$. By the definition of outer measure, for every $k = 1, 2, \dots$ there exists $E_j^k \in \mathcal{R}(X)$ such that

$$E \subset \bigcup_{j=1}^\infty E_j^k \quad \text{and} \quad \sum_{j=1}^\infty \mu(E_j^k) < \tilde{\mu}(E) + \frac{1}{2^k}. \tag{5.1}$$

Put $G = \bigcap_{k=1}^\infty \bigcup_{j=1}^\infty E_j^k$. Then G is a G_δ-set and $E \subset G$. Thus $\tilde{\mu}(E) \leq \mu^*(G) = \tilde{\mu}(G)$. On the other hand, by (5.1) $\mu^*(G) \leq \tilde{\mu}(E) + 1/2^k$, $k = 1, 2, \dots$. Thus $\tilde{\mu}(G) \leq \tilde{\mu}(E)$. Therefore $\tilde{\mu}(G) = \tilde{\mu}(E)$. If $N = G - E$, then $\tilde{\mu}(N) = \tilde{\mu}(G) - \tilde{\mu}(E) = 0$. Thus the theorem is proved when $\tilde{\mu}(E) < \infty$.

In general, by the σ-finiteness X is written as a countable union $\bigcup_{n=1}^\infty X_n$, where X_n are sets in $\mathcal{R}(X)$ with finite measures. For each n there exists a G_δ-set G_n such that $X_n \cap E \subset G_n \subset X_n$ and $N_n = G_n - (X_n \cap E)$ is μ-null. If $G = \bigcup G_n$ and $N = G - E$, then $E = G - N$. Now $G \in \sigma(\mathcal{R})(X)$ and $N \subset \bigcup N_n$, which is μ-null. Therefore the proof is complete. □

Let $\mathcal{M} \subset \mathcal{P}(X)$ be a σ-field and μ be a nonnegative measure on $\mathcal{M}$. When μ is a finite measure, the triple $(X, \mathcal{M}, \mu)$ is called a *finite measure space*. In particular, if $\mu(X) = 1$, $(X, \mathcal{M}, \mu)$ is called a *probability space*. Similarly, if μ is σ-finite, $(X, \mathcal{M}, \mu)$ is called a *σ-finite measure space*.

DEFINITION 5.3. A measure space $(X, \mathcal{M}, \mu)$ is said to be *complete* if $\mathcal{M}$ contains every μ-null set.

Let μ be a σ-finite measure on a finitely additive ring of sets $\mathcal{R}$. Let $\overline{\mu} = \mu^*|_{\sigma(\mathcal{R})}$. Then $(X, \sigma(\mathcal{R}), \overline{\mu})$ is an extension of μ to the σ-field generated by $\mathcal{R}$, and $(X, \mathcal{M}_\mu, \tilde{\mu})$ is a completion. That is, Theorems 5.3 and 5.4 imply that for a given σ-finite measure on a ring of sets there exists an extension to the complete measure space.

In the following we assume that every σ-finite measure space is complete, and we write simply μ for $\tilde{\mu}$.

Note that a finitely additive set function on a ring of sets is not necessarily extendable to a measure on a σ-field.

THEOREM 5.5 (The Hopf-Kolmogorov extension theorem). *Let μ be a nonnegative finitely additive set function on a ring of sets $\mathcal{R}(X)$. A necessary and sufficient condition for μ to be extendable to a measure on $\sigma(\mathcal{R})(X)$ is that μ is countably additive on $\mathcal{R}(X)$.*

Furthermore, if $(X, \mathcal{R}(X), \mu)$ is σ-finite, the extension is unique.

PROOF. The sufficiency part is already proved. The necessity part is obvious. We shall prove the uniqueness.

Let us denote the extension $\mu^*|_{\sigma(\mathcal{R})(X)}$ of μ by the same notation μ, and let ν be any extension of μ to a measure on $\sigma(\mathcal{R})(X)$. Let $E \in \sigma(\mathcal{R})(X)$. Then by the definition of μ^*

$$\nu(E) \leq \mu(E). \tag{5.2}$$

Indeed, if $\{E_j : j = 1, 2, \ldots\}$ is a family in $\mathcal{R}(X)$ which covers E, then

$$\nu(E) \leq \nu(\bigcup_j E_j) \leq \sum_j \nu(E_j) = \sum_j \mu(E_j).$$

Taking the infimum over $\{E_j\}$, we get (5.2) by the definition of μ^*.

To get the reverse inequality to (5.2), we assume that

$$\nu(E) < \mu(E)$$

for a set E in $\sigma(\mathcal{R})(X)$. By σ-finiteness of μ there exists an increasing sequence $\{X_j\}$ in $\mathcal{R}(X)$ such that $\mu(X_j) < \infty \, (j = 1, 2, \ldots)$ and $\lim X_j = X$. Since $\mu(E) = \lim_{j\to\infty}\mu(E \cap X_j)$ and $\nu(E) = \lim_{j\to\infty}\nu(E \cap X_j)$, we have

$$\nu(E \cap X_n) < \mu(E \cap X_n) \tag{5.3}$$

for at least one n. On the other hand,

$$\mu(X_n) = \mu(X_n \cap E) + \mu(X_n - E), \tag{5.4}$$

$$\nu(X_n) = \nu(X_n \cap E) + \nu(X_n - E). \tag{5.5}$$

Since $\mu = \nu$ on $\mathcal{R}(X)$, $\nu(X_n) = \mu(X_n)$. Therefore by (5.3), (5.4) and (5.5) we have $\mu(X_n - E) < \nu(X_n - E)$, which contradicts (5.2). □

5.3. Integration in a measure space

Let $(X, \mathcal{M}, \mu)$ be a complete σ-finite measure space. An extended real valued function f in X is said to be *measurable* if

$$\{x \in X : f(x) > a\} \in \mathcal{M} \quad \text{for every } a \in \mathbb{R}.$$

The integral $\int f d\mu$ for a measurable function f is defined quite similarly to the integration on the Euclidean spaces which we studied in Chapter 3.

Let $s(x) = \sum_{j=1}^{k} a_j \chi_{E_j}(x)$ be a simple function, where $a_j > 0$ and $E_j \in \mathcal{M}$. Then the integral of s is defined by

$$\int_X s d\mu = \sum_{j=1}^{k} a_j \mu(E_j).$$

For nonnegative measurable function f let $\{s_j\}$ be a sequence of simple functions such that

$$0 \leq s_1(x) \leq s_2(x) \leq \cdots \leq s_j(x) \to f(x). \tag{5.6}$$

Then the integral of f over the space X is defined by

$$\int_X f d\mu = \lim_{j\to\infty} \int_X s_j d\mu.$$

One example of a sequence $\{s_j\}$ of simple functions which satisfies condition (5.6) is given in the following way. Let $\{X_j\}$ be an increasing sequence of sets in $\mathcal{M}$ of finite measures which converges to X. Define simple functions s_j by

$$s_j(x) = \begin{cases} (k-1)2^j & (x \in X_j \text{ and } (k-1)2^{-j} \le f(x) < k\cdot 2^{-j}), \\ & \qquad k = 1, 2, \ldots, j\cdot 2^j, \\ j & (x \in X_j \text{ and } f(x) \ge j), \\ 0 & (x \notin X_j). \end{cases}$$

The definitions of the integral of s and f do not depend on the expression of the simple function s and the choice of the increasing sequences $\{s_j\}$ of simple functions, respectively. This is proved quite similarly as in Chapter 3.

For a real valued function the integral is defined as a difference of integrals of positive part and negative part of the function. For a complex valued function the integral is defined as the sum of the real part and the imaginary part.

The definition of the space $L^1(X, \mathcal{M}, \mu)$ is similar to that of $L^1(\mathbb{R}^d, \mathcal{M}(\mathbb{R}^d), m)$. Let $\mathcal{L}$ be the set of all measurable functions f such that $\int_X |f| d\mu < \infty$. For $f, g \in \mathcal{L}$ the definition

$$f \sim g \quad \Leftrightarrow \quad f = g \text{ a.e.}$$

yields an equivalence relation in $\mathcal{L}$. Let $L^1 = \mathcal{L}/\sim$ be the family of all equivalence classes in $\mathcal{L}$. To clarify situations, L^1 is sometimes written as $L^1(X, \mathcal{M}, \mu)$, $L^1(X)$, $L^1(\mu)$, etc.

Let $\dot{f}$ be a class in L^1. Then for $f', f'' \in \dot{f}$ we have $f' = f''$ a.e. Thus

$$\int_X f' d\mu = \int_X f'' d\mu,$$

which enables us to define the integral of $\dot{f} \in L^1$ by

$$\int_X \dot{f} d\mu = \int_X f' d\mu.$$

In the following we do not distinguish functions of $\mathcal{L}$ and classes of L^1 when there will be no confusion.

B. Levi's theorem, Fatou's lemma and Lebesgue's convergence theorems hold on the σ-finite complete measure space $(X, \mathcal{M}, \mu)$. All theorems on Lebesgue integration stated in Chapter 3 are valid on the space $(X, \mathcal{M}, \mu)$ except Dini's and Lusin's theorems in § 3.5. The representation by Riemann-Stieltjes integrals still holds in the present situation (see Theorem 3.30).

The proofs of these theorems, except Lusin's and Dini's, remain valid when $(\mathbb{R}^d, \mathcal{M}(\mathbb{R}^d), m)$ is replaced by $(X, \mathcal{M}, \mu)$.

5.4. Product measures

Let $(X, \mathcal{X}, \mu)$ and $(Y, \mathcal{Y}, \nu)$ be two nonnegative σ-finite complete measure spaces.

We denote by $\mathcal{R}(X) \otimes \mathcal{R}(Y)$ the collection of all finite unions of sets of the form $E \times F$, where $E \in \mathcal{X}$ and $F \in \mathcal{Y}$.

LEMMA 5.1. *$\mathcal{R}(X) \otimes \mathcal{R}(Y)$ is a Boolean algebra.*

PROOF. We have $X \times Y \in \mathcal{R}(X) \otimes \mathcal{R}(Y)$, since $X \in \mathcal{X}$ and $Y \in \mathcal{Y}$.

First, we show that if $A \in \mathcal{R}(X) \otimes \mathcal{R}(Y)$, then $A^c \in \mathcal{R}(X) \otimes \mathcal{R}(Y)$. Indeed, if A is a product set $E \times F$ where $E \in \mathcal{X}$ and $F \in \mathcal{Y}$, then we have

$$A^c = (E \times F)^c = (E^c \times Y) \cup (X \times F^c) \in \mathcal{R}(X) \otimes \mathcal{R}(Y). \tag{5.7}$$

In general, a set A in $\mathcal{R}(X) \otimes \mathcal{R}(Y)$ is written as $E_1 \times F_1 \cup E_2 \times F_2 \cup \cdots \cup E_k \times F_k$, where $E_j \in \mathcal{R}(X)$ and $F_j \in \mathcal{R}(Y)$. For simplicity we consider the case $k = 2$. The general case is proved by induction. Since $A^c = (E_1 \times F_1)^c \cap (E_2 \times F_2)^c$, we have, by (5.7) and the distributive law,

$$\begin{aligned} A^c &= [(E_1^c \times Y) \cup (X \times F_1^c)] \cap [(E_2^c \times Y) \cup (X \times F_2^c)] \\ &= [(E_1^c \times Y) \cap (E_2^c \times Y)] \cup [(E_1^c \times Y) \cap (X \times F_2^c)] \\ &\quad \cup [(X \times F_1^c) \cap (E_2^c \times Y)] \cup [(X \times F_1^c) \cap (X \times F_2^c)] \\ &= (E_1^c \cap E_2^c \times Y) \cup (E_1^c \times F_2^c) \cup (E_2^c \times F_1^c) \cup (X \times F_1^c \cap F_2^c). \end{aligned}$$

The last term belongs to $\mathcal{R}(X) \otimes \mathcal{R}(Y)$ by the definition. Thus $A_1 \cup A_2 \in \mathcal{R}(X) \otimes \mathcal{R}(Y)$.

Finally, if $A_1, A_2 \in \mathcal{R}(X) \otimes \mathcal{R}(Y)$, then by the above argument $A_1^c \cup A_2$ belongs to $\mathcal{R}(X) \otimes \mathcal{R}(Y)$, and thus $A_1 - A_2 = (A_1^c \cup A_2)^c \in \mathcal{R}(X) \otimes \mathcal{R}(Y)$. □

Define a set function λ for a product set $E \times Y \in \mathcal{R}(X) \otimes \mathcal{R}(Y)$ by

$$\lambda(E \times F) = \mu(E)\nu(F).$$

Our next aim is to show that λ can be extended to a measure in $\mathcal{R}(X) \otimes \mathcal{R}(Y)$.

First we observe that every set of $\mathcal{R}(X) \otimes \mathcal{R}(Y)$ is a finite union of mutually disjoint product sets. Let us consider the case where a set is a union of two product sets $[E_1 \times F_1] \cup [E_2 \times F_2]$ with $E_j \in \mathcal{X}$ and $F_j \in \mathcal{Y} (j = 1, 2)$. Since $E_1 = (E_1 - E_2) \cup (E_1 \cap E_2)$, we have

$$\begin{aligned} &[E_1 \times F_1] \cup [E_2 \times F_2] \\ &= \{[(E_1 - E_2) \times F_1] \cup [(E_1 \cap E_2) \times F_1]\} \cup [E_2 \times F_2]. \end{aligned}$$

Remark that $F_1 = (F_1 - F_2) \cup (F_1 \cap F_2)$ and that $(E_1 \cap E_2) \times (F_1 \cap F_2) \subset E_2 \times F_2$. Then the last union becomes

$$[(E_1 - E_2) \times F_1] \cup [(E_1 \cap E_2) \times (F_1 - F_2)] \cup [E_2 \times F_2],$$

which is a union of mutually disjoint product sets. Now the general case is proved by induction.

For a set A which is a disjoint union of product sets $A = \bigcup_{j=1}^k (E_j \times F_j)$, define

$$\lambda(A) = \sum_{j=1}^k \mu(E_j)\nu(F_j). \tag{5.8}$$

The right-hand side of (5.8) does not depend on the expression of A. The following proof is essentially similar to the proof given in § 2.1. Suppose that

$$A = \bigcup_{j=1}^k (E_j \times F_j) = \bigcup_{p=1}^q (G_p \times H_p)$$

are two expressions of A as disjoint unions of product sets. Since the set $E_j \times F_j$ is written as a union $\bigcup_{p=1}^q [(E_j \cap G_p) \times (F_j \cap H_p)]$, the family $\{(E_j \cap G_p) \times (F_j \cap H_p)\}$

is a partition of $E_j \times F_j$. Therefore we have

$$\chi_{E_j}(x)\chi_{F_j}(y) = \sum_{p=1}^{q} \chi_{E_j \cap G_p}(x)\chi_{F_j \cap H_p}(y).$$

Integrate both sides with respect to x and then with respect to y, to obtain

$$\mu(E_j)\nu(F_j) = \sum_{p=1}^{q} \mu(E_j \cap G_p)\nu(F_j \cap H_p). \tag{5.9}$$

In the same way we have $\mu(G_p)\nu(H_p) = \sum_{j=1}^{k} \mu(E_j \cap G_p)\nu(F_j \cap H_p)$. Summing over j and changing the order of summation, we have from (5.9)

$$\begin{aligned}\sum_{j=1}^{k} \mu(E_j)\nu(F_j) &= \sum_{j=1}^{k}\sum_{p=1}^{q} \mu(E_j \cap G_p)\nu(F_j \cap H_q) \\ &= \sum_{p=1}^{q}\sum_{j=1}^{k} \mu(E_j \cap G_p)\nu(F_j \cap H_q),\end{aligned}$$

which is equal to $\sum_{p=1}^{q}\mu(G_p)\nu(H_q)$. This implies that the right-hand side of (5.8) does not depend on the expression of A as disjoint unions.

LEMMA 5.2. *The set function λ is a measure on $\mathcal{R}(X) \otimes \mathcal{R}(Y)$.*

PROOF. The proof given above shows that λ is a finitely additive set function in $\mathcal{R}(X) \otimes \mathcal{R}(Y)$. Let $A \in \mathcal{R}(X) \otimes \mathcal{R}(Y)$ and suppose that A is the countable disjoint union of sets A_j in $\mathcal{R}(X) \otimes \mathcal{R}(Y)$ $(j = 1, 2, \dots)$. Our object is to prove that

$$\lambda(A) = \sum_{j=1}^{\infty} \lambda(A_j). \tag{5.10}$$

If $A = B_1 \cup \cdots \cup B_N$ and $B_1, \dots, B_N$ are disjoint product sets, then equation (5.10) follows from

$$\lambda(B_i) = \sum_{j=1}^{\infty} \lambda(A_j \cap B_i), \quad i = 1, 2, \dots, N. \tag{5.11}$$

For, if we assume (5.11), then we have $\lambda(A) = \sum_{i=1}^{N} \lambda(B_i) = \sum_{i=1}^{N} \sum_{j=1}^{\infty} \lambda(A_j \cap B_i)$, from which we obtain (5.10) using $\sum_{i=1}^{N} \lambda(A_j \cap B_i) = \lambda(A_j)$.

Therefore to prove (5.10) we can suppose that A is of the form $A = E \times F$. Furthermore, since A_j is a disjoint finite union of product sets, we can suppose that $A_j = E_j \times F_j$. Thus we can assume that $E \times F = \bigcup_{j=1}^{\infty} E_j \times F_j$, where the sets $E_j \times F_j$ are mutually disjoint.

By our assumption we have

$$\chi_E(x)\chi_F(y) = \sum_{j=1}^{\infty} \chi_{E_j}(x)\chi_{F_j}(y). \tag{5.12}$$

Integrating both sides of (5.12) with respect to x and then with respect to y, we get

$$\lambda(A) = \mu(E)\nu(F) = \sum_{j=1}^{\infty} \mu(E_j)\nu(F_j) = \sum_{j=1}^{\infty} \lambda(A_j),$$

which is (5.10). Thus the proof is complete. □

The measure λ can be extended to a measure $\tilde{\lambda}$ on the smallest λ-complete σ-field which imcludes $\mathcal{R}(X \times Y)$ by the Hopf-Kolmogorov theorem. We denote such a σ-field by $\mathcal{X} \otimes \mathcal{Y}$ or $\mathcal{X} \times \mathcal{Y}$ and the extended measure $\tilde{\lambda}$ by $\mu \otimes \nu$ or $\mu \times \nu$. The measure space $(X \times Y, \mathcal{X} \otimes \mathcal{Y}, \mu \otimes \nu)$ is called the *product measure space* of $(X, \mathcal{X}, \mu)$ and $(Y, \mathcal{Y}, \nu)$. By a simple observation the product measure $\mu \otimes \nu$ is σ-finite, since μ and ν are σ-finite.

The integral on the product space is defined as an integral with respect to the measure λ.

5.5. Integration on product spaces

Let $(X \times Y, \mathcal{X} \otimes \mathcal{Y}, \mu \otimes \nu)$ be the product measure spaces of σ-finite measure spaces $(X, \mathcal{X}, \mu)$ and $(Y, \mathcal{Y}, \nu)$.

THEOREM 5.6 (Fubini's theorem). *If $f(x,y) \in L^1(X \times Y, \mathcal{X} \otimes \mathcal{Y}, \mu \otimes \nu)$, then*

(i) $f(\cdot, y) \in L^1(X, \mathcal{X}, \mu)$ *for a.e.* y,
$f(x, \cdot) \in L^1(Y, \mathcal{Y}, \nu)$ *for a.e.* x.

(ii) *We have*

$$\int_{X \times Y} f(x,y) d(\mu \otimes \nu)(x,y) = \int_X \left\{ \int_Y f(x,y) d\nu(y) \right\} d\mu(x)$$
$$= \int_Y \left\{ \int_X f(x,y) d\mu(x) \right\} d\nu(y).$$

The first part of (i) asserts that for almost every y the function $f(x,y)$ is measurable as a function of x and integrable with respect to x.

PROOF. We can assume that $f \geq 0$.

First Step. *If the theorem is true for f and g, then it is truc for $f \perp g$.*

If $0 \leq f_1 \leq f_2 \leq \cdots$ and if the theorem holds for each f_j, $j = 1, 2, \ldots$, then it holds for $f = \lim f_j$.

The first part is obvious. To prove the second part observe that, by our assumption, $f_j(x,y)$ is measurable with respect to x for almost every y. Thus $\lim_{j \to \infty} f_j(x,y)$ is measurable in x for a.e. y.

Apply B. Levi's theorem to $\{f_j\}$ on the measure space $(X \times Y, \mathcal{X} \otimes \mathcal{Y}, \mu \otimes \nu)$, and then use the assumption (ii) for f_j. Then we have

$$\int_{X \times Y} f(x,y) d(\mu \otimes \nu)(x,y) = \lim_{j \to \infty} \int_{X \times Y} f_j(x,y) d(\mu \otimes \nu)(x,y)$$
$$= \lim_{j \to \infty} \int_Y \left\{ \int_X f_j(x,y) d\mu(x) \right\} d\nu(y).$$

By repeated applications of B. Levi's theorem the last term is equal to

$$\int_Y \lim_{j \to \infty} \left\{ \int_X f_j(x,y) d\mu(x) \right\} d\nu(y) = \int_Y \left\{ \int_X \lim_{j \to \infty} f_j(x,y) d\mu(x) \right\} d\nu(y)$$
$$= \int_Y \left\{ \int_X f(x,y) d\mu(x) \right\} d\nu(y),$$

which implies the second equality of (ii).

Now, by (ii) we have $\int_Y\{\int_X f(x,y)d\mu(x)\}d\nu(y) < \infty$. Since $f \geq 0$ by our assumption, $\int_X f(x,y)d\mu(x) < \infty$ for ν-a.e. y. Thus $f(\cdot,y) \in L^1(X,\mathcal{X},\mu)$ for ν-a.e. y. A proof of the second part of (i) is similar.

Second Step. *The theorem holds for the characteristic function of a countable union of sets in $\mathcal{R}(X \times Y)$.*

Indeed, the thorem obviously holds for the characteristic function $\chi_{E\times F}(x,y) = \chi_E(x)\chi_F(y)$, where E and F are sets of finite measure in $\mathcal{R}(X)$ and $\mathcal{R}(Y)$ respectively. Therefore it holds for the finite sum of $\chi_{E_j}(x)\chi_{F_j}(y)$ by the first part of the first step, and thus for characteristic functions of sets in $\mathcal{R}(X)\otimes\mathcal{R}(Y)$. The characteristic function of a countable union of sets in $\mathcal{R}(X)\otimes\mathcal{R}(Y)$ can be expressed as a limit of a nondecreasing sequence of finite sums of characteristic functions of sets in $\mathcal{R}(X)\otimes\mathcal{R}(Y)$, for which the theorem holds by the first part of the first step. Thus the second step follows from the second part of the first step.

Third Step. *The theorem holds for G_δ-sets of finite measure.*

Indeed, let A be a G_δ-set, that is, $A = \bigcap_{j=1}^\infty G_j$ where each G_j is a countable union of sets in $\mathcal{R}(X)\otimes\mathcal{R}(Y)$, and assume that $\mu\otimes\nu(A) < \infty$. By the definition of outer measure we can choose G_1 so that $(\mu\times\nu)(G_1) < \infty$. Observe that $\bigcap_{j=1}^k G_j$ is also a countable union of sets in $\mathcal{R}(X)\otimes\mathcal{R}(Y)$ for each k. The theorem holds for the functions χ_{G_1} and $f_k = \chi_{G_1} - \chi_{\bigcap_{j=1}^k G_j}$ by the preceding steps. Thus by the first step the theorem holds for $\lim_{k\to\infty} f_k = \chi_{G_1} - \chi_A$ and thus for χ_A.

Fourth Step. *The theorem holds for $\mu\otimes\nu$-null sets.*

Indeed, let N be a null set. By definition there exists a G_δ-set G such that $N \subset G$ and G is μ-null, that is,

$$0 = (\mu\otimes\nu)(G) = \int_{X\times Y} \chi_G(x,y)d(\mu\otimes\nu)(x,y).$$

Since the theorem holds for G_δ-sets, the last equation is reduced to

$$0 = \int_X\left\{\int_Y \chi_G(x,y)d\nu(y)\right\}d\mu(x).$$

Thus for a.e. x we have

$$\int_Y \chi_G(x,y)d\nu(y) = 0,$$

which implies that for almost every x, $\chi_G(x,y) = 0$ for a.e. y. Therefore for almost every x we have $\chi_N(x,y) = 0$ for a.e. y. Thus, for almost every x, $\chi_N(x,y)$ is ν-measurable and

$$\int_X\left\{\int_Y \chi_N(x,y)d\nu(y)\right\}d\mu(x) = 0.$$

This implies that the theorem holds for χ_N.

Fifth Step. *The theorem holds for any characteristic functions of measurable sets in $\mathcal{X}\otimes\mathcal{Y}$ of finite measure.*

Indeed, let A be a measurable set of finite measure. Then there exists a G_δ-set G such that

$$A \subset G, \quad \text{and} \quad N = G - A \text{ is null.}$$

Therefore the theorem holds for $\chi_A = \chi_G - \chi_N$.

Sixth Step. *The theorem holds for nonnegative functions in $L^1(X\times Y)$.*

In fact, any nonnegative integrable function in $X \times Y$ is a pointwise limit of a nondecreasing sequence of nonnegative simple measurable functions. The assertion is obtained by the fifth and first steps.

Therefore the theorem is valid for general integrable functions. □

REMARK 5.1. In Fubini's theorem the assumption that $f \in L^1(X \times Y)$ is important. Indeed, there exists a measurable function f in $[0,1] \times [0,1]$ such that both integrals

$$\int_0^1 \left\{ \int_0^1 f(x,y) dm(x) \right\} dm(y)$$

and

$$\int_0^1 \left\{ \int_0^1 f(x,y) dm(y) \right\} dm(x)$$

exist, but the values are different.

To construct such a function put $g_j(x) = j(j+1)\,(1/(j+1) \le x < 1/j)$, $=$ $0\,(\text{otherwise})$ for $j = 1, 2, \ldots$. Then we have $\int_0^1 g_j dm = 1$. Let

$$f(x,y) = \sum_{j=1}^{\infty} [g_{j+1}(x) - g_j(x)] g_j(y).$$

For $0 < y < 1$ choose an integer k so that $1/(k+1) \le y < 1/k$. Then, since $g_j(y) = 0$ for $j \ne k$,

$$\int_0^1 f(x,y) dm(x) = g_k(y) \int_0^1 [g_{k+1}(x) - g_k(x)] dm(x) = 0.$$

Thus

$$\int_0^1 \left\{ \int_0^1 f(x,y) dm(x) \right\} dm(y) = 0. \tag{5.13}$$

On the other hand, when $1/2 \le x < 1$, $g_j(x) = 0$ for $j > 1$, so that $f(x,y) = -g_1(x) g_1(y)$. Thus

$$\int_0^1 f(x,y) dm(y) = -g_1(x) \int_0^1 g_1(y) dm(y) = -g_1(x) \quad (1/2 \le x < 1).$$

When $1/(k+1) \le x < 1/k\,(k \ge 2)$, $g_j(x) = 0$ for $j \ne k$, so that $f(x,y) = -g_k(x) g_k(y) + g_k(x) g_{k-1}(y)$. Thus

$$\begin{aligned} \int_0^1 f(x,y) dm(y) &= -g_k(x) \int_0^1 g_k(y) dm(y) + g_k(x) \int_0^1 g_{k-1}(y) dm(y) \\ &= 0 \quad (1/(k+1) \le x < 1/k). \end{aligned}$$

Thus

$$\int_0^1 \left\{ \int_0^1 f(x,y) dm(y) \right\} dm(x) = -\int_{1/2}^1 g_1(x) dm(x) = -1. \tag{5.14}$$

Therefore we get

$$\int_0^1 \left\{ \int_0^1 f(x,y) dm(x) \right\} dm(y) \ne \int_0^1 \left\{ \int_0^1 f(x,y) dm(y) \right\} dm(x).$$

The reader may check by a direct computation that the function f defined above is not in $L^1([0,1]\times[0,1])$.

From Fubini's theorem we get another useful theorem.

THEOREM 5.7 (Tonelli's theorem). *Let f be a nonnegative $\mathcal{X}\otimes\mathcal{Y}$-measurable function. Then the existence of any one of the three integrals*

$$\int f d(\mu\otimes\nu), \quad \int\left\{\int f d\mu\right\}d\nu \quad and \quad \int\left\{\int f d\nu\right\}d\mu$$

implies the existence and equality of all these integrals.

PROOF. Since the measure spaces are assumed to be σ-finite, there exists an increasing sequence $\{Z_n\}$ in $\mathcal{X}\otimes\mathcal{Y}$ such that $\lim_{n\to\infty} Z_n = X\times Y$ and $(\mu\otimes\nu)(Z_n) < \infty$ for each n. Let

$$f_n(x,y) = \begin{cases} f(x,y), & (x,y)\in Z_n \text{ and } f(x,y)\le n, \\ 0, & \text{otherwise.}\end{cases}$$

Then $f_n \in L^1(X\times Y)$ for each n. By Fubini's theorem (ii)

$$\int f_n d\mu\otimes\nu = \int\left\{\int f_n d\mu\right\}d\nu.$$

Since $0\le f_1\le f_2\le\cdots\le f_n\to f$, applying B. Levi's theorem repeatedly we get

$$\begin{aligned}\int \lim f_n d\mu\otimes\nu &= \lim\int f_n d\mu\otimes\nu = \lim\int\left\{\int f_n d\mu\right\}d\nu \\ &= \int \lim\left\{\int f_n d\mu\right\}d\nu = \int\left\{\int \lim f_n d\mu\right\}\nu.\end{aligned}$$

The rest of the proof is similar. □

5.6. Jordan and Hahn decomposition theorems

Let μ be a real valued measure. Then it can be expressed as follows:

$$\mu = \mu_+ - \mu_-,$$

where μ_+ and μ_- are nonnegative measures. We call such an expression a Jordan decomposition. We illustrate it by examples.

EXAMPLE 5.4. Let $(X,\mathcal{M},\mu)$ be a nonnegative measure space. Let f be a real valued function in $L^1(X,\mathcal{M},\mu)$. Then the set function μ_f defined by

$$\mu_f(E) = \int_E f d\mu$$

is a real valued measure. If

$$\mu_+(E) = \int_E f_+ d\mu \quad \text{and} \quad \mu_-(E) = \int_E f_- d\mu,$$

then μ_+ and μ_- are nonnegative measures and

$$\mu_f = \mu_+ - \mu_-.$$

EXAMPLE 5.5. Let f be a bounded real valued function in $\mathbb{R}$ which is of bounded variation and continuous on the right. Let P_f and N_f be the positive and negative variations of f respectively.

Let μ_f, μ_P and μ_N be the measures associated with the functions f, P_f and N_f respectively. Then μ_P and μ_N are nonnegative, and we have

$$\mu_f = \mu_P - \mu_N. \tag{5.15}$$

In fact, note that

$$f(x) = f(c) + P_f([c, x]) - N_f([c, x])$$

for any $c \in \mathbb{R}$ (see § 3.8). Let

$$P_f(x) = \lim_{c\to-\infty} P_f([c, x]) \quad \text{and} \quad N_f(x) = \lim_{c\to-\infty} N_f([c, x])$$

be the positive and negative variations of f on $(-\infty, x]$. Then P_f and N_f are nondecreasing and continuous on the right. By the definition of μ_f we have

$$\begin{aligned}\mu_f([a, b)) &= f(b) - f(a) = [P_f(b+0) - P_f(a)] - [N_f(b+0) - N_f(a)] \\ &= \mu_P([a, b)) - \mu_N([a, b)),\end{aligned}$$

which implies (5.15)

For the rest of this section, $(X, \mathcal{M}, \mu)$ is assumed to be a real measure space. Thus $\mu(E) \in \mathbb{R}$ for all $E \in \mathcal{M}$. The values $\pm\infty$ are excluded from the range of μ. We remark that the case of complex valued measure space reduces to the real valued case.

A subset P in $\mathcal{M}$ is called a *positive set* if

$$\mu(E) \geq 0 \quad \text{for all measurable subsets } E \text{ of } P.$$

Similarly a set N is said to be *negative* if $\mu(E) \leq 0$ for all measurable subsets E of N.

In Example 5.4 the sets $P = \{x : f(x) \geq 0\}$ and $N = \{x : f(x) < 0\}$ are positive and negative sets respectively. Furthermore, $X = P \cup N$ and $P \cap N = \emptyset$.

THEOREM 5.8 (The Hahn decomposition theorem). *Let μ be a real measure on $\mathcal{M}$. Then there exists a decomposition of X such that*

$$X = P \cup N \quad \textit{and} \quad P \cap N = \emptyset,$$

where P and N are positive and negative sets with respect to μ, respectively.

LEMMA 5.3. *For every set E in $\mathcal{M}$ there exists a positive set P such that $\mu(P) \geq \mu(E)$.*

PROOF. If E is a positive set, choose $P = E$. Otherwise, put

$$\delta_1 = \min[1, -\inf\{\mu(F) : F \subset E,\ F \in \mathcal{M}\}].$$

Then, since E is not a positive set, we have $0 < \delta_1 \leq 1$. Pick $F_1 \subset E$ so that $\mu(F_1) < -\delta_1/2$. If $E - F_1$ is a positive set, the proof is accomplished with $P = E - F_1$. If it is not positive, put

$$\delta_2 = \min[1, -\inf\{\mu(F) : F \subset E - F_1,\ F \in \mathcal{M}\}]$$

and pick $F_2 \subset E - F_1$ so that $\mu(F_2) < -\delta_2/2$. Continuing the process, we get a sequence of positive numbers $\{\delta_j\}$ and sets $\{F_j\}$ in $\mathcal{M}$ such that

$$\delta_j = \min[1, -\inf\{\mu(F) : F \subset E - F_1 - \cdots - F_{j-1},\ F \in \mathcal{M}\}],$$

$$F_j \subset E - F_1 - \cdots - F_{j-1}, \quad \text{and} \quad \mu(F_j) < -\delta_j/2.$$

If this process finishes at the j-th step, the lemma is valid with $P = E - F_1 - \cdots - F_j$.

Suppose that the process continues infinitely. Since the F_j are mutually disjoint,

$$\begin{aligned}(5.16) \qquad \mu(E) &= \mu(E - \bigcup_{j=1}^{\infty} F_j) + \mu(\bigcup_{j=1}^{\infty} F_j) = \mu(E - \bigcup_{j=1}^{\infty} F_j) + \sum_{j=1}^{\infty} \mu(F_j) \\ &< \mu(E - \bigcup_{j=1}^{\infty} F_j) - \frac{1}{2}\sum_{j=1}^{\infty} \delta_j.\end{aligned}$$

Since the left-hand side and the first term on the right-hand side is finite, $\sum \delta_j < \infty$. In particular, $\delta_j \to 0$.

Let

$$P = E - \bigcup_{j=1}^{\infty} F_j.$$

Then $\mu(P) \geq \mu(E)$ by (5.16). The set P is positive. In fact, if $F \subset P$, then $F \subset E - F_1 - \cdots - F_{j-1}$ for any j. Thus by the definition of δ_j, we have $-\mu(F) \leq \delta_j$ for all j. Thus $\mu(F) \geq 0$. □

Proof of Theorem 5.8. Put $a = \sup\{\mu(E) : E \in \mathcal{M}\}$. By Lemma 5.3, for every n there is a positive set P_n such that $a \geq \mu(P_n) > a - 1/n$. If $P = \bigcup_{n=1}^{\infty} P_n$, then P is a positive set. Furthermore, $\mu(P) = a$, since $a \geq \mu(P) \geq \mu(P_n) \to a$.

Let $N = X - P$. Then N is negative. Indeed, for $F \subset N$ we have $a \geq \mu(P \cup F) = \mu(P) + \mu(F) = a + \mu(F)$. Thus $\mu(F) \leq 0$. The proof is complete. □

Definition 5.4. Let μ be a real valued measure in a σ-field $\mathcal{M}$. For $E \in \mathcal{M}$ define

$$\mu_+(E) = \sup\{\mu(F) : F \subset E,\ F \in \mathcal{M}\},$$

$$\mu_-(E) = -\inf\{\mu(F) : F \subset E,\ F \in \mathcal{M}\},$$

and put

$$|\mu| = \mu_+ + \mu_-.$$

μ_+, μ_- and $|\mu|$ are called the *positive, negative* and *total variations* of μ respectively.

Theorem 5.9 (The Jordan decomposition theorem). *Let μ be a real valued measure in $\mathcal{M}$. Then, μ_+, μ_- and $|\mu|$ are finite measure in $\mathcal{M}$ and*

$$(5.17) \qquad \mu = \mu_+ - \mu_-.$$

If $X = P \cup N$ is a Hahn decomposition with positive set P and negative set N, then

$$(5.18) \qquad \mu_+(E) = \mu(E \cap P) \ \text{ and } \ \mu_-(E) = -\mu(E \cap N)$$

for $E \in \mathcal{M}$.

(5.17) is called the *Jordan decomposition* of μ.

PROOF. Let $X = P \cup N$ be a Jordan decomposition. By the definition of μ_+ we have

$$\mu_+(E) \geq \mu(E \cap P) \quad \text{for all } E \in \mathcal{M}. \tag{5.19}$$

Fix a set $E \in \mathcal{M}$. Then for any $F \subset E$ we have

$$\mu(F) = \mu(F \cap P) + \mu(F \cap N).$$

Since P is a positive set, $\mu(F \cap P) \leq \mu(E \cap P)$. Also, the second term is ≤ 0. Thus $\mu(F) \leq \mu(E \cap P)$ for every $F \subset E$, and so $\mu_+(E) \leq \mu(E \cap P)$. Combining this with (5.19), we get

$$\mu_+(E) = \mu(E \cap P).$$

In the same way we obtain

$$\mu_-(E) = -\mu(E \cap N).$$

Thus μ_+ and μ_- are nonnegative measures, and $\mu = \mu_+ - \mu_-$. □

THEOREM 5.10. *Let μ be a real valued measure in $\mathcal{M}$. For a set $E \in \mathcal{M}$ we have*

$$|\mu|(E) = \sup\{\sum_{j=1}^{k} |\mu(E_j)| : \{E_j\}_{j=1}^{k} \text{ is a finite partition of } E\}. \tag{5.20}$$

PROOF. By the definition of total variation and Theorem 5.9 we have

$$|\mu|(E) = \mu^+(E) + \mu^-(E) = |\mu(E \cap P)| + |\mu(E - P)|,$$

where $X = P \cup N$ is a Hahn decomposition. Thus the left-hand side of (5.20) is bounded by the right-hand side.

On the other hand, for any partition $\{E_j\}_{j=1}^{k}$ of E we have

$$|\mu|(E) = \sum |\mu|(E_j) \geq \sum |\mu(E_j)|.$$

Thus the left-hand side of (5.20) is bounded below by the right-hand side. Thus we get (5.20). □

For a complex valued measure μ, define

$$||\mu|| = \sup \left\{ \sum_{j=1}^{k} |\mu(E_j)| : \{E_j\}_{j=1}^{k} \text{ is a finite partition of } X \right\}$$

and call it the *norm of the measure* μ.

If μ is a real measure, then $|\mu|(X) = ||\mu||$.

5.7. Absolutely continuous measures

Let $\mathcal{M}$ be a σ-field of subsets of X.

DEFINITION 5.5. Let μ and ν be real valued measures in $\mathcal{M}$.

(i) ν is said to be *absolutely continuous* with respect to μ if every $|\mu|$-null set is a $|\nu|$-null set. This fact is denoted by $\nu \ll \mu$.

(ii) ν is said to be *singular* with respect to μ if there exists a $|\mu|$-null set N such that $X - N$ is a $|\nu|$-null set. This fact is denoted by $\mu \perp \nu$.

(iii) ν is said to be *discrete* if $\nu \neq 0$ and there exists a countable set D such that

$$\nu(E) = \nu(E \cap D) \quad \text{for all } E \in \mathcal{M}.$$

(iv) ν is said to be *continuous* if

$$\nu(\{x\}) = 0 \text{ for all } x \in X.$$

REMARK 5.2. Let μ and ν be real valued measures in $\mathcal{M}$.

(i) $\nu \ll \mu$ if and only if for every $\varepsilon > 0$ there exists $\delta > 0$ such that

$$|\mu|(E) < \delta \ \Rightarrow \ |\nu|(E) < \varepsilon.$$

(ii) $\nu \perp \mu$ if and only if for every $\varepsilon > 0$ there exists $E \in \mathcal{M}$ such that

$$|mu|(E) < \varepsilon \text{ and } |\nu|(E^c) < \varepsilon.$$

(iii) $\nu \perp \mu \ \Leftrightarrow \ \mu \perp \nu$.

The "if" part of (i) is obvious. Suppose that $\nu \ll \mu$. If the "only if" part is not true, there exist an $\varepsilon > 0$ and sets E_j $(j = 1, 2, \dots)$ such that $|\mu|(E_j) < 2^{-j}$ but $|\nu|(E_j) > \varepsilon$. Put $E = \limsup E_j$. Then $|\mu|(E) = 0$ and $|\nu|(E) \geq \limsup |\nu|(E_j) \geq \varepsilon$ (see Problrm 2.2.), which contradicts our assumption.

The "only if " part of (ii) is obvious. To prove the "if" part, for each $j = 1, 2, \dots$ choose a set E_j so that $|\mu|(E_j) < 2^{-j}$ and $|\nu|(E_j^c) < 2^{-j}$. Put $E = \limsup_{j\to\infty} E_j$. Then $|\mu|(E) = 0$. On the other hand, by Fatou's lemma $|\nu|(E^c) = |\nu|(\liminf_{j\to\infty} E_j^c) \leq \liminf_{j\to\infty} |\nu|(E_j^c) = 0$. Thus ν is singular with respect to μ.

EXAMPLE 5.6. A complex measure ν is, by definition, absolutely continuous with respect to μ if both the real and complex parts of ν are absolutely continuous.

Let $f \in L^1(X, \mathcal{M}, \mu)$ and define the measure μ_f by $\mu_f(E) = \int_E f(x)d\mu$. Then μ_f is a complex valued measure. If $\mu(E) = 0$, then $\mu_f(E) = 0$. Thus $\mu_f \ll \mu$

EXAMPLE 5.7. Let f be the Cantor singular function in $[0,1]$ associated with the Cantor ternary set C. Let μ be the measure defined by f, so that,

$$\mu([a,b)) = f(b) - f(b) \quad \text{for } [a,b) \subset [0,1].$$

In particular $\mu([1/3, 2/3)) = 0$. In the same way, $\mu(I) = 0$ for every middle-third interval I. Thus $\mu([0,1] - C) = 0$. On the other hand, $m(C) = 0$. Therefore $m \perp \mu$.

EXAMPLE 5.8. Let $a \in \mathbb{R}^d$. The measure defined by

$$\delta_a(E) = \begin{cases} 1 & (a \in E), \\ 0 & (a \notin E) \end{cases}$$

is called the *Dirac measure* with support at a. δ_a is a discrete measure.

For any continuous function f we have

$$\int_{\mathbb{R}^d} f d\delta_a = f(a).$$

EXAMPLE 5.9. The function Y in $\mathbb{R}$ defined by

$$Y(t) = \begin{cases} 1 & (t \geq 0), \\ 0 & (t < 0) \end{cases}$$

is called the *Heaviside function.*

Let $\{a_j\}_{j=-\infty}^{\infty}$ be a sequence such that $\sum |a_j| < \infty$, and let $\{t_j\}$ be an increasing sequence. Put

$$\varphi(t) = \sum_{j=-\infty}^{\infty} a_j Y(t - t_j).$$

Then φ is a function of bounded variation. If μ is the measure defined by φ, then

$$\mu(E) = \mu(E \cap \{t_j\}) = \sum_{t_j \in E} a_j,$$

which implies that $\mu = \sum a_j \delta_{a_j}$ and μ is a discrete measure supported on $\{t_j\}$.

THEOREM 5.11 (The Lebesgue decomposition theorem). *Let $(X, \mathcal{M}, \mu)$ be a σ-finite nonnegative measure space and ν be a complex valued measure in $\mathcal{M}$. Then ν can be uniquely decomposed into two measures as follows :*

$$\nu = \nu_a + \nu_s, \tag{5.21}$$

where $\nu_a \ll \mu$ and $\nu_s \perp \mu$.

Furthermore, there exists a unique $f \in L^1(d\mu)$ such that

$$\nu_a(E) = \int_E f d\mu \quad \text{for } E \in \mathcal{M}. \tag{5.22}$$

LEMMA 5.4. *Suppose that μ and ν are finite nonnegative measures in a σ-field $\mathcal{M}$. Then for a given $\varepsilon > 0$ the set X can be decomposed as follows:*

(i) *$X = H \cup \bigcup_{k=1}^{\infty} E_k$, where H and E_k $(k = 1, 2, \dots)$ are mutually disjoint sets in $\mathcal{M}$,*

(ii) *$\mu(H) = 0$ and*

$$\varepsilon(k-1)\mu(F) \le \nu(F) \le \varepsilon k \mu(F) \quad \textit{for } F \subset E_k,\ F \in \mathcal{M}.$$

PROOF OF THE LEMMA. Let $X = P_k \cup N_k$ be a Hahn decomposition with respect to the real valued measure $\varepsilon k \mu - \nu$. Then by the definition

$$\varepsilon k \mu(F) - \nu(F) \ge 0 \quad \text{for } F \subset P_k,\ F \in \mathcal{M}, \tag{5.23}$$

$$\varepsilon k \mu(G) - \nu(G) \le 0 \quad \text{for } G \subset N_k,\ G \in \mathcal{M}. \tag{5.24}$$

Put $H = \limsup_{j \to \infty} N_j$. Then for any positive integer k we have

$$\begin{aligned} H = \bigcap_{k=1}^{\infty} \bigcup_{j=k}^{\infty} N_j &\subset \bigcup_{j=k}^{\infty} N_j \\ &= N_k \cup (N_{k+1} - N_k) \cup (N_{k+2} - N_{k+1} - N_k) \cup \cdots, \end{aligned}$$

which implies that

$$\mu(H) \le \mu(N_k) + \mu(N_{k+1} - N_k) + \mu(N_{k+2} - N_{k+1} - N_k) + \cdots.$$

By (5.24) $\mu(A) \le \frac{1}{\varepsilon j}\nu(A) \le \frac{1}{\varepsilon k}\nu(A)$ for $A \subset N_j$ $(j \ge k)$. Therefore we have

$$\begin{aligned} \mu(H) &\le \frac{1}{\varepsilon k}\Big\{\nu(N_k) + \nu(N_{k+1} - N_k) + \nu(N_{k+2} - N_{k+1} - N_k) + \cdots\Big\} \\ &= \frac{1}{\varepsilon k}\nu(\bigcup_{j=k}^{\infty} N_j) \le \frac{1}{\varepsilon k}\nu(X) \end{aligned}$$

for every k, from which it follows that $\mu(H) = 0$.

By the definitions of H and P_j we have

$$X - H = \Big(\bigcap_{k=1}^{\infty} \bigcup_{j=k}^{\infty} N_j\Big)^c = \bigcup_{k=1}^{\infty} \bigcap_{j=k}^{\infty} P_j = \bigcup_{k=1}^{\infty} E_k,$$

where

$$E_1 = \bigcap_{j=1}^{\infty} P_j \quad \text{and} \quad E_k = \bigcap_{j=k}^{\infty} P_j - \bigcap_{j=k-1}^{\infty} P_j \ (k \geq 2).$$

Therefore we get the expression $X = H \cup \bigcup_{k=1}^{\infty} E_k$ with mutually disjoint sets. The second inequality of (ii) is obvious when $k = 1$ by the definition of P_1 and by the fact that $E_1 \subset P_1$. Let $k \geq 2$. By the definition of the set E_k we have

$$E_k = \bigcap_{j=k}^{\infty} P_j - \left[\bigcap_{j=k}^{\infty} P_j \cap P_{k-1}\right] = \bigcap_{j=k}^{\infty} P_j \cap P_{k-1}^c = \bigcap_{j=k}^{\infty} P_j \cap N_{k-1}.$$

Therefore if $A \subset E_k$, then $A \subset P_k$ and $A \subset N_{k-1}$. The former relation implies

$$\varepsilon k\mu(A) - \nu(A) \geq 0$$

by (5.23), and the latter implies

$$\varepsilon(k-1)\mu(A) - \nu(A) \leq 0$$

by (5.24). Therefore (ii) is proved. □

PROOF OF THEOREM 5.11. We can assume that ν is a real valued measure by decomposing ν into its real and imaginary parts. Furthermore, we can suppose that $\nu \geq 0$ by the Jordan decomposition theorem.

First assume that μ is a finite measure.

First Step. *Decomposition of the set X.*

By Lemma 5.4, X is decomposed as follows: For each $p = 1, 2, \ldots$

(a) $X = H^p \cup \bigcup_{k=1}^{\infty} E_k^p$, where H^p and E_k^p $(k = 1, 2, \ldots)$ are mutually disjoint sets in $\mathcal{M}$,

(b) $\mu(H^p) = 0$, and

$$(k-1) \cdot 2^{-p}\mu(F) \leq \nu(F) \leq k \cdot 2^{-p}\mu(F) \quad (F \subset E_k^p,\ F \in \mathcal{M}).$$

We have, furthermore,

(c) if $i > 2j + 1$ or $i < 2j - 2$, then $\mu(E_i^{p+1} \cap E_j^p) = 0$.

To prove (c), apply the first inequality of (b) to the set $F = E_i^{p+1} \cap E_j^p$ as a subset of E_j^p, and then the second inequality of (b) as a subset of E_i^{p+1}. Then we have

$$\begin{aligned}(5.25) \qquad (j-1) \cdot 2^{-p}\mu(E_i^{p+1} \cap E_j^p) &\leq \nu(E_i^{p+1} \cap E_j^p) \\ &\leq i \cdot 2^{-p-1}\mu(E_i^{p+1} \cap E_j^p).\end{aligned}$$

If $i < 2j - 2$, then $i \cdot 2^{-p-1} < (j-1) \cdot 2^{-p}$. Substituting the last inequality into the right-hand side of (5.25), we get $\mu(E_i^{p+1} \cap E_j^p) = 0$. In the same way, apply the first inequality of (b) to the set $E_i^{p+1} \cap E_j^p$ as a subset of E_i^{p+1}, and then the second inequality of (b) as a subset of E_j^p, to get

$$\begin{aligned}(5.26) \qquad (i-1) \cdot 2^{-p-1}\mu(E_i^{p+1} \cap E_j^p) &\leq \nu(E_i^{p+1} \cap E_j^p) \\ &\leq j \cdot 2^{-p}\mu(E_i^{p+1} \cap E_j^p).\end{aligned}$$

If $i > 2j + 1$, then $j \cdot 2^{-p} < (i-1) \cdot 2^{-p-1}$. Substituting this in (5.26), we get $\mu(E_i^{p+1} \cap E_j^p) = 0$. Thus (c) is obtained.

Put $N_j^p = \bigcup_{\{i : i>2j+1 \text{ or } i<2j-2\}} E_i^{p+1} \cap E_j^p$ and define

$$N = \bigcup_{p=1}^{\infty} H^p \cup \bigcup_{j,p=1}^{\infty} N_j^p.$$

Then we have

(d) $\mu(N) = 0$,

and

(e) $X = N \cup \bigcup_{j=1}^{\infty}(E_j^p - N)$ for each $p = 0, \pm 1, \pm 2, \ldots$,

where the sets on the right-hand side are mutually disjoint.

Second Step. *Definitions of the singular measure ν_s and the absolutely continuous measure ν_a.*

Define the measure ν_s by $\nu_s(E) = \nu(E \cap N)$. Since the set N is μ-null, ν_s is singular with respect to μ.

Next we shall construct a function f in $L^1(X, d\mu)$ such that

$$\nu(E) = \nu_s(E) + \int_E f d\mu \quad (E \in \mathcal{M}). \tag{5.27}$$

Then the proof will be accomplished.

Fix an integer p, and put

$$f_p(x) = \begin{cases} (j-1) \cdot 2^{-p} & (x \in E_j^p - N), \\ 0 & (x \in N). \end{cases}$$

The function f_p is measurable and $f_p \geq 0$. Furthermore, we have

$$|f_{p+1}(x) - f_p(x)| \leq 2^{-p}. \tag{5.28}$$

In fact, if $x \in N$, then $f_p(x) = f_{p+1}(x) = 0$. If $x \in E_j^p - N$, then, by the definition of N, $x \in E_{2j-2}^{p+1} \cup E_{2j-1}^{p+1} \cup E_{2j}^{p+1} \cup E_{2j+1}^{p+1}$. Thus $f_{p+1}(x) = (2j-2) \cdot 2^{-p-1}$, $(2j-1) \cdot 2^{-p-1}, 2j \cdot 2^{-p-1}$, or $(2j+1) \cdot 2^{-p-1}$. Thus we have (5.28).

By (5.28) the sequence $\{f_p(x)\}$ converges uniformly to a function $f(x)$. We shall show that the function f satisfies (5.27).

Fix a set $E \in \mathcal{M}$. By (e) and (b) we have

$$\begin{aligned} \nu(E) &= \nu(E \cap N) + \sum_{j=1}^{\infty} \nu(E \cap (E_j^p - N)) \\ &\geq \nu_s(E) + \sum_{j=1}^{\infty} (j-1) \cdot 2^{-p} \mu(E \cap (E_j^p - N)), \end{aligned}$$

which is, by the definition of f_p, equal to $\nu_s(E) + \int_E f_p d\mu$.

Since $\{f_p\}$ converges uniformly to f as $p \to \infty$, and μ is a finite measure,

$$\nu(E) \geq \nu_s(E) + \int_E f d\mu. \tag{5.29}$$

On the other hand, we have, again by (e) and (b),

$$\begin{aligned} \nu(E) &\leq \nu_s(E) + \sum_{j=1}^{\infty} j \cdot 2^{-p} \mu(E \cap (E_j^p - N)) \\ &= \nu_s(E) + \int_E f_p d\mu + 2^{-p} \mu(E). \end{aligned}$$

Thus

$$\nu(E) \le \nu_s(E) + \int_E f d\mu. \tag{5.30}$$

Therefore (5.29) and (5.30) yield (5.27).

Third Step. *The case where μ is σ-finite.*

Let $X = \bigcup_{n=1}^{\infty} X_n$ be a disjoint union with X_n of finite measure. Put $\mu^n(E) = \mu(E \cap X_n)$ and $\nu^n(E) = \nu(E \cap X_n)$. Since μ^n and ν^n are finite measures, we have a decomposition $\nu^n = \nu_a^n + \nu_s^n$ by the second step. Put $\nu_a = \sum_{n=1}^{\infty} \nu_a^n$ and $\nu_s = \sum_{n=1}^{\infty} \nu_s^n$. Then ν_a is clearly absolutely continuous with respect to μ, and ν_s is singular.

On the other hand, there exists $f_n \in L^1(X_n, d\mu^n)$ such that

$$\nu_a^n(E) = \int_E f_n d\mu^n \quad (E \subset X_n,\ E \in \mathcal{M}).$$

Thus the total variation of ν_a is estimated by

$$|\nu_a|(X_n) = |\nu_a^n|(X_n) = \int_{X_n} |f_n| d\mu^n.$$

Let f be the function which coincides with f_n in X_n. Then $f \in L^1(X, \mu)$. In fact by B. Levi's theorem we have

$$\int_X |f| d\mu = \int_X \sum_{n=1}^{\infty} |f| \cdot \chi_{X_n} d\mu = \sum_{n=1}^{\infty} \int_{X_n} |f_n| d\mu$$
$$= \sum_{n=1}^{\infty} |\nu_a|(X_n) = |\nu_a|(X) < \infty.$$

Therefore the function f is integrable on X, and

$$\nu_a(E) = \int_E f d\mu.$$

Fourth Step. *Uniqueness of the function f.*

This is a direct consequence of the following lemma.

LEMMA 5.5. *Let $f, g \in L^1(X, d\mu)$. If*

$$\int_E f d\mu = \int_E g d\mu \quad \textit{for all } E \in \mathcal{M},$$

then

$$f = g \quad \mu\text{-}a.e.$$

PROOF OF LEMMA 5.5. We can assume that f and g are real valued functions. Put $h = f - g$. Then $h \in L^1$ and $\int_E h d\mu = 0\,(E \in \mathcal{M})$. Let $E = \{x : h(x) \ge 0\}$. Then $\int_E h d\mu = \int h_+ d\mu = 0$. Therefore $h_+ = 0$ a.e. (see Problem 3.4). In the same way we have $h_- = 0$ a.e. □

Fifth Step. *Uniqueness of the decomposition* (5.21).

Let

$$\nu = \nu_a + \nu_s = \nu_a' + \nu_s'$$

be two decompositions of ν into absolutely continuous measures and singular measures. Then $\nu_a - \nu_a' = \nu_s' - \nu_s$. The right-hand side is a singular measure, so it vanishes outside a null set. On the other hand, the left-hand side vanishes on any null set. Thus it vanishes identically. □

We have the following from Theorem 5.11.

THEOREM 5.12 (The Radon–Nikodým theorem). *Let μ be a σ-finite measure and ν be a finite measure. If $\nu \ll \mu$, then there exists a unique function $f \in L^1(X, d\mu)$ such that*

$$\nu(E) = \int_E f d\mu \quad \text{for } E \in \mathcal{M}.$$

DEFINITION 5.6. The function f given by the Radon–Nikodým theorem is denoted by

$$\frac{d\nu}{d\mu} = f$$

and called the *Radon–Nikodým derivative* of ν with respect to μ.

COROLLARY 5.1. *Let μ be a bounded Borel measure on $\mathbb{R}^d$. Then μ can be decomposed as follows:*

$$\mu = \mu_a + \mu_s + \mu_d,$$

where $\mu_a \ll m$, $\mu_s \perp m$, μ_s is continuous and μ_d is discrete. The decomposition is unique.

PROOF. We can assume that μ is a real valued measure. Put $D = \{x : \mu(\{x\}) \neq 0\}$ and suppose that D is nonempty. Then D is countable and the measure μ_d defined by $\mu_d(E) = \mu(E \cap D)$ is discrete. In fact, if $D_n = \{x : |\mu(\{x\})| > 1/n\}$, then D_n $(n = 1, 2, \dots)$ is a finite set, since for any finite subset $\{x_1, x_2, \dots, x_k\} \subset D_n$ we have

$$\infty > \|\mu\| \geq |\mu_d|(D_n) \geq |\mu_d|(\{x_1, x_2, \dots, x_k\}) \geq \sum_{j=1}^{k} |\mu(\{x_j\})| > \frac{k}{n}.$$

Thus k is bounded by $n\|\mu\|$. Therefore $D = \bigcup_{n=1}^{\infty} D_n$ is countable.

Applying the Lebesgue decomposition theorem to the continuous measure $\mu - \mu_d$, we get the corollary. The proof of uniqueness is an easy exercise and is left to the reader. □

EXAMPLE 5.10. Let f be a bounded nondecreasing function in $\mathbb{R}$. Furthermore we assume that f is continuous on the right. Let μ be the measure which satisfies $\mu([a, b)) = f(b) - f(a)$ for $a < b$. By Theorem 5.11 and Cororally 5.1 μ can be decomposed as follows:

$$\mu(E) = \int_E g dx + \mu_s(E) + \mu_d(E), \tag{5.31}$$

where g is the Radon-Nikodým derivative, μ_s is a continuous singular measure, and μ_d is a discrete measure.

By the Lebesgue differentiation theorem the derivative f' of f in the ordinary sense exists a.e. and $f' \geq 0$ almost everywhere. We shall show that

$$g = f', \tag{5.32}$$

and if $\{a_n\}$ is the set of points of discontinuity of f, then

$$\mu_d = \sum_{n=1}^{\infty}[f(a_n) - f(a_n - 0)]\delta_{a_n}, \tag{5.33}$$

where δ_a is the Dirac measure supported on $\{a\}$.

In fact, applying Fatou's lemma we get

$$\begin{aligned}\int_a^b f'(t)dt &= \int_a^b \liminf_{h\to 0} \frac{f(t+h)-f(t)}{h}dt \\ &\le \liminf_{h\to 0}\frac{1}{h}\int_a^b [f(t+h)-f(t)]dt \\ &= \liminf_{h\to 0}\frac{1}{h}\left(\int_b^{b+h} f(t)dt - \int_a^{a+h} f(t)dt\right).\end{aligned}$$

Since f is nondecreasing, we have $\int_b^{b+h} f(t)dt \le hf(b+h)$, and $\int_a^{a+h} f(t)dt \ge hf(a)$. Thus by the continuity on the right we conclude that

$$\int_a^b f'(t)dt \le f(b) - f(a) = \mu_f([a,b)). \tag{5.34}$$

Therefore $f' \in L^1(\mathbb{R})$, since f is bounded and $f' \ge 0$.

If the measure $\nu(E) = \mu(E) - \int_E f'dt$ is singular, (5.32) follows from the uniqueness of the Lebesgue decomposition theorem. Now, by the Lebesgue decomposition theorem $\nu(E) = \int_E \varphi dt + \nu_s(E)$, where ν_s is singular and φ is an integrable function. By (5.34), $\nu([a,b)) \ge 0$. Thus $\varphi \ge 0$ a.e., and $\nu_s \ge 0$. Therefore

$$\lim_{h\to +0} \frac{\nu([t,t+h))}{h} \ge \varphi(t) \ge 0 \text{ a.e.}$$

On the other hand,

$$\lim_{h\to +0}\frac{\nu([x,x+h))}{h} = \lim_{h\to +0}\frac{\mu([x,x+h))}{h} - \frac{d}{dx}\int_{-\infty}^x f'dt = 0 \text{ a.e.}$$

Thus $\varphi = 0$ a.e., and so $g = f'$.

To prove (5.33) let $D = \{a_j\}_{j=-\infty}^{\infty}$ be the set of points of discontinuitty of f. Then $\mu(\{a_n\}) = f(a_n) - f(a_n - 0)$. Thus

$$\mu_d(E) = \mu(E\cap D) = \sum_{\{j;a_j\in E\}} [f(a_j) - f(a_j - 0)] = \sum_j [f(a_j) - f(a_j-0)]\delta_{a_j}(E).$$

5.8. Measures on metric spaces

Sometimes Lebesgue measure gives us rather meager information on sets. Some curves in the plane are null in the sense of Lebesgue measure but of infinite one-dimensional measure. The fractal set which is called the Koch curve is such an example.

The two dimensional Lebesgue measure of the set obtained in each step is zero, but the one dimensional measures of the sets are

$$4,\quad \frac{1}{3}\cdot 4^2,\quad \left(\frac{1}{3}\right)^2\cdot 4^3,\quad \ldots\ldots,\quad \left(\frac{1}{3}\right)^n \cdot 4^{n+1},\quad \ldots,$$

which diverge to infinity (see Figure 5.1).

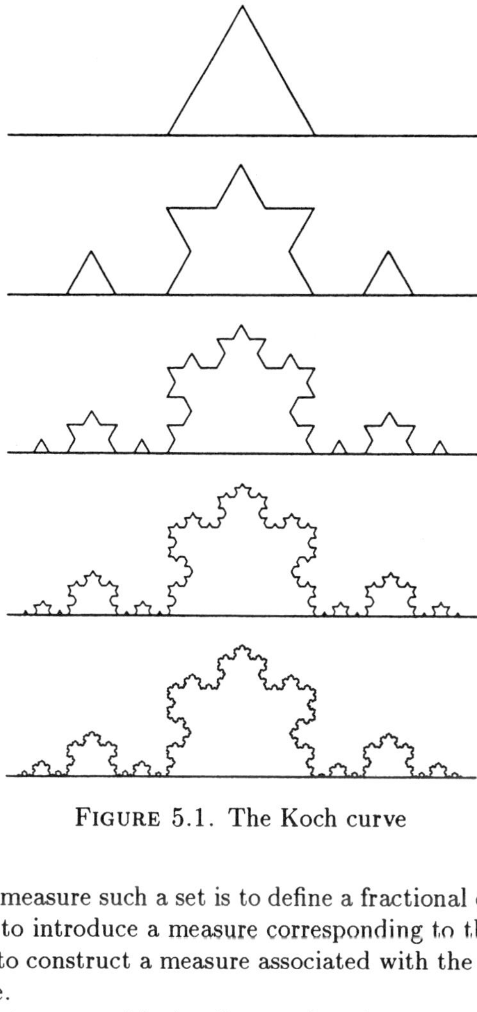

FIGURE 5.1. The Koch curve

One method to measure such a set is to define a fractional dimension which fits in with the set and to introduce a measure corresponding to the dimension.

We shall begin to construct a measure associated with the notion of dimension in an abstract space.

Let X be a metric space with the distance function ρ. For subsets A and B of X let $\rho(A, B)$ be the distance between A and B, that is,

$$\rho(A, B) = \inf\{\rho(a, b) : a \in A,\ b \in B\}.$$

The outer measure μ^* on $\mathcal{P}(X)$ is called a *metric outer measure* if

$$\rho(A, B) > 0 \ \Rightarrow\ \mu^*(A \cup B) = \mu^*(A) + \mu^*(B).$$

For example, the Lebesgue outer measure m^* is a metric outer measure (see also Remark 5.3 below). Indeed, let A and B be any subsets of $\mathbb{R}^d$. If the distance $\mathrm{dist}(A, B)$ is positive, the set $O_A = \{x : \mathrm{dist}(x, A) < \mathrm{dist}(A, B)/2\}$ is open and contains A. In the same way, define $O_B = \{x : \mathrm{dist}(x, B) < \mathrm{dist}(A, B)/2\}$. Then obviously $O_A \cap O_B = \emptyset$. Thus for any open set $O \supset A \cup B$ we have

$$\begin{aligned} m(O) &\geq m((O_A \cup O_B) \cap O) = m(O_A \cap O) + m(O_B \cap O) \\ &\geq m^*(A) + m^*(B), \end{aligned}$$

which implies that $m^*(A \cup B) \geq m^*(A) + m^*(B)$. The reverse inequality holds in general.

LEMMA 5.6. *Let μ^* be a metric outer measure on (X,ρ). Let E be a subset of X and O be an open subset such that $E \subset O$. If $E_n = \{x \in E : \rho(x, O^c) \geq 1/n\}$ $(n = 1, 2, \dots)$, then*

$$\lim_{n\to\infty} \mu^*(E_n) = \mu^*(E).$$

PROOF. Put $D_0 = E_1$ and $D_n = E_{n+1} - E_n$ $(n = 1, 2, \dots)$ (see Figure 5.2). If $D_{n+1} \neq \emptyset$ and $E_n \neq \emptyset$, then $\rho(E_n, D_{n+1}) \geq 1/n(n+1) > 0$. In fact, let $a \in E_n$ and $b \in D_{n+1}$. By the definition of E_n we have $\rho(a,x) \geq 1/n$ for any $x \in O^c$. Thus $\rho(a,b) \geq \rho(a,x) - \rho(b,x) \geq 1/n - \rho(b,x)$ for all $x \in O^c$. On the other hand, since $b \notin E_{n+1}$, $\inf\{\rho(b,x) : x \in O^c\} \leq 1/(n+1)$. Thus $\rho(a,b) \geq 1/n - 1/(n+1) = 1/n(n+1)$.

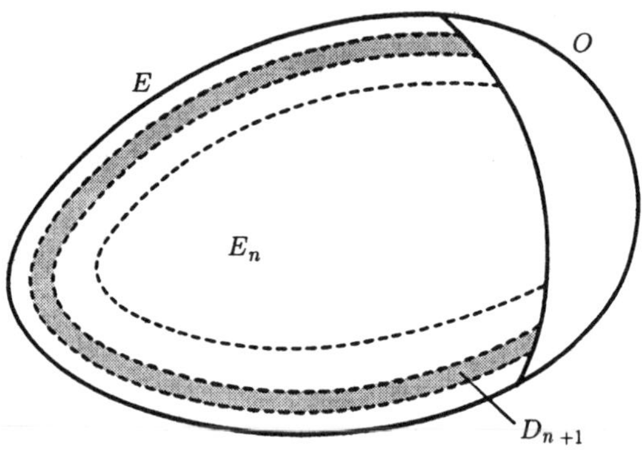

FIGURE 5.2. Decomposition of E

Therefore we have $\rho(D_{n+1}, D_{n-1}) \geq \rho(D_{n+1}, E_n) > 0$. Since $E_{2n+1} \supset D_2 \cup D_4 \cup \cdots \cup D_{2n}$, by the property of metric outer measure we get

$$\mu^*(E_{2n+1}) \geq \sum_{j=1}^{n} \mu^*(D_{2j}).$$

In the wame way,

$$\mu^*(E_{2n}) \geq \sum_{j=1}^{n} \mu^*(D_{2j-1}).$$

When $\sum_{j=1}^{\infty} \mu^*(D_{2j}) = \infty$ or $\sum_{j=1}^{\infty} \mu^*(D_{2j-1}) = \infty$, the lemma holds obviously. Assume that the both sums are finite. Since $E = E_{2n} \cup \bigcup_{j=n}^{\infty} D_{2j} \cup \bigcup_{j=n}^{\infty} D_{2j+1}$,

$$\mu^*(E) \leq \mu^*(E_{2n}) + \sum_{j=n}^{\infty} \mu^*(D_{2j}) + \sum_{j=n}^{\infty} \mu^*(D_{2j+1}).$$

Thus $\mu^*(E) \leq \liminf \mu^*(E_{2n})$.

On the other hand, since $E_n \subset E$, we have $\limsup \mu^*(E_n) \leq \mu(E)$, which completes the proof. □

Let $\mathcal{B}_0(X)$ be the smallest σ-field containing all open subsets of X.

THEOREM 5.13. *Let μ^* be a metric outer measure on the metric space (X,ρ). Then every open set is μ^*-measurable. Thus $\mathcal{B}_0(X) \subset \mathcal{M}_{\mu^*}$.*

PROOF. Let O be an open set of X. Let A be any subset of X and put $E = A \cap O$. We apply Lemma 5.6 to the sets O and E. With the notations in Lemma 5.6 we have $A \supset E_n$ and $\rho(E_n, A \cap O^c) \geq 1/n > 0$. Thus

$$\mu^*(A) \geq \mu^*(E_n \cup (A \cap O^c)) = \mu^*(E_n) + \mu^*(A \cap O^c).$$

By Lemma 5.6 the right-hand side converges to $\mu^*(A \cap O) + \mu^*(A - O)$ as $n \to \infty$. Thus O satisfies the Carathéodory condition

$$\mu^*(A) \geq \mu^*(A \cap O) + \mu^*(A - O)$$

for any subset A of X. Thus $O \in \mathcal{M}_{\mu^*}$. □

REMARK 5.3. The converse of Theorem 5.13 is also true. That is, let μ^* be an outer measure on X. If $\mathcal{B}_0(X) \subset \mathcal{M}_{\mu^*}$, then μ^* is a metric outer measure.

In fact, let E and F be subsets of X, and suppose $\rho(E, F) = \delta > 0$. If $U = \{x \in E : \rho(x, E) < \delta/2\}$, then U is an open set containing E, and $F \cap U = \emptyset$. By assumption $U \in \mathcal{M}_{\mu^*}$, or, equivalently, U satisfies the Carathéodory condition. Thus for $A = E \cup F$ we have

$$\mu^*(E \cup F) = \mu^*(A \cap U) + \mu^*(A - U) = \mu^*(E) + \mu^*(F).$$

Therefore μ^* is a metric outer measure.

For a subset A of X let

$$\operatorname{diam}(A) = \sup\{\rho(x, y) : x, y \in A\}$$

be the diameter of A. If A is empty, then $\operatorname{diam}(A) = 0$ by definition. A countable family $\{U_j\}$ is said to be a *δ-covering* of A, if $\{U_j\}$ is a covering of A and $\operatorname{diam}(U_j) \leq \delta$.

Fix $s > 0$. For $\delta > 0$ denote

$$\mathcal{H}_\delta^s(A) = \inf\{\sum_{j=1}^{\infty} [\operatorname{diam}(U_j)]^s : \{U_j\} \text{ is a } \delta\text{-covering of } A\}.$$

If there is no countable δ-covering of A, we put $\mathcal{H}_\delta^s(A) = \infty$. Clearly, $\mathcal{H}_\delta^s(A)$ increases when δ decreases. Define

$$\mathcal{H}^s(A) = \lim_{\delta \to 0} \mathcal{H}_\delta^s(A).$$

THEOREM 5.14. *Let $s > 0$. Then*

(i) *$\mathcal{H}^s$ is a metric outer measure,*
(ii) *$\mathcal{H}^s$ is regular in the following sense: For any subset E there exists a sequence of open sets $\{O_n\}$ such that*

$$\mathcal{H}^s(E) = \mathcal{H}^s(\bigcap_{n=1}^{\infty} O_n) \quad \text{and} \quad E \subset O_n.$$

PROOF. (i) Let A_n $(n = 1, 2, \ldots)$ be subsets of X. First we shall prove

$$\mathcal{H}^s(\bigcup_{n=1}^{\infty} A_n) \leq \sum_{n=1}^{\infty} \mathcal{H}^s(A_n). \tag{5.35}$$

It is obvious that $\mathcal{H}^s$ satisfies the other two properties of outer measure.

We assume that $\mathcal{H}^s(A_n) < \infty$ for all n. Otherwise (5.35) is trivial. For a given $\varepsilon > 0$, choose a δ-covering $\{U_n^j : j = 1, 2, \dots\}$ of A_n such that

$$\sum_{j=1}^{\infty} [\mathrm{diam}(U_n^j)]^s < \mathcal{H}_\delta^s(A_n) + \frac{\varepsilon}{2^n}.$$

Since $\{U_n^j : n, j = 1, 2, \dots\}$ is a δ-covering of $\bigcup_{n=1}^{\infty} A_n$,

$$\mathcal{H}_\delta^s\Big(\bigcup_{n=1}^{\infty} A_n\Big) \le \sum_{n=1}^{\infty}\sum_{j=1}^{\infty} [\mathrm{diam}(U_n^j)]^s < \sum_{n=1}^{\infty} \mathcal{H}_\delta^s(A_n) + \sum_{n=1}^{\infty} \frac{\varepsilon}{2^n}.$$

The last inequalities hold for any $\varepsilon > 0$. Thus $\mathcal{H}_\delta^s(\bigcup_{n=1}^{\infty} A_n) \le \sum_{n=1}^{\infty} \mathcal{H}_\delta^s(A_n)$. Letting $\delta \to 0$, we get (5.35).

To prove that $\mathcal{H}^s$ is a metric outer measure, let E and F be subsets of X such that $\rho(E, F) = \delta > 0$. If $\{W_j\}$ is a $\delta/3$-covering of $E \cup F$, put

$$\mathcal{U} = \{W_j : E \cap W_j \ne \emptyset\} \quad \text{and} \quad \mathcal{V} = \{W_j : F \cap W_j \ne \emptyset\}.$$

Then $\mathcal{U}$ and $\mathcal{V}$ are $\delta/3$-coverings of E and F respectively, and $\mathcal{U} \cap \mathcal{V} = \emptyset$. Thus it follows that

$$\begin{aligned}\sum_j [\mathrm{diam}(W_j)]^s &= \sum_{W_k \in \mathcal{U}} [\mathrm{diam}(W_k)]^s + \sum_{W_l \in \mathcal{V}} [\mathrm{diam}(W_l)]^s \\ &\ge \mathcal{H}_{\delta/3}^s(E) + \mathcal{H}_{\delta/3}^s(F).\end{aligned}$$

When we take the infimum on the $\delta/3$-coverings of $E \cup F$, the left-hand side is replaced by $\mathcal{H}_{\delta/3}^s(E \cup F)$. Letting $\delta \to 0$, we obtain $\mathcal{H}^s(E \cup F) \ge \mathcal{H}^s(E) + \mathcal{H}^s(F)$.

To prove (ii) we may suppose that $\mathcal{H}^s(E) < \infty$. Since $\mathcal{H}_{1/n}^s(E) < \infty$ for all $n = 1, 2, \dots,$ there exists a $1/n$-covering $\{U_j^n : j = 1, 2, \dots\}$ of E such that

$$\sum_{j=1}^{\infty} [\mathrm{diam}(U_j^n)]^s < \mathcal{H}_{1/n}^s(E) + \frac{1}{n}. \tag{5.36}$$

If $O_j^n = \{x : \rho(x, U_j^n) < (1/n)\mathrm{diam}(U_j^n)\}$, then the family $\{O_j^n\}_{j=1}^{\infty}$ is an open covering of E, since $U_j^n \subset O_j^n$ $(j = 1, 2, \dots)$. Furthermore we have

$$\mathrm{diam}(O_j^n) < \left(1 + \frac{2}{n}\right)\mathrm{diam}(U_j^n) < \left(1 + \frac{2}{n}\right)\frac{1}{n}, \quad n = 1, 2, \dots.$$

The union $O_n = \bigcup_{j=1}^{\infty} O_j^n$ is an open set which includes E and satisfies condition (ii). In fact, for each n we have

$$\begin{aligned}\mathcal{H}_{(1+2/n)/n}^s\Big(\bigcap_{k=1}^{\infty} O_k\Big) &\le \mathcal{H}_{(1+2/n)/n}^s(O_n) \\ &\le \sum_{j=1}^{\infty} [\mathrm{diam}(O_j^n)]^s \le \sum_{j=1}^{\infty} \left(1 + \frac{2}{n}\right)^s [\mathrm{diam}(U_j^n)]^s.\end{aligned}$$

By (5.36) the last sum is bounded by

$$\le \left(1 + \frac{2}{n}\right)^s \left(\mathcal{H}_{1/n}^s(E) + \frac{1}{n}\right) \le \left(1 + \frac{2}{n}\right)^s \left(\mathcal{H}^s(E) + \frac{1}{n}\right).$$

Therefore we get $\mathcal{H}^s(\bigcap_{k=1}^{\infty} O_k) \le \mathcal{H}^s(E)$ for all n. The reverse inequality is obvious. □

DEFINITION 5.7. The value $\mathcal{H}^s(A)$ is called the *s-dimensional Hausdorff measure* of A.

There exists a unique number s such that

$$\mathcal{H}^s(E) < \infty \Rightarrow \mathcal{H}^t(E) = 0 \text{ for every } t > s,$$

and

$$\mathcal{H}^s(E) > 0 \Rightarrow \mathcal{H}^u(E) = \infty \text{ for every } s > u.$$

This implies that Hausdorff measure $\mathcal{H}^t(E)$ is essentially meaningful only at $t = s$.

To prove the existence of such a number, suppose that $\mathcal{H}^s(E) < \infty$ and $t > s$. By definition we have, for $\delta > 0$,

$$\mathcal{H}^t_\delta(E) = \inf\{\sum_{j=1}^{\infty}[\operatorname{diam}(U_j)]^t : \{U_j\} \text{ is a } \delta\text{-covering of } E\}.$$

Since $[\operatorname{diam}(U_j)]^t \leq [\operatorname{diam}(U_j)]^s \cdot \delta^{t-s}$, we have

$$\mathcal{H}^t_\delta(E) \leq \mathcal{H}^s_\delta(E)\delta^{t-s} \leq \mathcal{H}^s(E)\delta^{t-s}.$$

Thus $\mathcal{H}^t_\delta(E) \to 0$ as $\delta \to 0$.

If $s > u$, observe that $[\operatorname{diam}(U_j)]^u \geq [\operatorname{diam}(U_j)]^s\delta^{u-s}$. Then the second assertion follows in a similar way.

DEFINITION 5.8. Define

$$\dim_H(E) = \inf\{s > 0 : \mathcal{H}^s(E) = 0\}.$$

The number $\dim_H(E)$ is callled the *Hausdorff dimension* of E.

EXAMPLE 5.11. Let O be a nonempty bounded open set of $\mathbb{R}^d$. Then

$$\dim_H(O) = d \quad \text{and} \quad \mathcal{H}^d(O) = cm(O),$$

where $c > 0$ is a constant.

Indeed, let Q be a cube of side length 1. For $\delta = \sqrt{d}2^{-k}$ we have

$$\mathcal{H}^d_\delta(Q) \leq (\sqrt{d}2^{-k})^d \times 2^{dk} = \sqrt{d}^d,$$

taking the δ-covering of Q consisting of 2^{dk} cubes of side length 2^{-k}.

On the other hand, for any δ-covering $\{U_j\}$ of Q there exist cubes Q_j of side length $\leq \operatorname{diam}(U_j)$ such that $U_j \subset Q_j$. Thus $Q \subset \bigcup_j Q_j$. Thus

$$\sum_j \operatorname{diam}(U_j)^d \geq \sum_j m(Q_j) \geq m(Q) = 1.$$

Therefore we have $1 \leq \mathcal{H}^d(Q) \leq \sqrt{d}^d$.

The last inequality implies that the Hausdorff dimension of Q is d.

Put $\mathcal{H}^d(Q) = c$, which is between 1 and $\sqrt{d}^d$. Since an open set is a disjoint countable union of dyadic cubes $\bigcup_j Q_j$, we have $\mathcal{H}^d(O) = \sum_j \mathcal{H}^d(Q_j) = c\sum_j m(Q_j) = cm(O)$.

EXAMPLE 5.12. The Hausdorff dimension of the Cantor ternary set C is $\log 2/\log 3 = 0.6309\ldots$, which we denote by s, and the Hausdorff measure $\mathcal{H}^s(C) = 1$.

In fact, $C = \bigcap_{n=1}^{\infty} C_n$, where C_n are the union of 2^n closed intervals of length 3^{-n} which are complements of the "middle-third" intervals. Taking a 3^{-n}-covering of C, we have $\mathcal{H}^s_{3^{-n}}(C) \leq \mathcal{H}^s_{3^{-n}}(C_n) \leq 2^n 3^{-sn} = 1$. Thus $\mathcal{H}^s(C) \leq 1$.

To prove the reverse inequality let $\{U_j\}$ be a δ-covering of C. We may assume that the U_j are open intervals. Let $\varepsilon > 0$ and put $V_j = (1+\varepsilon)U_j$; that is, V_j is the interval concentric with U_j and with $1+\varepsilon$ times the side length. Choose a finite covering $\{V_j : j = 1, 2, \ldots, k\}$ of C, which is possible by the compactness of C. Choose n so that each interval in C_n is included in an interval V_j, which is also possible by the finite intersection property of compact sets. First, suppose that each V_j contains at most one interval of C_n. Then we have

$$\begin{aligned}\sum_{j=1}^{\infty}[\mathrm{diam}(U_j)]^s &= (1+\varepsilon)^{-s}\sum_{j=1}^{\infty}[\mathrm{diam}(V_j)]^s \\ &\geq (1+\varepsilon)^{-s}2^n 3^{-ns} = (1+\varepsilon)^{-s}.\end{aligned}$$

Thus $\mathcal{H}^s_\delta(C) \geq (1+\varepsilon)^{-s}$. Let $\delta \to 0$. Since $\varepsilon > 0$ is an arbitrarily given number, we get $\mathcal{H}^s(C) \geq 1$.

If V_j contains more than two intervals of length 3^{-n} from set C_n, let K be the biggest "middle third" open interval which is removed up to the n-th step in the construction of the Cantor ternary set and which is located between intervals of length 3^{-n} contained in V_j. Divide the interval V_j into three intervals, say, $V_j = J \cup K \cup L$. Then we have

$$\mathrm{diam}(J), \mathrm{diam}(L) \leq \mathrm{diam}(K).$$

Since $0 < s < 1$, we have

$$\begin{aligned}[\mathrm{diam}(V_j)]^s &= [\mathrm{diam}(J) + \mathrm{diam}(K) + \mathrm{diam}(L)]^s \\ &\geq \left[\frac{3}{2}(\mathrm{diam}(J) + \mathrm{diam}(L))\right]^s \\ &\geq 3^s\frac{1}{2}([\mathrm{diam}(J)]^s + [\mathrm{diam}(L)]^s) \\ &= [\mathrm{diam}(J)]^s + [\mathrm{diam}(L)]^s.\end{aligned}$$

We apply such a procedure to J if J containes more than two intervals of length 3^{-n} composing the set C_n, and then to L. By such a procedure, we can suppose, finally, that V_j contains only one interval of length 3^{-n} without losing generality.

The Hausdorff dimension of the Koch curve is $\log 4/\log 3$. For a proof the reader is referred to [Fa].

5.9. Problems

5.1. Let X be a set and $\mathcal{X}$ a σ-field of subsets of X. Then the pair $(X, \mathcal{X})$ is called a *meaurable space*. Let $(X, \mathcal{X})$ and $(Y, \mathcal{Y})$ be two measurable spaces and let φ be a mapping of X to Y. Prove the following.

(i) The family $\varphi^{-1}(\mathcal{Y})$ defined by $\{\varphi^{-1}(B) : B \in \mathcal{Y}\}$ is a σ-field of X.
(ii) The family $\{B \subset Y : \varphi^{-1}(B) \in \mathcal{X}\}$ is a σ-field of Y.
(iii) Is the family $\varphi(\mathcal{X})$ a σ-field of Y?

5.2. Let $(X, \mathcal{X})$ and $(Y, \mathcal{Y})$ be two measurable spaces. A mapping T of X to Y is said to be a *measurable transform* if $T^{-1}(B) = \{x \in X : T(x) \in B\} \in \mathcal{X}$ for all $B \in \mathcal{Y}$.

Suppose that $T : X \mapsto Y$ and $g : Y \mapsto (\mathbb{R}, \mathcal{M}(\mathbb{R}))$ are measurable transforms. Then prove that $g(T(\cdot)) : X \mapsto \mathbb{R}$ is a measurable transform.

5.3. Let $(X, \mathcal{X}, \mu)$ be a measure space. Let $\nu(E) = \mu(E \cap A)$. Then prove that ν is a measure on $(X, \mathcal{M})$.

5.4. For subsets E and F of a set X the *symmetric difference* is defined by

$$E \Delta F = (E - F) \cup (F - E).$$

Let $(X, \mathcal{X}, \mu)$ be a finite measure space. For $E, F \in \mathcal{X}$ put $\rho(E, F) = \mu(E \Delta F)$. Prove the following:

(i) $\rho(E, F) = 0 \;\Leftrightarrow\; E - F$ and $F - E$ are null.
(ii) $\rho(E, F) = \rho(F, E)$.
(iii) $\rho(E, F) \leq \rho(E, G) + \rho(G, F)$ for $E, F, G \in \mathcal{X}$.

5.5. For a subset E of the natural numbers $\mathbb{N}$ define ν by $\nu(E) = \text{card}(E)$. Prove that ν is a measure on the σ-field $\mathcal{P}(\mathbb{N})$ and, for $f \in L^1(\mathbb{N})$,

$$\int_{\mathbb{N}} f(n) d\nu(n) = \sum_{n=1}^{\infty} f(n).$$

5.6. Fix an integer $k \geq 0$. Let $\mathcal{F}_k$ be the σ-field generated by the intervals $I_j = [j2^{-k}, (j+1)2^{-k})$, $j = 0, 1, \ldots, 2^k - 1$. For $f \in L^1([0,1), \mathcal{M}, m)$ give an $\mathcal{F}_k$-measurable function f_k such that

$$\int_E f dm = \int_E f_k dm \quad \text{for all } E \in \mathcal{F}_k.$$

Is such an f_k unique?

5.7. Let μ be a σ-finite measure on $(X, \mathcal{M})$. Prove directly that μ can be decomposed uniquely as $\mu = \mu_c + \mu_d$ and $\mu_c \perp \mu_d$, where μ_c is a continuous measure and μ_d is a discrete measure.

5.8. Let μ_1, μ_2 and μ_3 be σ-finite measures on a σ-finite measure space. If $\mu_1 \ll \mu_2$ and $\mu_2 \ll \mu_3$, then prove that $\mu_1 \ll \mu_3$ and

$$\frac{d\mu_1}{d\mu_3} = \frac{d\mu_1}{d\mu_2} \frac{d\mu_2}{d\mu_3} \quad \mu_3\text{-a.e.}$$

5.9. Let D be a domain in $\mathbb{R}^n$ and $0 < \alpha \leq 1$. A mapping f of D to $\mathbb{R}^m$ is said to satisfy a *Lipschitz condition of order* α with the Lipschitz constant C if

$$|f(x) - f(y)| \leq C|x - y|^\alpha \quad (x, y \in D).$$

Prove that $\dim_H f(D) \leq (1/\alpha) \dim_H D$.

5.10. Compute the Hausdorff dimension $\dim_H(D)$ of the disc $D = \{(x_1, x_2, 0) : x_1^2 + x_2^2 < 1\}$ in the three dimensional Euclidean space $\mathbb{R}^3$.

CHAPTER 6

Lebesgue Spaces and Continuous Functions

The function spaces which are mainly studied in this chapter are the Lebesgue space L^p and the space of continuous functions C_0. Not only they are important for their role in analysis, but also they have interesting properties in themselves and offer basic concepts to functional analysis.

In §1 some fundamental notions of functional analysis are explained.

In §2 the space $L^p, 1 \le p < \infty$, is discussed, where the inequality has a special role.

By a simple observation, to a linear functional on a finite dimensional vector space there corresponds a vector, and a linear functional is represented by an inner product. There exists a similar relation between the bounded linear functionals on L^p and the elements of the space $L^{p'}$, if $1 \le p < \infty$ and $1/p + 1/p' = 1$. More precisely, to a bounded linear functional T there corresponds a function $u \in L^{p'}$ such that

$$T(f) = \int f\bar{u}d\mu \quad \text{for } f \in L^p$$

(see §6.3). When $p = 2$, the space L^2 is a typical Hilbert space and has various particularly interesting properties.

When $p = \infty$, such a direct representation no longer exists in general. However, if we replace the space L^∞ by the space of continuous functions C_0, we shall have an important expression of bounded linear functionals, called the F. Riesz representation theorem; that is, to every continuous linear functional on C_0 there corresponds a finite Borel measure, and every finite Borel measure defines a continuous linear functional, which will be proved in §§6.6 and 6.7.

We shall introduce in §6.4 a kind of multiplication in the space $L^1(\mathbb{R}^d)$, called convolution, which has an important role in Fourier analysis and will be studied in Chapter 8.

6.1. Elements of functional analysis

The set L^1 of integrable functions is a vector space, as we have already shown. We shall introduce some geometries on function spaces.

6.1.1. Banach spaces. Let $\mathbb{K}$ be the field $\mathbb{C}$ or $\mathbb{R}$.

DEFINITION 6.1. Let V be a vector space on $\mathbb{K}$. The mapping $\|\ \|$ of V to $\mathbb{R}$ is called a *seminorm* if

(i) $0 \le \|f\| < \infty \quad (f \in V)$,
(ii) $\|af\| = |a|\|f\| \quad (f \in V,\ a \in \mathbb{K})$,
(iii) $\|f + g\| \le \|f\| + \|g\| \quad (f, g \in V)$.

Furthermore, if

(iv) $\|f\| = 0 \Leftrightarrow f = 0$,

then $\| \ \|$ is called a *norm*.

A vector space with a norm $(V, \| \ \|)$ is called a *normed space*.

If

$$\rho(f, g) = \|f - g\| \quad (f, g \in V),$$

then it is an easy exercise to see that $\rho(\cdot, \cdot)$ satisfies the three conditions for a metric. Thus any normed space is a metric space with the metric $\rho(\cdot, \cdot)$.

DEFINITION 6.2. A complete normed space is called a *Banach*[1] *space*.

Let $(V, \| \ \|_V)$ and $(W, \| \ \|_W)$ be two Banach spaces. A mapping T of V to W is said to be *linear* if

$$T(af + bg) = aTf + bTg \quad (f, g \in V \text{ and } a, b \in \mathbb{K}).$$

PROPOSITION 6.1. *For a linear mapping T of V to W the following are equivalent:*

(i) *T is continuous.*
(ii) *T is continuous at 0.*
(iii) *There exists a constant $C > 0$ such that*

$$\|Tf\|_W \le C\|f\|_V \quad \text{for } f \in V.$$

PROOF. The implication (i) $\Rightarrow$ (ii) is trivial. If T is continuous at 0, there exists $\delta > 0$ such that $\|Tf\|_W < 1$ for $f \in V$ with $\|f\|_V \le \delta$. Thus for every $g \neq 0$ we have

$$\|Tg\|_W = \left\| \frac{\|g\|_V}{\delta} T\left(\frac{\delta}{\|g\|_V} g \right) \right\|_W \le \frac{1}{\delta}\|g\|_V,$$

since $\|(\delta/\|g\|_V)g\|_V = \delta$. Thus (iii) holds with $C = 1/\delta$.

Let $f_0 \in V$. Under the condition (iii), $\|T(f - f_0)\|_W \le C\|(f - f_0)\|_V$. Thus f is continuous at $f_0 \in V$. □

When (iii) holds for some constant $C < \infty$, T is said to be *bounded*. By Proposition 6.1, T is bounded if and only if T is continuous.

We denote by $B(V, W)$ the collection of all bounded linear operators from V to W. For $T \in B(V, W)$ put

$$\|T\| = \sup\left\{ \frac{\|Tf\|_W}{\|f\|_V} : f \neq 0 \right\} = \sup_{\|f\|_V = 1} \|Tf\|_W,$$

and call it the norm of T. $\|T\|$ is the smallest number C which satisfies the inequality in (iii).

For $S, T \in B(V, W)$ and $a, b \in \mathbb{K}$ put

$$(aS + bT)f = aSf + bTf \quad (f \in V).$$

Then $aS + bT$ is again a bounded linear operator in $B(V, W)$. Thus $B(V, W)$ is a vector space. Furthermore, $\|T\|$ is a norm in $B(V, W)$. In fact, $\|(S + T)f\|_W \le \|Sf\|_W + \|Tf\|_W \le \|S\|\|f\|_V + \|T\|\|f\|_V$, which implies $\|S + T\| \le \|S\| + \|T\|$. Other properties of norm are easily obtained.

---

[1] S. Banach, 1892–1945. Polish mathematician.

THEOREM 6.1. *$B(V, W)$ is a Banach space.*

PROOF. To show the completeness of the space $B(V, W)$, let $\{T_n\}$ be a sequence in $B(V, W)$ such that $\|T_m - T_n\| \to 0$ as $m, n \to \infty$. Since for every $f \in V$

$$\|T_m f - T_n f\|_W \leq \|T_m - T_n\| \|f\|_V \to 0 \quad \text{as } m, n \to \infty, \tag{6.1}$$

$\{T_n f\}$ is a Cauchy sequence in W. Let Tf be the limit of $T_m f$. Since each T_m is linear, T is also linear. Furthermore, we have $\|T\| = \lim \|T_n\|$. Thus T is bounded. Therefore $T \in B(V, W)$.

Finally, by (6.1) we have $\|Tf - T_n f\| \leq (\lim_{m\to\infty} \|T_m - T_n\|)\|f\|$ for every $f \in V$. Thus $\|T - T_n\| \leq \lim_{m\to\infty} \|T_m - T_n\| \to 0$ as $n \to \infty$. Thus $B(V, W)$ is complete. □

In particular, when $W = \mathbb{K}$ the space $B(V, \mathbb{K})$ is denoted by V' and a linear operator in V' is called *a linear functional.*

A sequence $\{f_n\}$ in V is said to converge to f in V or to *converge strongly* to f if

$$\|f_n - f\|_V \to 0 \quad \text{as } n \to \infty.$$

If $Tf_n \to Tf$ for every $T \in V'$, then $\{f_n\}$ is said to *converge weakly to* f, and we write

$$f_n \to f \text{ weakly} \quad \text{or} \quad f_n \to f \text{ in the } \sigma(V, V')\text{-topology}.$$

V' is a Banach space, so the set of all continuous linear functionals on V', which is denoted by $V'' = B(V', \mathbb{K})$, is a Banach space, too.

A sequence $\{T_n\}$ in V' converges to T weakly or in the $\sigma(V', V'')$-topology, if $F(T_n) \to F(T)$ for every $F \in V''$.

If $f \in V$, the mapping $V' \ni T \mapsto Tf \in \mathbb{K}$ is linear and the norm does not exceed $\|f\|_V$. Thus f is identified with an operator in V''.

If $T_n f \to Tf$ for every $f \in V$, the sequence $\{T_n\}$ of V' is said to *converge to* T *in the weak* topology* or in the $\sigma(V', V)$-topology.

6.1.2. Topological vector spaces. Let V be a vector space over the field $\mathbb{K}$ with a topology. If the following two mappings are continuous, V is said to be a *topological vector space:*

(i) $V \times V \ni (f, g) \mapsto f + g \in V$.
(ii) $\mathbb{K} \times V \ni (a, f) \mapsto af \in V$.

If a family of sets $\mathcal{U}(0)$ in V is a neighborhood system of 0, then (i) implies that the family $f_0 + \mathcal{U}(0) = \{f_0 + U : U \in \mathcal{U}(0)\}$ is a neighborhood system of f_0 for all $f_0 \in V$.

A family $\mathcal{V}(0)$ is called a fundamental system of neighborhoods of 0, if $\mathcal{V}(0) \subset \mathcal{U}(0)$ and for every $U_1 \in \mathcal{U}(0)$ there exists a $U_2 \in \mathcal{V}(0)$ such that $U_2 \subset U_1$.

A set C is said to be *convex* if $tf + (1-t)g \in C$ for all $f, g \in C$ and $0 < t < 1$. When V has a fundamental system of neighborhoods of 0 which consists of convex sets, V is said to be a *locally convex topological vector space.*

Let $\{p_\alpha : \alpha \in A\}$ be a family of seminorms on the vector space V. For any finite number of seminorms $p_{\alpha_1}, \dots, p_{\alpha_n}$ and $\varepsilon_1, \dots, \varepsilon_n > 0$, put

$$U(0 : p_{\alpha_1}, \dots, p_{\alpha_n}; \varepsilon_1, \dots, \varepsilon_n) = \{f \in V : p_{\alpha_j}(f) < \varepsilon_j,\ j = 1, \dots, n\}. \tag{6.2}$$

$U(0 : p_{\alpha_1}, \dots, p_{\alpha_n}; \varepsilon_1, \dots, \varepsilon_n)$ is clearly a convex set.

Let $\mathcal{V}(0)$ be the collection of all such sets $U(0 : p_{\alpha_1}, \dots, p_{\alpha_n}; \varepsilon_1, \dots, \varepsilon_n)$ and let $\mathcal{V}(f) = f + \mathcal{V}(0)$.

It is not difficult to show the following:

(V1) If $U \in \mathcal{V}(f)$, then $f \in U$.

(V2) If $U_1, U_2 \in \mathcal{V}(f)$, then there exists $U_3 \in \mathcal{V}(f)$ such that $U_3 \subset U_1 \cap U_2$.

(V3) If $U_1 \in \mathcal{V}(f)$ and $g \in U_1$, then there exists $U_2 \in \mathcal{V}(g)$ such that $U_2 \subset U_1$.

Let $\mathcal{O}(V)$ be the family of finite or infinite unions of sets in the family $\bigcup_{f \in V} \mathcal{V}(f)$. We assume that $\mathcal{O}(V) \ni \emptyset$.

PROPOSITION 6.2. *For a vector space V we have:*

(i) *$\mathcal{O}(V)$ satisfies the three conditions for open sets (cf. §1.3).*

(ii) *V is a topological vector space with the topology $\mathcal{O}(V)$.*

(iii) *$\mathcal{V}(0)$ is a fundamental system of neighborhood of 0 which consists of convex sets.*

Before the proof we mension some examples.

Let B be a Banach space over $\mathbb{K}$ and B' the space of all continuous linear functionals. For $F \in B''$ let

$$p_F(T) = |F(T)| \quad \text{for } T \in B'.$$

p_F is a seminorm on B'. For a finite number of $F_1, \dots, F_n \in B''$ and $\varepsilon_1, \dots, \varepsilon_n > 0$, the sets

$$U(0 : F_1, \dots, F_n; \varepsilon_1, \dots, \varepsilon_n) = \{T \in B' : p_{F_j}(f) < \varepsilon_j,\ j = 1, \dots, n\}$$

constitute a fundamental system of neighborhoods of 0 of B'. The topology defined by this system is called the *weak topology* or the *$\sigma(B', B'')$-topology.*

In a similar way, for $f \in B$ let

$$p_f(T) = |T(f)| \quad (T \in B').$$

p_f is also a seminorm on B'. For a finite number of $f_1, \dots, f_n \in B$ and $\varepsilon_1, \dots, \varepsilon_n > 0$, the sets

$$U(0 : f_1, \dots, f_n; \varepsilon_1, \dots, \varepsilon_n) = \{T \in B' : p_{f_j}(f) < \varepsilon_j,\ j = 1, \dots, n\}$$

again constitute a fundamental system of neighborhoods of 0 of B' and define a topology, which is called the *weak* topology* or *$\sigma(B', B)$-topology.* The weak* topology is the weakest topology in which every $T \in B'$ is continuous on B.

PROOF OF THE PROPOSITION. (i) The set V obviously belongs to $\mathcal{O}(V)$. Thus condition (O1) for open sets is satisfied (see Definition 1.4). Condition (O2) is trivial.

Let $O_1, O_2 \in \mathcal{O}(V)$. We shall show that $O_1 \cap O_2 \in \mathcal{O}(V)$. If $f \in O_1 \cap O_2$, then by the definition of $\mathcal{O}(V)$ there exist $U_j \subset O_j$, $j = 1, 2$, such that $f \in U_j \in \mathcal{V}(f_j)$ for some $f_j \in V$. By (V3) there exists $W_j \in \mathcal{V}(f)$ such that $W_j \subset U_j$. Thus $f \in W_1 \cap W_2 \subset O$. On the other hand, by (V2) there exists $W \in \mathcal{V}(f)$ such that $W \subset W_1 \cap W_2$. Thus $f \in W \subset O$. Therefore $O \in \mathcal{O}(V)$. Thus condition (O3) is satisfied.

(ii) Let $f_0, g_0 \in V$, and let O be a neighborhood of $f_0 + g_0$. Then there exists $U \in \mathcal{V}(0)$ such that $f_0 + g_0 + U \subset O$. Since U is of the form (6.2), it is easy to find a neighborhoof $W \in \mathcal{V}(0)$ such that $W + W \subset U$. Thus $(f_0 + W) + (g_0 + W) \subset f_0 + g_0 + U$. Thus the mapping $V \times V \ni (f, g) \mapsto f + g \in V$ is continuous at $f_0 + g_0$.

Next, we show that the mapping $\mathbb{K} \times V \ni (a, f) \mapsto af \in V$ is continuous at (a_0, f_0). Let $U \in \mathcal{V}(0)$. It suffices to find $\delta > 0$ and $U_0 \in \mathcal{V}(0)$ such that $(a + a_0)(f + f_0) \in a_0 f_0 + U$ for $|a| < \delta$ and $f \in U_0$. Choose $W \in \mathcal{V}(0)$ so that $W + W \subset U$. If $\delta > 0$ is sufficiently small, $af_0 \in W$ for all $|a| < \delta$.

Observe that for a seminorm p we have $p((a + a_0)f) \leq (\delta + |a_0|)p(f)$ if $|a| < \delta$. Thus $(a + a_0)f \in W$ if $|a| < \delta$ and $f \in (\delta + |a_0|)^{-1}W$, where $\lambda W = \{\lambda f : f \in W\}$. Thus if $|a| < \delta$ and $f \in (\delta + |a_0|)^{-1}W$, then $(a+a_0)(f+f_0) = a_0 f_0 + (a+a_0)f + af_0 \in a_0 f_0 + W + W \subset a_0 f_0 + U$. This proves the continuity of the mapping $(a, f) \mapsto af$.

(iii) is obvious. □

Hereafter, we call a vector space over the complex field $\mathbb{C}$ simply a vector space, unless otherwise mentioned.

6.1.3. Banach algebras.

DEFINITION 6.3. A vector space over the complex field is said to be a *Banach algebra* if

(i) B is a Banach space over $\mathbb{C}$,

(ii) B is a ring over $\mathbb{C}$, that is, a multiplication is defined on B which satisfies

$$u(vw) = (uv)w,$$
$$(u + v)w = uw + vw, \quad w(u + v) = wu + wv,$$
$$(au)v = u(av) = a(uv), \quad \text{for } u, v, w \in B \text{ and } a \in \mathbb{C},$$

(iii) $\|uv\| \leq \|u\|\|v\|$.

In particular, if

$$uv = vu \quad \text{for all } u, v \in B,$$

then B is called a commutative Banach algebra.

EXAMPLE 6.1. The space $C[0, 1]$ of continuous functions is a commutative Banach algebra with pointwise multiplication and the uniform norm $\|u\|_\infty = \sup_{x\in[0,1]} |u(x)|$. The algebra $C[0, 1]$ has the unit that is the function identically 1. It is easily shown that

$$\|uv\|_\infty \leq \|u\|_\infty \|v\|_\infty.$$

Set

$$C_0(\mathbb{R}^d) = \{f \in C(\mathbb{R}^d) : \lim_{|x|\to\infty} f(x) = 0\}.$$

The space $C_0(\mathbb{R}^d)$ is also a commutative Banach algebra with the uniform norm. It does not have a unit.

EXAMPLE 6.2. Let V be a Banach space. Then $B(V, V)$ is a Banach algebra. Indeed, for $S, T \in B(V, V)$ the multiplication is defined by $(ST)f = S(Tf)$ for $f \in V$. Since $\|S(T(f))\| \leq \|S\|\|Tf\| \leq \|S\|\|T\|\|f\|$, we have $\|ST\| \leq \|S\|\|T\|$.

The identity operator is the unit of $B(V, V)$. The algebra $B(V, V)$ is not commutative in general.

6.1.4. Hilbert spaces. Let H be a vector space over $\mathbb{C}$. A mapping $(\cdot, \cdot)$ of $H \times H$ to $\mathbb{C}$ is called an *inner product* if the following conditions are satisfied:

(i) $(f, f) \geq 0$, and $(f, f) = 0 \Leftrightarrow f = 0$,

(ii) $(f, g) = \overline{(g, f)}$,

(iii) $(af + bg, h) = a(f, h) + b(g, h) \quad (a, b \in \mathbb{C},\ f, g, \in H)$.

A vector space with an inner product is called a *pre-Hilbert space* or an *inner product space.*

Put

$$\|f\| = \sqrt{(f,f)} \quad \text{for } f \in H.$$

Then $\|\cdot\|$ is a norm, as is shown in Lemma 6.1 below.

DEFINITION 6.4. A *Hilbert*[2] *space* is a complete inner product space with the norm $\|\cdot\| = \sqrt{(\cdot,\cdot)}$.

LEMMA 6.1. *Let $(\cdot,\cdot)$ be an inner product on a vector space H over $\mathbb{C}$. Then $\|\cdot\| = \sqrt{(\cdot,\cdot)}$ is a norm.*

PROOF. By the condition (i) of the inner product, $\|f\| = \sqrt{(f,f)}$ vanishes if and only if $f = 0$. By (iii) and (ii) we have, for $a \in \mathbb{C}$,

$$\|af\|^2 = (af, af) = a\bar{a}(f,f) = |a|^2\|f\|^2,$$

from which it follows that $\|af\| = |a|\|f\|$.

Finally, let $f, g \in H$. Assume that $(f,g) \neq 0$, which implies that $(f,f) \neq 0$. Let z be any complex number. Then we have, for $t \in \mathbb{R}$,

$$(tf + zg, tf + zg) = t^2(f,f) + t\bar{z}(f,g) + tz(g,f) + z\bar{z}(g,g). \tag{6.3}$$

If $z = (f,g)/|(f,g)|$, the right-hand side equals

$$t^2(f,f) + 2t|(f,g)| + (g,g),$$

which is nonnegative for all t, since the left-hand side of (6.3) is nonnegative. Thus $|(f,g)|^2 \leq (f,f)(g,g)$, from which it follows that

$$|(f,g)| \leq \|f\|\|g\| \quad \text{for all } f, g \in H. \tag{6.4}$$

By (6.3) and (6.4) we have

$$\begin{aligned} \|f+g\|^2 &= (f+g, f+g) = (f,f) + (f,g) + (g,f) + (g,g) \\ &\leq \|f\|^2 + \|f\|\|g\| + \|g\|\|f\| + \|g\|^2 \\ &\leq (\|f\| + \|g\|)^2. \end{aligned}$$

Thus $\|f+g\| \leq \|f\| + \|g\|$.

This completes the proof. □

LEMMA 6.2 (The parallelogram law). *For $f, g \in H$,*

$$\|f+g\|^2 + \|f-g\|^2 = 2\|f\|^2 + 2\|g\|^2.$$

PROOF. We have $\|f \pm g\|^2 = (f,f) \pm (f,g) \pm (g,f) + (g,g)$ for $f, g \in H$. Adding both sides, we get the lemma. □

For the rest of this section, H stands for a Hilbert space.

Let $g \in H$. An element f of H is said to be *orthogonal* to g if $(f,g) = 0$, which is denoted by $f \perp g$.

For a subset $M \subset H$ the *orthogonal complement* of M is defined by

$$M^{\perp} = \{f \in H : (f,g) = 0 \text{ for all } g \in M\}.$$

By the definition we have $f \perp g$ for all $f \in M$ and $g \in M^{\perp}$.

[2] D. Hilbert, 1862–1943. German mathematician.

The set $M^\perp$ is a closed subspace. In fact, if $g, g' \in M^\perp$, then $(ag + bg', h) = a(g,h) + b(g',h) = 0$ for all $h \in M$, which implies that $af + bg \in M^\perp$. Thus $M^\perp$ is a subspace. Let $\{f_n\}$ be a Cauchy sequence in $M^\perp$ which converges to $f \in H$. Then for every $h \in M$ we have $|(f,h)| = |(f,h) - (f_n,h)| \leq \|f - f_n\|\|h\| \to 0$ as $n \to \infty$. Thus $(f,h) = 0$ for all $h \in M$, that is, $f \in M^\perp$. Thus $M^\perp$ is a closed subspace.

THEOREM 6.2. *Let M be a closed subspace of a Hilbert space H. Then:*

(i) $H = M \oplus M^\perp$, *that is, each element f of H can be uniquely expressed as*

$$f = Pf + Qf, \tag{6.5}$$

where $Pf \in M$ and $Qf \in M^\perp$.

(ii) *The mappings P and Q are linear operators from H to M and $M^\perp$, respectively.*

PROOF. To prove the uniqueness let $f = f_1 + f_2 = f_1' + f_2'$, where $f_1, f_1' \in M$ and $f_2, f_2' \in M^\perp$. Since $(f_1 - f_1') \perp (f_2 - f_2')$ we have

$$(f_1 - f_1', f_1 - f_1') = -(f_1 - f_1', f_2 - f_2') = 0.$$

Thus $f_1 - f_1' = 0$.

Let $f \in H$ and put $\delta = \inf\{\|f - f'\| : f' \in M\}$. Choose a sequence $\{f_n'\}$ in M so that $\|f - f_n'\| \to \delta$. Then $\{f_n'\}$ is a Cauchy sequence. In fact, by the parallelogram law

$$2(\|f_n' - f\|^2 + \|f_m' - f\|^2) = \|f_n' - f_m'\|^2 + \|f_n' + f_m' - 2f\|^2.$$

The last term equals $4\|(f_n' + f_m')/2 - f\|^2$, which is, by the definition of δ, not less than $4\delta^2$, since $(f_n' + f_m')/2 \in M$. On the other hand, the left-hand side converges to $4\delta^2$. Thus $\|f_n' - f_m'\|^2 \to 0$ as $n \to \infty$.

Put $\lim f_n' = Pf$ and $f - Pf = Qf$. Since M is a closed subspace, it follows that $Pf \in M$ and $\delta = \|f - Pf\| = \inf\{\|f - f'\| : f' \in M\}$.

Now we show that $Qf \in M^\perp$. For $h \in M$ put

$$\varphi(t) = \|Qf - th\|^2 = \|f - Pf - th\|^2.$$

Since $\varphi(t)$ is a polynomial of t of order 2 and attains its minimal at $t = 0$, the coefficient of t must be 0. That is, $(Qf, h) + (h, Qf) = 2\Re(Qf, h) = 0$. If h is replaced by ih, $\Im(Qf, h) = 0$ is obtained by a similar argument. Thus $(Qf, h) = 0$. Therefore $Qf \perp M$.

The mapping P is surjective. For, if $f \in M$, then $Pf = f$ by the uniqueness of the decomposition.

Applying (6.5) to $f, g \in H$ and $af + bg$ with $a, b \in \mathbb{C}$, we have

$$af + bg = a(Pf + Qf) + b(Pg + Qg).$$

On the other hand,

$$af + bg = P(af + bg) + Q(af + bg).$$

Thus $P(af + bg) - (aPf + bPg) = Q(af + bg) - (bPg + bQg)$. Since both sides belong to M and $M^\perp$, they must be 0. Thus the mappings P and Q are linear. □

The linear operator P is called the *orthogonal projection* of H to M.

THEOREM 6.3 (The Riesz representation theorem). [3] *For every bounded linear functional T of a Hilbert space H there exists $g \in H$ such that*

$$Tf = (f, g) \quad \text{for } f \in H.$$

The vector g is unique, and $\|T\| = \|g\|$.

PROOF. We begin with the proof of uniqueness. If there exists $g' \in H$ such that $(f, g) = (f, g')$ for all $f \in H$, then substituting $f = g - g'$ for the equation $(f, g - g') = 0$, we get $g - g' = 0$.

Let T be a nonzero continuous linear functional on H. Put $M = \ker(T)$, which is $\{f : Tf = 0\}$ by definition. Since T is continuous and linear, M is a proper closed subspace. Thus by Theorem 6.2, $M^\perp \neq \{0\}$. Let h be an element of $M^\perp$ with $\|h\| = 1$. Since $T[(Tf)h - (Th)f] = 0$ for every $f \in H$, we have $0 = ([(Tf)h - (Th)f], h) = Tf\|h\|^2 - (Th)(f, h)$. Thus $Tf = (f, g)$ with $g = \overline{Th}h$. □

Suppose $T \in B(H, H)$ and $g \in H$. Since the mapping $H \ni f \mapsto (Tf, g) \in \mathbb{C}$ is continuous and linear, there exists $h \in H$ such that

$$(Tf, g) = (f, h) \quad \text{for } f \in H \tag{6.6}$$

by the Riesz representation theorem. To each g there corresponds a unique such element h, which we denote by $h = T^*g$. By a simple argument T^* is a linear operator from H to H. Since $(Tf, g) = (f, T^*g)$ by definition, taking the supremum over f and g with $\|f\| = \|g\| = 1$ we get $\|T\| = \|T^*\|$.

T^* is called the *adjoint operator* of T.

T is called a *unitary operator* if T is invertible and $T^{-1} = T^*$.

6.2. Lebesgue spaces

6.2.1. Definition of Lebesgue spaces. Let $(X, \mathcal{M}, \mu)$ be a nonnegative measure space and $0 < p < \infty$. The space $\mathcal{L}^p(X) = \mathcal{L}^p(X, \mathcal{M}, \mu)$ is the set of $\mathcal{M}$-measurable functions in X for which the L^p-norm is finite, that is,

$$\|f\|_p = \left(\int_X |f|^p d\mu\right)^{1/p} < \infty.$$

If $p = \infty$, the L^∞-norm is defined by

$$\|f\|_\infty = \inf\{x : \mu(\{|f(x)| > a\}) = 0\}.$$

The norm $\|f\|_\infty$ is also called the *essential supremum* of f and denoted sometimes by ess sup $|f|$.

In the same manner as in the space $\mathcal{L}^1(\mathbb{R}^d)$, an equivalence relation on the spaces $\mathcal{L}^p(X)$ $(0 < p \leq \infty)$ is defined by

$$f \sim g \quad \Longleftrightarrow \quad f = g \quad \mu\text{-a.e.}$$

The space of all equivalence classes in $\mathcal{L}^p(X)$ (that is, the space $\mathcal{L}^p(X)/\sim$) is called the Lebesgue space $L^p(X)$.

If f' and f'' are in the same equivalence class $\dot{f}$, then $\|f'\|_p = \|f''\|_p$. For $\dot{f} \in L^p(X)$ the norm is defined by

$$\|\dot{f}\|_p = \|f'\|_p \quad (f' \in \dot{f}).$$

[3] F. Riesz, 1880–1956. Hungarian mathematician

Actually, if $1 \le p \le \infty$, then $\|\cdot\|_p$ is a norm in L^p, as is shown below. If $0 < p < 1$, then $\|\cdot\|_p$ does not always satisfy the conditions of a norm, but it is, habitually, called the L^p-norm.

As in the Euclidean space case, the equivalence class which contains the measurable function f is denoted simply by f. When there is no additional note, f denotes either the function itself or the equivalence class.

In the following, the measure space $(X, \mathcal{M}, \mu)$ is assumed to be nonnegative and σ-finite.

6.2.2. L^p-norm. A real valued function φ in $\mathbb{R}$ is said to be convex if

$$\varphi(\theta x + (1-\theta)y) \le \theta\varphi(x) + (1-\theta)\varphi(y) \tag{6.7}$$

for all $x, y \in \mathbb{R}$ and $0 < \theta < 1$.

LEMMA 6.3. *If φ is a convex function in $\mathbb{R}$, then*

$$\frac{\varphi(a) - \varphi(s)}{a - s} \le \frac{\varphi(t) - \varphi(a)}{t - a} \quad \textit{for } s < a < t.$$

PROOF. Since $a = \theta s + (1-\theta)t$ for some $\theta \in (0,1)$, by the convexity of φ we have

$$\varphi(a) \le \theta\varphi(s) + (1-\theta)\varphi(t).$$

Substituting θ into the inequality, we get the lemma. □

THEOREM 6.4. *Let φ be a nonnegative convex function in $\mathbb{R}$. Let ν be a nonnegative measure such that $\int_X d\nu = 1$. If u is a real valued function in $L^1(X, \nu)$, then*

$$\varphi\Big(\int u d\nu\Big) \le \int \varphi(u) d\nu \qquad \textit{(Jensen's inequality)}.$$

PROOF. Fix a real number a and put

$$M = \sup_{s<a} \frac{\varphi(a) - \varphi(s)}{a - s}.$$

Then

$$\varphi(t) \ge \varphi(a) - M(t-a) \tag{6.8}$$

for all t. If $t < a$ the inequality is a direct consequence of the definition of M, and if $t > a$ it follows from Lemma 6.3. The case $t = a$ is trivial.

Now, put $t = u(x)$ and $a = \int u d\nu$ in (6.8). Then we have

$$\varphi(u(x)) \ge \varphi\Big(\int u d\nu\Big) + M\left[u(x) - \int u d\nu\right].$$

Integrating both sides with respect to $d\nu$, we get Jensen's inequality. □

THEOREM 6.5. *Let $1 \le p \le \infty$ and $1/p + 1/p' = 1$. If $f \in L^p(X, \mu)$ and $g \in L^{p'}(X, \mu)$, then $fg \in L^1(X, \mu)$, and we have*

$$\int_X |fg| d\mu \le \|f\|_p \|g\|_{p'} \qquad \textit{(Hölder's inequality)}.$$

In particular, if $p = 2$, then for $f, g \in L^2(X, \mu)$ we have $fg \in L^1(X, \mu)$, and

$$\int_X |fg| d\mu \le \|f\|_2 \|g\|_2 \qquad \textit{(Schwarz's inequality)}.$$

PROOF. Let $p = 1$. If $f \in L^1$ and $g \in L^\infty$, then we have

$$\int_X |f(x)g(x)|d\mu \leq \int_X |f(x)|\|g\|_\infty d\mu = \|f\|_1\|g\|_\infty.$$

If $p = \infty$, just swap the roles of f and g.

Let $1 < p < \infty$. If $\|g\|_{p'} = 0$, then $g = 0$ a.e. Thus $|f(x)g(x)| = 0$ a.e., which implies $\int |fg|d\mu = 0$. Thus the inequality is trivial.

When $\|g\|_{p'} \neq 0$, choose $a > 0$ so that $a\|g\|_{p'} = 1$. To apply Jensen's inequality put $\varphi(t) = |t|^p$, $d\nu(x) = |ag(x)|^{p'} d\mu(x)$, and

$$u(x) = \begin{cases} |f(x)||ag(x)|^{-p'/p} & (g(x) \neq 0), \\ 0 & (g(x) = 0). \end{cases}$$

Note that $\int_X d\nu = 1$, $ud\nu = a|fg|d\mu$ and $\varphi(u)d\nu \leq |f|^p d\mu$. Then we have the inequality

$$\left(a \int_X |fg|d\mu\right)^p \leq \int_X |f|^p d\mu,$$

which in turn yields

$$\int_X |fg|d\mu \leq \frac{1}{a}\left(\int_X |f|^p d\mu\right)^{1/p} = \|f\|_p\|g\|_{p'}.$$

□

THEOREM 6.6.

(i) *Let* $1 \leq p \leq \infty$. *If* f, $g \in L^p(X)$, *then* $f + g \in L^p(X)$ *and*

$$\|f + g\|_p \leq \|f\|_p + \|g\|_p \qquad \textit{(Minkowski's inequality)}.$$

(ii) *Let* $0 < p < 1$. *If* f, $g \in L^p(X)$, *then* $f + g \in L^p(X)$ *and*

$$\|f + g\|_p^p \leq \|f\|_p^p + \|g\|_p^p.$$

PROOF. (i) Let $p = \infty$. For $f, g \in L^\infty(X)$ we have

$$|f(x) + g(x)| \leq |f(x)| + |g(x)| \leq \|f\|_\infty + \|f\|_\infty \text{ a.e.}$$

Thus $f + g \in L^\infty$ and $\|f + g\|_\infty \leq \|f\|_\infty + \|f\|_\infty$.

If $1 \leq p < \infty$, then for $f, g \in L^p(X)$ we have

$$|f(x) + g(x)|^p \leq 2^p(|f(x)|^p + |g(x)|^p) \in L^1(X).$$

Therefore $f + g \in L^p(X)$. Since $p - 1 = p/p'$, we have

$$\int_X |f + g|^p d\mu \leq \int_X |f + g|^{p-1}(|f| + |g|)d\mu$$
$$= \int_X |f + g|^{p/p'}|f|d\mu + \int_X |f + g|^{p/p'}|g|d\mu.$$

Apply Hölder's inequality to each term on the right-hand side. Then we obtain

$$\|f + g\|_p^p \leq \|f + g\|_p^{p-1}(\|f\|_p + \|g\|_p).$$

When $\|f + g\|_p \neq 0$, Minkowski's inequality follows from the last inequality. If $\|f + g\|_p = 0$, (i) is obvious.

(ii) If $0 < p < 1$, we have

$$(a+b)^p \leq a^p + b^p \quad \text{for } a, b \geq 0.$$

Substitute $f(x)$ and $g(x)$ for a and b, respectively, and integrate with respect to x. Then (ii) is obtained. □

THEOREM 6.7. *Let $(X, \mathcal{M}, \mu)$ be a σ-finite measure space and let $1 \leq p \leq \infty$. Then, for every measurable function f,*

$$\sup_{\|g\|_{p'}=1} \int_X |fg| d\mu = \|f\|_p. \tag{6.9}$$

PROOF. (6.9) is obvious if $\|f\|_p = 0$. Suppose that $0 < \|f\|_p < \infty$. Let $1/p + 1/p' = 1$. If $\|g\|_{p'} = 1$, by Hölder's inequality

$$\int_X |fg| d\mu \leq \|f\|_p. \tag{6.10}$$

In particular, if we put

$$g(x) = \overline{\text{sign} f(x)} |f(x)|^{p-1} \|f\|_p^{-p/p'},$$

then $\|g\|_{p'} = 1$ by a computation, and $\int_X fg d\mu = \|f\|_p$, where $\text{sign} z = z/|z|$ for $z \neq 0$, and $= 0$ for $z = 0$. Thus the supremum of (6.9) is attained by g, and the integral $\int_X fg d\mu$ equals $\|f\|_p$.

Suppose that $\|f\|_p = \infty$. Let $X = \bigcup_{n=1}^{\infty} X_n$, where $\{X_n\}$ is an increasing sequence of sets of finite measure. Put

$$f_n(x) = \begin{cases} f(x) & (x \in X_n \text{ and } |f(x)| < n), \\ 0 & (\text{otherwise}). \end{cases}$$

Then $f_n \in L^p$ and

$$\sup_{\|g\|_{p'}=1} \int_X |fg| d\mu \geq \sup_{\|g\|_{p'}=1} \int_X |f_n g| d\mu = \|f_n\|_p.$$

Thus $\sup_{\|g\|_{p'}=1} \int_X |fg| d\mu = \infty$, since $\|f_n\|_p \to \infty$ as $n \to \infty$. □

The space $L^p(X)$ is a vector space over the complex field, and $\|\cdot\|_p$ defines a norm on it when $1 \leq p \leq \infty$. Therefore $L^p(X)$ is a normed space. On the other hand, when $0 < p < 1$, $\|\cdot\|_p$ does not always satisfy, the condition for a norm. But, by Theorem 6.6 (ii),

$$\rho(f, g) = \|f - g\|_p^p \quad (f, g \in V)$$

defines a metric on L^p.

6.2.3. Completeness of the spaces L^p.

THEOREM 6.8.

(i) *If $1 \leq p \leq \infty$, the space L^p is a Banach space.*
(ii) *If $0 < p < 1$, the space L^p is a complete metric space.*

PROOF. (i) Let $1 \leq p \leq \infty$. We have already shown that L^p is a normed space. To prove the completeness, let $\{f_n\}$ be a Cauchy sequence in L^p. By definition, for every k there exists an integer n_k such that

$$\|f_{n_k} - f_n\|_p < 2^{-k} \quad (n \geq n_k).$$

We can suppose that $n_1 < n_2 < \cdots$. If we put

$$g(x) = |f_{n_1}(x)| + \sum_{k=1}^{\infty} |f_{n_{k+1}}(x) - f_{n_k}(x)|, \tag{6.11}$$

then the series on the right-hand side converges almost everywhere and $g \in L^p$. In fact, by Minkowski's inequality

$$\|g\|_p \leq \|f_{n_1}\|_p + \sum_{k=1}^{\infty} \|f_{n_{k+1}} - f_{n_k}\|_p < \infty.$$

Thus $g \in L^p$, which implies that $g(x) < \infty$ a.e. Thus the series on the right-hand side of (6.11) converges almost everywhere. Therefore if

$$f(x) = f_{n_1}(x) + \sum_{k=1}^{\infty} [f_{n_{k+1}}(x) - f_{n_k}(x)],$$

then the series converges almost everywhere and $f \in L^p$.

Now we shall show that $f_n \to f$ in L^p. Since

$$f(x) = f_{n_j}(x) + \sum_{k=j}^{\infty} [f_{n_{k+1}}(x) - f_{n_k}(x)]$$

for every j, we have

$$\|f - f_{n_j}\|_p \leq \sum_{k=j}^{\infty} \|f_{n_{k+1}} - f_{n_k}\|_p \leq \sum_{k=j}^{\infty} \frac{1}{2^k} = 2^{-j+1}.$$

Therefore

$$\|f - f_n\|_p \leq \|f - f_{n_j}\|_p + \|f_{n_j} - f_n\|_p \leq 2^{-j+1} + 2^{-j}$$

for all $n \geq n_j$, which implies that $\|f - f_n\|_p \to 0$ as $n \to \infty$.

(ii) If $0 < p < 1$, replace the norm $\|\cdot\|_p$ by the metric $\|\cdot\|_p^p$ in the above proof. Then the all arguments in (i) can be carried out. The details are left to the reader. □

Let $p = 2$. Then the integral

$$(f, g) = \int_X f\bar{g}\,d\mu \quad \text{for } f, g \in L^2(X)$$

always exists by Schwarz's inequality, and (f, g) possesses the all properties of an inner product. Furthermore, $(f, f) = \|f\|_2^2$. Therefore $L^2(X)$ is a Hilbert space with the inner product $(\cdot, \cdot)$.

6.3. Linear functionals on L^p

THEOREM 6.9 (The Riesz representation theorem). *Let $(X, \mathcal{M}, \mu)$ be a σ-finite measure space. Let $1 \leq p < \infty$ and $1/p + 1/p' = 1$.*

(i) *For each continuous linear functional T on $L^p(X)$ there exists a unique function $f = f_T$ in $L^{p'}(X)$ such that*

$$Tu = \int_X fu\,d\mu, \quad u \in L^p(X). \tag{6.12}$$

(ii) *Conversely, if* $f \in L^{p'}(X)$, *then the mapping* T_f *defined by*

$$T_f u = \int_X f u d\mu, \quad u \in L^p(X)$$

is a continuous linear functional on $L^p(X)$, *and*

$$\|T_f\| = \|f\|_{L^{p'}}. \tag{6.13}$$

Therefore, the mapping

$$(L^p(X))' \ni T \mapsto f_T \in L^{p'}(X)$$

is surjective and isometric.

Since L^2 is a Hilbert space, the theorem reduces to Theorem 6.3 when $p = 2$.

PROOF. First Step. *Proof of* (ii).

If $f \in L^{p'}(X)$, then by Hölder's inequality

$$\left| \int_X f u d\mu \right| \le \|f\|_{p'} \|u\|_p \quad \text{for } u \in L^p(X).$$

Thus $Tu = \int_X f u d\mu$ is a bounded linear functional and $\|T\| \le \|f\|_{p'}$.

On the other hand we have $\sup_{\|u\|_p = 1} \left| \int_X f u d\mu \right| = \|f\|_{p'}$ by Theorem 6.7. Thus $\|T\| = \|f\|_{p'}$.

The uniqueness follows from the equation (6.13). Therefore to complete the proof we must show the existence part of (i).

Second Step. *Existence of* f_T*; the case where* $\mu(X) < \infty$.

Put

$$\nu(E) = T(\chi_E) \quad \text{for } E \in \mathcal{M}.$$

Then, ν is a complex valued measure. First we remark that $|\nu(E)| \le \|T\| \|\chi_E\|_p < \infty$. If $\{E_j\}$ is a sequence of mutually disjoint subsets, then by definition

$$\left| \nu\Big(\bigcup_{j=1}^{\infty} E_j\Big) - \nu\Big(\bigcup_{j=1}^{k} E_j\Big) \right| = |T(\chi_{\bigcup_{j=1}^{\infty} E_j}) - T(\chi_{\bigcup_{j=1}^{k} E_j})| = |T(\chi_{\bigcup_{j=k+1}^{\infty} E_j})|.$$

The last term is bounded by $\|T\| (\mu(\bigcup_{j=k+1}^{\infty} E_j))^{1/p}$, which converges to 0 as $k \to \infty$. On the other hand, we have $\nu(\bigcup_{j=1}^{k} E_j) = T(\chi_{\bigcup_{j=1}^{k} E_j}) = \sum_{j=1}^{k} T(\chi_{E_j}) = \sum_{j=1}^{k} \nu(E_j)$ by the linearity of T. Therefore we obtain

$$\nu\Big(\bigcup_{j=1}^{\infty} E_j\Big) = \sum_{j=1}^{\infty} \nu(E_j),$$

which implies that ν is a complex measure.

Observe that if $\mu(N) = 0$, then $|\nu(N)| \le \|T\| (\mu(N))^{1/p} = 0$. Thus the measure ν is absolutely continuous with respect to μ. Thus by the Radon-Nikodým theorem there exists $f \in L^1(X)$ such that

$$T(\chi_E) = \nu(E) = \int \chi_E f d\mu \quad \text{for } E \in \mathcal{M}.$$

Therefore we have, for any simple function $s = \sum a_j \chi_{E_j}$,

$$Ts = \sum a_j T(\chi_{E_j}) = \sum a_j \int \chi_{E_j} f d\mu = \int f s d\mu.$$

Next, we shall show that the last relation holds for all functions in $L^p(X)$.

Let u be a bounded function and let $\{s_j\}$ be a sequence of simple functions such that $|s_j(x)| \leq \|u\|_\infty$ and $\lim_j s_j(x) = u(x)$ a.e. Then by Lebesgue's bounded convergence theorem $\|u_k - s_j\|_p \to 0$ as $j \to \infty$. Thus

$$|T(u) - T(s_j)| \leq \|T\| \|u - s_j\|_p \to 0 \text{ as } j \to \infty.$$

On the other hand, since $|f(x)s_j(x)| \leq \|u\|_\infty |f(x)| \in L^1(X)$, the dominated convergence theorem yields

$$\lim_{j\to\infty} \int_X f s_j d\mu = \int_X f u d\mu.$$

Therefore we obtain

$$T(u) = \int_X f u d\mu \quad \text{for } u \in L^\infty(X). \tag{6.14}$$

Now we observe that

$$\int_X |fu| d\mu = \int_X f[|u|\overline{\operatorname{sign}(f)}] d\mu \leq \|T\| \|u \overline{\operatorname{sign}(f)}\|_p \leq \|T\| \|u\|_p.$$

For a general function u put $u_k(x) = u(x)$ $(|u(x)| \leq k)$ and $= 0$ $(|u(x)| > k)$ for $k = 1, 2, \ldots$. Then each u_k is bounded and $\int_X |fu_k| d\mu \leq \|T\| \|u_k\|_p$. Thus by B. Levi's theorem we get

$$\int_X |fu| d\mu \leq \|T\| \|u\|_p \quad \text{for } u \in L^p(X). \tag{6.15}$$

Taking the supremum over u with $\|u\|_p = 1$, we obtain that $f \in L^{p'}(X)$ and $\|f\|_{p'} \leq \|T\|$.

From the formula $Tu_k = \int f u_k d\mu$, equation (6.12) follows by approximation. In fact, we have $|Tu_k - Tu| \leq \|T\| \|u - u_k\|_p \to 0$ as $k \to \infty$, and $|\int f u_k d\mu - \int f u d\mu| \leq \|f\|_{p'} \|u_k - u\|_p \to 0$.

Third Step. *Existence of f_T; the case where $(X, \mathcal{M}, \mu)$ is σ-finite.*

Choose an increasing sequence $\{X_n\}$ of measurable sets of finite measure in $\mathcal{M}$ such that $X = \bigcup_{n=1}^\infty X_n$.

By the second step, for each n there exists $f_n \in L^{p'}(X_n)$ such that

$$T(u\chi_n) = \int_{X_n} f_n u d\mu = \int_X f_n u \chi_n d\mu$$

for $u \in L^p(X)$, where χ_n is the characteristic function of the set X_n. Put $f(x) = f_n(x)$ $(x \in X_n)$. By the uniqueness of f_n, $f_m(x) = f_n(x)$ a.e. x in X_m, if $m \leq n$. Thus f is well-defined except on a null set. Now, the rest of the proof is to show that $f \in L^{p'}(X)$ and the equation (6.12) is valid.

For $u \in L^p(X)$ let $u_n(x) = u(x)\chi_n(x)$. Then we have

$$T(u_n) = T(u\chi_n) = \int_X f_n u \chi_n d\mu = \int_{X_n} f u d\mu. \tag{6.16}$$

Replacing u by $|u|\overline{\operatorname{sign}(f)}$ in (6.16), we get

$$\int_{X_n} |fu| d\mu = T(|u|\overline{\operatorname{sign}(f)}\chi_n) \leq \|T\| \|u_n\|_p \leq \|T\| \|u\|_p.$$

Letting $n \to \infty$, we get (6.15). We have $f \in L^p(X)$ just as in the second step. The equation (6.12) follows from (6.16). □

6.4. Convolution

In this and the next sections we study some properties of the Lebesgue spaces on $\mathbb{R}^d$ which are closely related with the translation operation $\mathbb{R}^d$.

For measurable functions f, g on $\mathbb{R}^d$ the integral

$$f * g(x) = \int_{\mathbb{R}^d} f(x-y)g(y)dm(y)$$

is called the *convolution* of f and g. Here we assume the integrability of the integrand.

THEOREM 6.10. *If $f, g \in L^1(\mathbb{R}^d)$, the convolution $f * g(x)$ exists almost everywhere and $f * g \in L^1(\mathbb{R}^d)$. Furthermore, we have*

$$\|f * g\|_1 \leq \|f\|_1 \|g\|_1.$$

LEMMA 6.4. *If f is a measurable function in $\mathbb{R}^d$, then $F(x,y) = f(x-y)$ is measurable in $\mathbb{R}^d \times \mathbb{R}^d = \mathbb{R}^{2d}$.*

PROOF OF THE LEMMA. If f is the characteristic function χ_O of an open set O in $\mathbb{R}^d$, then $\{(x,y) \in \mathbb{R}^d \times \mathbb{R}^d : F(x,y) > a\}$ is an open set in $\mathbb{R}^{2d}$ for every $a \in \mathbb{R}$. Thus F is measurable. This implies that Lemma 6.4 holds for characteristic functions of G_δ-sets.

Let N be a null set in $\mathbb{R}^d$. We know that there is a null G_δ-set G which contains N. By Fubini's theorem we have

$$\begin{aligned}\int_{\mathbb{R}^{2d}} \chi_G(x-y)dm(x,y) &= \int_{\mathbb{R}^d} \left\{ \int_{\mathbb{R}^d} \chi_G(x-y)dm(x) \right\} dm(y) \\ &= \int_{\mathbb{R}^d} m(G)dm(y) = 0.\end{aligned}$$

Thus $\chi_G(x-y) = 0$ for a.e. (x,y) in $\mathbb{R}^{2d}$, which reduces to $\chi_N(x-y) = 0$ for a.e. (x,y). Therefore $\chi_N(x-y)$ is a measurable function of (x,y).

Now, any measurable set E in $\mathbb{R}^d$ can be written as a difference $E = G - N$, where G is a G_δ-set and N is a null set contained in G. Thus $\chi_E(x-y) = \chi_G(x-y) - \chi_N(x-y)$ is a measurable function of (x,y). Thus the lemma holds for any measurable simple functions. Therefore the lemma is valid for all measurable functions, since every measurable function is a limit of a sequence of measurable simple functions. □

PROOF OF THEOREM 6.10. By the definition of convolution we have

$$\int_{\mathbb{R}^d} |f * g(x)| dm(x) \leq \int_{\mathbb{R}^d} \left\{ \int_{\mathbb{R}^d} |f(x-y)g(y)| dm(y) \right\} dm(x).$$

The right-hand side equals, by Tonelli's theorem,

$$\int_{\mathbb{R}^d} \left\{ \int_{\mathbb{R}^d} |f(x-y)| dm(x) \right\} |g(y)| dm(y) = \int_{\mathbb{R}^d} |f(x)| dm(x) \left\{ \int_{\mathbb{R}^d} |g(y)| dm(y) \right\}.$$

Since the last term is finite, $f * g(x)$ exists almost everywhere, and $f * g(x) \in L^1(\mathbb{R}^d)$. □

THEOREM 6.11. *The space $L^1(\mathbb{R}^d)$ is a commutative Banach algebra with multiplication defined by convolution.*

PROOF. We have already shown that $L^1(\mathbb{R}^d)$ is a Banach space. To prove that it is a ring, we remark first that if $u, u \in L^1$, then $u*v \in L^1$ and $\|u*v\|_1 \leq \|u\|_1\|v\|_1$ by Theorem 6.10.

Let $u, v, w \in L^1$ and observe that

$$(u * v) * w(x) = \int (u * v)(x - y)w(y)dm(y)$$
$$= \int \left(\int u(x - y - z)v(z)dm(z) \right) w(y)dm(y).$$

Replacing the variable z of the inner integral by $z - y$ and then changing the order of the integrals, we find that the last double integral is equal to

$$\int \left(\int v(z - y)w(y)dm(y) \right) u(x - z)dm(z) = u * (v * w)(x).$$

Thus the convolution satisfies the associative law.

The distributive law follows from the linearity of the integration. The commutativity is shown by changing variables. In fact,

$$u * v(x) = \int u(x - y)v(y)dm(y) = \int u(z)v(x - z)dm(z) = v * u(x).$$

Thus the theorem is established. □

THEOREM 6.12. *Let* $1 \leq p \leq \infty$. *If* $u \in L^p(\mathbb{R}^d)$ *and* $k \in L^1(\mathbb{R}^d)$, *then* $u * k \in L^p(\mathbb{R}^d)$ *and*

$$\|u * k\|_p \leq \|k\|_1 \|u\|_p. \tag{6.17}$$

PROOF. A proof for the case $p = 1$ is given by Theorem 6.10. Let $1 < p < \infty$ and $1/p + 1/p' = 1$. We have

$$|u * k(x)| \leq \int |u(x - y)||k(y)|dm(y). \tag{6.18}$$

Since $|k(y)| = |k(y)|^{1/p+1/p'}$, by Hölder's inequality the right-hand does not exceed

$$\left(\int |u(x - y)|^p |k(y)|dm(y) \right)^{1/p} \left(\int |k(y)|dm(y) \right)^{1/p'}.$$

Thus, integrating the pth power of (6.18) with respect to x, we get

$$\int |u * k(x)|^p dm(x) \leq \iint |u(x - y)|^p |k(y)|dm(y)dm(x)\|k\|_1^{p/p'}.$$

The last term equals, by Fubini's theorem,

$$\int \left\{ \int |u(x - y)|^p |k(y)|dm(x) \right\} dm(y)\|k\|_1^{p/p'} = \|u\|_p^p \|k\|_1 \|k\|_1^{1+p/p'},$$

from which (6.17) follows.

If $p = \infty$, we have, by (6.18),

$$|u * k(x)| \leq \int \|u\|_\infty |k(y)\ dm(y) = \|u\|_\infty \|k\|_1 \text{ a.e.}$$

Thus $\|u * k\|_\infty \leq \|k\|_1 \|u\|_\infty$. □

6.5. Approximate identity

6.5.1. Approximate identities and summability kernels. As we have shown in the previous section, $L^1(\mathbb{R}^d)$ is a commutative Banach algebra with the convolution multiplicatin. This algebra does not have a unit, that is, there is no integrable function u which satisfies

$$f * u = f \quad \text{for all } f \in L^1(\mathbb{R}^d)$$

(see Problem 8.4). But there exists a sequence $\{u_n\}$ in $L^1(\mathbb{R}^d)$ such that

$$\|f * u_n - f\|_1 \to 0 \text{ as } n \to \infty$$

for all $f \in L^1(\mathbb{R}^d)$. Such a sequence $\{u_n\}$ is called an *approximate identity.*

DEFINITION 6.5. A family $\{k_t(x) : t > 0\}$ in $L^1(\mathbb{R}^d)$ is called a *summability kernel* if

(S1) $\int_{\mathbb{R}^d} k_t(x) = 1$,
(S2) there exists a constant $C > 0$ independent of t such that $\|k_t\|_1 \leq C$,
(S3) for every $\delta > 0$, $\int_{|x|>\delta} |k_t(x)|dm(x) \to 0$ as $t \to 0$.

For a function $k(x)$ on $\mathbb{R}^d$, denote

$$k_t(x) = t^{-d}k(t^{-1}x), \quad t > 0.$$

PROPOSITION 6.3. *If $k \in L^1(\mathbb{R}^d)$ and $\int_{\mathbb{R}^d} kdm = 1$, then $\{k_t : t > 0\}$ is a summability kernel.*

PROOF. Changing variables, we get the properties (S1) and (S2) for k_t. To prove (S3) we remark that

$$\int_{|x|>\delta} |k_t(x)|dm(x) = \int_{\{x:|x|>\delta/t\}} |k(x)|dm(x).$$

Since the integral domain of the last term converges to $\emptyset$ as $t \to 0$, it converges to 0 by the dominated convergence theorem. □

THEOREM 6.13. *Let $\{k_t : t > 0\}$ be a summability kernel.*

(i) *If $1 < p < \infty$, then for every $f \in L^p(\mathbb{R}^d)$*

$$\|f * k_t - f\|_p \to 0 \quad \text{as } t \to 0.$$

(ii) *For every $f \in C_0(\mathbb{R}^d)$*

$$\|f * k_t - f\|_\infty \to 0 \quad \text{as } t \to 0.$$

For a function f on $\mathbb{R}^d$ the *translation operator* τ_y ($y \in \mathbb{R}^d$) is defined by

$$\tau_y f(x) = f(x - y) \quad (x \in \mathbb{R}^d).$$

LEMMA 6.5.

(i) *Let $1 \leq p < \infty$. If $f \in L^p(\mathbb{R}^d)$, then $\|\tau_y f - f\|_p \to 0$ as $y \to 0$.*
(ii) *If $f \in C_0(\mathbb{R}^d)$, then $\|\tau_y f - f\|_\infty \to 0$ as $y \to 0$.*

PROOF OF THE LEMMA. (i) Let $f \in L^p(\mathbb{R}^d)$ and $\varepsilon > 0$. Then there exists a simple function $s(x) = \sum_{j=1}^N a_j \chi_{I_j}(x)$ such that $I_1, \ldots, I_N$ are intervals and $\|f - s\|_p < \varepsilon/3$. If $|y|$ is sufficiently small, we have $\|\tau_y s - s\|_p < \varepsilon/3$. We have

$$\|\tau_y f - f\|_p \leq \|\tau_y f - \tau_y s\|_p + \|\tau_y s - s\|_p + \|s - f\|_p.$$

Since $\|\tau_y f - \tau_y s\|_p = \|k - s\|_p < \varepsilon/3$, the right-hand side of the last inequality is less than $\varepsilon/3 + \varepsilon/3 + \varepsilon/3 = \varepsilon$, which implies (i).

(ii) Let $f \in C_0(\mathbb{R}^d)$. Then f is uniformly continuous; that is, for a given $\varepsilon > 0$ there exists $\delta > 0$ such that $|f(x-y) - f(x)| < \varepsilon\,(|y| < \delta)$ for all $x \in \mathbb{R}^d$ (see Problem 1.6). □

PROOF OF THEOREM 6.13. (i) Let $f \in L^p(\mathbb{R}^d)$. Since $\int k_t(y)dm(y) = 1$, we have

$$(f * k_t)(x) - f(x) = \int [f(x-y) - f(x)]k_t(y)dm(y).$$

By Hölder's inequality with indices $1/p + 1/p' = 1$ the last integral is, in absolute value, bounded by

$$\left(\int |f(x-y) - f(x)|^p |k_y(y)| dm(y)\right)^{1/p} \left(\int |k_y(y)| dm(y)\right)^{1/p'}.$$

Therefore we have

$$\int |f * k_t(x) - f(x)|^p dm(x)$$
$$\leq \int dm(x) \int |f(x-y) - f(x)|^p |k_t(y)| dm(y) \left(\int |k_t(y)| dm(y)\right)^{p/p'}.$$

Changing the order of integration and applying condition (S2), we see that the last term is bounded by

$$C^{p/p'} \int |k_t(y)| dm(y) \int |\tau_y f(x) - f(x)|^p dm(x)$$
$$\leq C^{p/p'} \left(\int_{|y|>\delta} + \int_{|y|\leq\delta}\right) |k_t(y)| \|\tau_y f - f\|_p^p dm(y)$$
$$= C^{p/p'}(I + J), \quad \text{say}.$$

For a given $\varepsilon > 0$ choose $\delta > 0$ so that $\|\tau_y f - f\|_p < \varepsilon\,(|y| < \delta)$, which is possible by Lemma 6.5. Then we have

$$J \leq \varepsilon \int_{|y|\leq\delta} |k_t(y)| dm(y) \leq \varepsilon \int_{\mathbb{R}^d} |k_t(y)| dm(y) = \varepsilon \|k\|_1.$$

Since $\|\tau_y f - f\|_p \leq \|\tau_y f\|_p + \|f\|_p = 2\|f\|_p$,

$$I \leq (2\|f\|_p)^p \int_{|y|>\delta} |k_t(y)| dm(y).$$

By condition (S3), $I \to 0$ as $t \to 0$. Therefore we obtain

$$\limsup_{t\to 0} \|f * k_t - f\|_p^p \leq C^p \varepsilon.$$

(ii) Let $f \in C_0(\mathbb{R}^d)$. As in case (i), we have

$$|f * k_t(x) - f(x)| \leq \int |f(x-y) - f(x)||k_t(y)| dm(y)$$
$$= \left(\int_{|y|>\delta} + \int_{|y|\leq\delta}\right) |k_t(y)||\tau_y f(x) - f(x)| dm(y).$$

For a given $\varepsilon > 0$ choose $\delta > 0$ so that $\|\tau_y f - f\|_\infty < \varepsilon\,(|y| < \delta)$. The rest of the proof proceeds as in case (i). □

6.5.2. Examples of summability kernels. If $k \in L^1(\mathbb{R}^d)$ and if $\int k dm = 1$, then $k_t(x) = t^{-d}k(t^{-1}x)\ (t > 0)$ is an approximate identity in $L^1(\mathbb{R}^d)$; this is a consequence of Theorem 6.13 and Proposition 6.3. We shall give some useful examples of summability kernels, which are applied especially in the theory of distributions and Fourier analysis.

For a function f in $\mathbb{R}^d$ the closure of the set $\{x \in \mathbb{R}^d : f(x) \neq 0\}$ is called the *support* of f and denoted by $\operatorname{supp} f$.

By definition a function f on $\mathbb{R}^d$ belongs to $C^{(n)}(\mathbb{R}^d)$ if the derivative

$$\left(\frac{\partial}{\partial x_1}\right)^{\alpha_1} \cdots \left(\frac{\partial}{\partial x_d}\right)^{\alpha_d} f$$

exists and is continuous for every $\alpha_1 + \cdots + \alpha_d \leq n$.

We use the following notation:

$$C^\infty(\mathbb{R}^d) = \bigcap_{n=0}^{\infty} C^{(n)}(\mathbb{R}^d),$$

and

$$C_c^\infty(\mathbb{R}^d) = \{f \in C^\infty(\mathbb{R}^d) : \operatorname{supp} f \text{ is compact}\}.$$

EXAMPLE 6.3 (Mollifier). Put $k(x) = \exp[-1/(1-|x|^2)]\ (|x| < 1),\ = 0\ (|x| \geq 1)$. Then $k \in C_c^\infty(\mathbb{R}^d)$. In fact, the support of k is compact, since $\operatorname{supp} k = \{x \in \mathbb{R}^d : |x| \leq 1\}$. If $u(t) = e^{-1/t}\ (t > 0)$ and $= 0\ (t \leq 0)$, then u is infinitely differentiable everywhere. For example, $u(t)$ is differentiable at $t = 0$ and $u'(0) = 0$. For u'' the situation is similar. The reader may proceed in general by induction. Since $v(x) = 1 - |x|^2 = 1 - x_1^2 - \cdots - x_d^2$ is infinitely differentiable, so is the composition $k(x) = u(v(x))$.

Pick a constant c so that $c\int k(x)dm(x) = 1$, and put $\varphi_t(x) = ct^{-d}k(t^{-1}x)$. Then $\varphi_t(x)$ is an approximate identity which belongs to $C_c^\infty(\mathbb{R}^d)$. $\varphi_t(x)$ is called a *mollifier.*

THEOREM 6.14. *Let $1 \leq p < \infty$. The space $C_c^\infty(\mathbb{R}^d)$ is dense in $L^p(\mathbb{R}^d)$ and in $C_0(\mathbb{R}^d)$.*

PROOF. Let $f \in L^p(\mathbb{R}^d)$ and $\varepsilon > 0$ be given. For $N > 0$ put $f_N(x) = f(x)\ (|x| \leq N)$ and $= 0\ (|x| > N)$. Choose N so that

$$\|f - f_N\|_p < \varepsilon/2.$$

By Theorem 6.13 (i),

$$\|f_N - f_N * \varphi_t\|_p < \varepsilon/2$$

for a small $t > 0$. Thus

$$\|f - f_N * \varphi_t\|_p \leq \|f - f_N\|_p + \|f_N - f_N * \varphi_t\|_p < \varepsilon.$$

Since $\operatorname{supp} f_N * \varphi_t \subset \{x : |x| \leq N + t\}$ and $\varphi_t(x) \in C_c^\infty(\mathbb{R}^d)$, it follows that $f_N * \varphi_t \in C_c^\infty(\mathbb{R}^d)$ (see Example 3.3 or Problem 3.14).

In the case $C_0^\infty(\mathbb{R}^d)$, apply the above argument with $p = \infty$. □

We shall give one more application of mollifiers, which will be used in the next chapter.

LEMMA 6.6. *Let K be a compact set in $\mathbb{R}^d$ and O be any open set containing K. Then there exists u in $C_c^\infty(\mathbb{R}^d)$ such that*

$$0 \le u \le 1, \quad \operatorname{supp} u \subset O, \quad \textit{and } u(x) = 1 \ (x \in K).$$

PROOF. Let $K_\delta = \{x : \operatorname{dist}(K, x) \le \delta\}$, where $\delta > 0$ is chosen so small that $K_{2\delta} \subset O$. Since the set K_δ is bounded and closed, K_δ is compact. Observe that if $u(x) = \int \chi_{K_\delta}(y)\varphi_\delta(x-y)dy$, then $u(x) = 0$ outside $K_{2\delta}$ and

$$u(x) = \int \chi_{K_\delta}(x-y)\varphi_\delta(y)dy = 1$$

in K. Furthermore, u is infinitely differentiable. □

Next, we mention the Poisson and the Gauss-Weierstrass kernels, which will be treated more precisely in Chapter 8 with applications to Fourier analysis.

EXAMPLE 6.4 (The Poisson kernel in $\mathbb{R} \times (0, \infty)$). Let

$$P(x) = \frac{1}{\pi}\frac{1}{1+x^2},$$

$x \in \mathbb{R}$. We have $P(x) \in L^1(\mathbb{R})$. Changing the variable by $x = \tan\theta$, we have

$$\int_{-\infty}^{\infty} P(x)dx = \int_{-\pi/2}^{\pi/2} \frac{1}{\pi}d\theta = 1.$$

The approximate identity

$$P_t(x) = t^{-1}P(t^{-1}x) = \frac{1}{\pi}\frac{t}{x^2+t^2}, \qquad t > 0,$$

is called the *Poisson*[4] *kernel.*

EXAMPLE 6.5 (The Poisson kernel in $\mathbb{R}^d \times (0, \infty)$). Let

$$P(x) = c_d \frac{1}{(1+|x|^2)^{(d+1)/2}},$$

where $c_d = \Gamma((d+1)/2)/\pi^{(d+1)/2}$. Then the Poisson kernel in the domain $\mathbb{R}^{d+1}_+ = \mathbb{R}^d \times (0, \infty)$ is defined by

$$P_t(x) = t^{-d}P(t^{-1}x) = c_d \frac{t}{(|x|^2+t^2)^{(d+1)/2}}, \qquad (x, t) \in \mathbb{R}^d.$$

We shall check that

$$\int_{\mathbb{R}^d} P(x)dm = 1. \tag{6.19}$$

Using the polar coordinate $x = r\omega$ with $r \ge 0$ and $\omega \in S^{d-1}$, we get

$$\int_{\mathbb{R}^d} P(x)dx = \int_0^\infty \int_{S^{d-1}} c_d r^{d-1}(1+r^2)^{-(d+1)/2} dr d\sigma(\omega),$$

where $S^{d-1} = \{x \in \mathbb{R}^d : |x| = 1\}$ is the unit shere and $d\sigma(\omega)$ is the area element on S^{d-1}. Let $\omega_{d-1} = 2\pi^{d/2}/\Gamma(d/2)$ be the area of the surface S^{d-1} (see Appendix

---

[4]S. D. Poisson, 1781–1840.

B.2). Then by a change of variable

$$\begin{aligned}\int_{\mathbb{R}^d} P(x)dx &= \frac{c_d\omega_{d-1}}{2}\int_0^1 t^{-1/2}(1-t)^{(d-2)/2}dt\\ &= \frac{c_d\omega_{d-1}}{2}\frac{\Gamma\left(\frac{1}{2}\right)\Gamma\left(\frac{d}{2}\right)}{\Gamma\left(\frac{d+1}{2}\right)} = \frac{\Gamma\left(\frac{1}{2}\right)}{\sqrt{\pi}} = 1\end{aligned}$$

(see Appendix B.1 for formulas on the gamma functions).

$P_t(x)$ is harmonic in $(x,t) \in \mathbb{R}^{d+1}_+$. In fact, it satisfies the Laplace equation:

$$\Delta_{(x,t)}P_t(x) = \left(\frac{\partial^2}{\partial x_1^2} + \cdots + \frac{\partial^2}{\partial x_d^2} + \frac{\partial^2}{\partial t^2}\right)P_t(x) = 0 \quad \text{for } (x,t) \in \mathbb{R}^{d+1}_+.$$

Let $f \in L^p(\mathbb{R}^d)\,(1 \le p < \infty)$, and put $P_t * f(x) = f(x,t)$. Then $f(x,t)$ is also harmonic in $\mathbb{R}^{d+1}_+$ (see Example 3.4) and $\|f(\cdot,t) - f(\cdot)\|_p \to 0$ as $t \to 0$.

EXAMPLE 6.6 (The Gauss-Weierstrass kernel). Let

$$w(x) = (2\sqrt{\pi})^{-d}\exp(-|x|^2/4), \quad x \in \mathbb{R}^d,$$

and put

$$W_t(x) = w_{\sqrt{t}}(x) = \frac{1}{(2\sqrt{\pi t})^d}e^{-|x|^2/4t},$$

which is called the *Gauss*[5]*-Weierstrass kernel.*

We have the formula

$$\int_{-\infty}^{\infty} e^{-s^2/4}ds = 2\sqrt{\pi}, \tag{6.20}$$

from which it follows that

$$\int_{\mathbb{R}^d} W_t(x)dm = \prod_{j=1}^d \frac{1}{2\sqrt{\pi}}\int_{-\infty}^{\infty} e^{-x_j^2/4t}dx_j = 1.$$

By Theorem 6.13, $W_t(x)$ is an approximate identity in $L^p(\mathbb{R}^d)$, $1 \le p < \infty$. We remark that

$$\left(\frac{\partial}{\partial t} - \frac{\partial^2}{\partial x_1^2} - \cdots + \frac{\partial^2}{\partial x_d^2}\right)W_t(x) = 0 \quad \text{for } x \in \mathbb{R}^d \text{ and } t > 0.$$

6.6. Space of continuous functions and Borel measures

To a continuous linear functional on the space $L^p(X)\,(1 \le p < \infty)$), there corresponds a unique function in $L^{p'}(X)$ by the Riesz representation theorem, where $1/p + 1/p' = 1$. When $p = \infty$, this relation does not hold in general, but if L^∞ is replaced by the space $C_0(X)$ of continuous functions, we get another representation by Borel measures.

---

[5] C. F. Gauss, 1777-1855. German mathematician, physicist and astronomer.

6.6.1. Locally compact Hausdorff spaces. A Hausdorff topological space X is said to be *locally compact* if every point x in X has at least one neighborhood U whose closure $\overline{U}$ is compact.

A compact space is a locally compact space, since the whole space is a compact neighborhood. The Euclidean spaces are other examples of locally compact Hausdorff spaces.

THEOREM 6.15. *Let K be a compact subset of a locally compact Hausdorff topological space X. If $x \notin K$, there exist a neighborhood V of x and an open set O such that*

$$K \subset O, \quad O \cap V = \emptyset, \text{ and } \overline{O} \text{ is compact.}$$

PROOF. Since X is a Hausdorff space, for every $p \in K$ there exist a neighborhood V_x^p of x and a neighborhood U_p of p such that $V_x^p \cap U_p = \emptyset$. Here, we can assume that $\overline{U}_p$ is compact. Indeed, since p has a neighborhood W_p whose closure is compact, the set $U_p \cap W_p$ has compact closure.

Since K is compact and $\{U_p : p \in K\}$ is an open covering of K, we can choose finite points $p_1, \cdots, p_N$ of K such that $K \subset U_{p_1} \cup \cdots \cup U_{p_N}$. The sets

$$V = V_x^{p_1} \cap \cdots \cap V_x^{p_N} \text{ and } O = U_{p_1} \cup \cdots \cup U_{p_N}$$

have the properties desired. □

REMARK 6.1. Theorem 6.15 implies that any compact set in a Hausdorff space is closed.

THEOREM 6.16. *Let X be a locally compact Hausdorff space. Let G be an open subset which contains a compact set K. Then there exists an open set O such that*

$$K \subset O \subset \overline{O} \subset G, \text{ and } \overline{O} \text{ is compact.}$$

PROOF. If $G = X$, the theorem follows from Theorem 6.15. Suppose $G \neq X$. By Theorem 6.15, for every $x \notin G$ there exist a neighborhood V_x of x and an open set O_x such that

$$K \subset O_x, \quad \overline{O}_x \cap V_x = \emptyset, \text{ and } \overline{O}_x \text{ is compact.}$$

Let us consider the family $\{\overline{O}_x \cap G^c : x \in G^c\}$ of compact sets and observe that $\bigcap_{x \in G^c} \overline{O}_x \cap G^c = \emptyset$. Then by the finite intersection property $\overline{O}_{x_1} \cap \cdots \cap \overline{O}_{x_N} \cap G^c = \emptyset$ for a finite number of $x_1, \ldots, x_N$ in G^c. Thus if $O = O_{x_1} \cap \cdots \cap O_{x_N}$, then $\overline{O} \cap G^c = \emptyset$, that is, $\overline{O} \subset G$. □

COROLLARY 6.1. *Any locally compact Hausdorff space has a fundamental system of neighborhoods which consists of open sets with compact closures.*

PROOF. Let $p \in X$, let G be any open neighborhood of p, and apply Theorem 6.16. □

6.6.2. Continuous functions on a locally compact space. Let $C(X)$ be the set of all continuous functions on X.

A function u on a non-compact space X is said to converge to 0 at ∞ if for every $\varepsilon > 0$ there exists a compact set K such that $|u(x)| < \varepsilon$ for $x \notin K$, which is written $u(x) \to 0\ (x \to \infty)$ or $\lim_{x \to \infty} u(x) = 0$. Put

$$C_0(X) = \{u : u \in C(X) \text{ and } \lim_{x \to \infty} u(x) = 0\},$$

and

$$C_c(X) = \{u : u \in C(X) \text{ and } \operatorname{supp} u \text{ is compact}\}.$$

For a nonempty open set O and $u \in C_c(X)$ we use the notation

$$u \prec O,$$

when $0 \le u \le 1$ and $\operatorname{supp} u \subset O$. Similarly, for a compact set K

$$K \prec u$$

means that $0 \le u \le 1$ and $u(x) = 1\,(x \in K)$. For a compace set K and an open set O such that $K \subset O$, $K \prec u \prec O$ means that these two conditions are simultaneously satisfied.

THEOREM 6.17 (Partition of unity). *Let $O_1, O_2, \dots, O_n$ be open sets of a locally compact Hausdorff space X and let K be a compact subset such that $K \subset O_1 \cup \cdots \cup O_n$. Then there exist $u_j \in C_c(X)$ such that*

$$u_j \prec O_j \quad (j = 1, 2, \dots, n),$$

and

$$u_1(x) + u_2(x) + \cdots + u_n(x) = 1 \quad \text{on } K. \tag{6.21}$$

PROOF. First we shall show that there exist compact sets $K_j\,(j = 1, \dots, n)$ such that $K_j \subset O_j$ and $K \subset \bigcup_{j=1}^n K_j$. In fact, for any $x \in K$ we can choose j and an open neighborhood V_x of x so that $V_x \subset \overline{V_x} \subset O_j$. Furthermore we can suppose that $\overline{V_j}$ is compact. Since $\{V_x : x \in K\}$ is an open covering of K, it has a finite subcovering $\{V_{x_i} : i = 1, \dots, N\}$. If $K_j = \bigcup\{\overline{V}_{x_i} : \overline{V}_{x_i} \subset O_j\}$, then the K_j are compact and we obviously have $K \subset \bigcup_{j=1}^n K_j$.

For a compact set K_j and an open set O_j there exists, by Urysohn's lemma (see Appendix A.2), $v_j \in C_c(X)$ such that $K_j \prec v_j \prec O_j$. The functions

$$u_1 = v_1,\ u_2 = (1 - v_1)v_2,\ \dots,\ u_n = (1 - v_1)(1 - v_2)\cdots(1 - v_{n-1})v_n$$

satisfy the conditions of the theorem. Actually, we have $\operatorname{supp} u_j \subset O_j$. If $x \in K$, then $v_j(x) = 1$ for some j. Therefore we have

$$u_1 + u_2 + \cdots + u_n = 1 - (1 - v_1)(1 - v_2)\cdots(1 - v_n) = 1.$$

□

REMARK 6.2. If X is d-dimensional Euclidean space, we can choose u_j from $C_c^\infty(\mathbb{R}^d)$ by Lemma 6.6.

6.6.3. Borel measures. The smallest σ-field which contains every compact set in X is called the *Borel field* and denoted by $\mathcal{B}(X)$. An element of $\mathcal{B}(X)$ is called a Borel set.

EXAMPLE 6.7. If u is a real valued function in $C_c(X)$, the set $\{x : u(x) > a\}$ is a Borel set for every $a \in \mathbb{R}$.

Indeed, for $a \ge 0$ we have $\{x : u(x) > a\} = \bigcup_{n=1}^\infty \{x : u(x) \ge a + 1/n\}$. Each set on the right hand side is a closed subset of $\operatorname{supp} u$ which is compact. Thus the set $\{x : u(x) > a\}$, which is a countable union of compact sets, belongs to $\mathcal{B}(X)$. If $a < 0$, then observe that $\{x : u(x) > a\}^c = \{x : u(x) \le a\}$ is a compact set.

We denote by $\mathcal{B}_0(X)$ the smallest σ-field which contains every open set in X. Since the complement of a compact set is open, the family $\mathcal{B}_0(X)$ contains all compact sets. Thus $\mathcal{B}(X) \subset \mathcal{B}_0(X)$.

The space X is said to be σ-compact, if there is a countable family of compact sets K_n such that $X = \bigcup_{n=1}^{\infty} K_n$.

Suppose that X is σ-compact. If O is an open subset, then $O^c = \bigcup_{n=1}^{\infty} O^c \cap K_n$, which is a countable union of compact sets. Thus $O^c \in \mathcal{B}(X)$ or, equivalently, $O \in \mathcal{B}(X)$. Therefore we have $\mathcal{B}(X) = \mathcal{B}_0(X)$.

DEFINITION 6.6. Let μ be a nonnegative measure on the Borel field $\mathcal{B}(X)$.

(i) μ is called a *Borel measure* if $\mu(K) < \infty$ for every compact set K.
(ii) If $\mu(E) = \inf\{\mu(O) : E \subset O \text{ and } O \text{ is open}\}$ for every Borel set E, then μ is said to be *outer regular*.
(iii) If $\mu(E) = \sup\{\mu(K) : K \subset E \text{ and } K \text{ is compact}\}$, then μ is said to be *inner regular*.

An inner and outer regular measure is called a *regular measure*. A regular Borel measure is called a *Radon measure*.

A real valued measure is a Radon measure if its positive and negative parts are both Radon measures. A complex valued measure is a Radon measure if its real and imaginary parts are Radon measures.

Lebesgue measure on $\mathbb{R}^d$ is a Radon measure.

6.6.4. Positive linear functionals and measures. Let T be a linear functional on $C_c(X)$. T is said to be *positive* if $0 \le Tu < \infty$ for any nonnegative function u in $C_c(X)$.

THEOREM 6.18 (The Riesz representation theorem). *Let T be a positive linear functional on $C_c(X)$. Then there exists a unique nonnegative Radon measure μ such that*

$$Tu = \int_X u\,d\mu, \quad u \in C_c(X). \tag{6.22}$$

The main part of the proof consists of the construction of a measure, which is accomplished by three lemmas.

For a nonempty open set O define

$$\mu(O) = \sup\{Tu : u \prec O\},$$

and $\mu(\emptyset) = 0$. For any subset A put

$$\mu^*(A) = \inf\{\mu(O) : A \subset O,\ O \text{ is open}\}.$$

Since T is positive, if $O_1 \subset O_2$, then $\mu(O_1) \le \mu(O_2)$. Therefore if $A \subset B$, then $\mu^*(A) \le \mu^*(B)$.

LEMMA 6.7. *μ^* is an outer measure on $\mathcal{P}(X)$.*

PROOF. We shall prove that

$$\mu^*\Big(\bigcup_{j=1}^{\infty} A_j\Big) \le \sum_{j=1}^{\infty} \mu^*(A_j) \tag{6.23}$$

for any countable sequence $\{A_j\}$ of subsets in X.

First Step. *For any open sets O_1 and O_2 we have*

$$\mu(O_1 \cup O_2) \le \mu(O_1) + \mu(O_2). \tag{6.24}$$

In fact, let u be an arbitrary function in $C_c(X)$ such that $u \prec O_1 \cup O_2$. By Theorem 6.17 there exist $v_j \in C_c(X)$ $(j = 1, 2)$ such that $v_1 + v_2 = 1$ on $\operatorname{supp} u$ and $v_j \prec O_j$. Thus if $u_j = uv_j$, then $u_j \in C_c(X)$ $(j = 1, 2)$, and

$$u = u_1 + u_2, \quad \text{and} \quad u_j \prec O_j.$$

Thus

$$Tu = Tu_1 + Tu_2 \le \mu(O_1) + \mu(O_2),$$

from which (6.24) follows.

Second Step. *Proof of* (6.23).

We may suppose that $\mu^*(A_j) < \infty$, $j = 1, 2, \dots$. For a given $\varepsilon > 0$ there exist open sets O_j such that

$$\mu(O_j) < \mu^*(A_j) + \varepsilon 2^{-j}, \quad A_j \subset O_j.$$

Let u be an arbitrary function in $C_c(X)$ such that $u \prec \bigcup_{j=1}^\infty O_j$. Since $\operatorname{supp} u$ is compact, $\operatorname{supp} u \subset \bigcup_{j=1}^n O_j$ for some n. By the first step

$$0 \le Tu \le \mu(\bigcup_{j=1}^n O_j) \le \sum_{j=1}^n \mu(O_j) < \sum_{j=1}^\infty \mu^*(A_j) + \varepsilon.$$

Taking the supremum over u, we get $\mu^*(\bigcup_{j=1}^\infty A_j) \le \sum_{j=1}^\infty \mu^*(A_j) + \varepsilon$. □

Let $\mathcal{M}_{\mu^*}$ be the family of all μ^*-measurable sets, that is, the sets which satisfy the Carathéodory condition with respect to the outer measure μ^*.

LEMMA 6.8. *We have $\mathcal{B}_0(X) \subset \mathcal{M}_{\mu^*}$. Thus every Borel set is μ^*-measurable.*

PROOF. Our goal is to prove that any open set G satisfies the Carathéodory condition:

$$\mu^*(A) \ge \mu^*(A - G) + \mu^*(A \cap G) \quad \text{for } A \in \mathcal{P}(X) \tag{6.25}$$

(see § 5.2).

Assume that $\mu^*(A) < \infty$ and G is an open set. For a given $\varepsilon > 0$ choose an open set O so that

$$A \subset O \quad \text{and} \quad \mu(O) < \mu^*(A) + \varepsilon.$$

Let u be a function in $C_c(X)$ such that $u \prec O \cap G$, and put $K = \operatorname{supp} u$. Since $K \subset G$, by Theorem 6.16 there exists an open set G_0 such that

$$K \subset G_0 \subset \overline{G_0} \subset G.$$

Thus $O - G \subset O - \overline{G_0}$. If $v \prec O - \overline{G_0}$, then $u + v \prec O$, and therefore we have

$$Tu + Tv = T(u + v) \le \mu(O) < \mu^*(A) + \varepsilon.$$

Taking the supremum over u and then over v, we get

$$\mu^*(A \cap G) + \mu^*(A - G) \le \mu(O) \le \mu^*(A) + \varepsilon,$$

which implies (6.25). □

Let $\tilde{\mu}$ be the extension of μ to $\mathcal{B}(X)$, that is, $\tilde{\mu} = \mu^* \,|_{\mathcal{B}(X)}$. In the sequel we use the same notation μ for $\tilde{\mu}$ on $\mathcal{B}(X)$.

LEMMA 6.9. *The measure μ is a regular Borel measure.*

PROOF. The outer regularity of μ follows directly from the definition of μ^*. In order to prove the inner regularity, let $\mathcal{M}_0$ be the family of the Borel sets E which satisfy

$$\mu(E) = \sup\{\mu(K) : K \subset E,\ \ K \text{ is compact}\}.$$

If $\mathcal{M}_0$ is a σ-field which contains all compact sets, then we have $\mathcal{M}_0 = \mathcal{B}(X)$ by the minimality of $\mathcal{B}(X)$, and thus the lemma is verified.

It is evident that $\mathcal{M}_0$ contains all compact subsets. We shall show that $\mathcal{M}_0$ is a σ-field. Let $E, F \in \mathcal{M}_0$. For $\varepsilon > 0$ let K be a compact set and O be an open set such that $K \subset E$, $\mu(K) + \varepsilon/2 > \mu(E)$ and $F \subset O$, $\mu(F) + \varepsilon/2 > \mu(O)$. Since

$$E - F \subset (E - K) \cup (K - O) \cup (O - F),$$

we have

$$\mu(E - F) \le \mu(E - K) + \mu(K - O) + \mu(O - F) < \mu(K - O) + \varepsilon.$$

If we observe that $K - O$ is a compact subset of $E - F$, the last inequality shows that the set $E - F$ is inner regular. Thus $E - F \in \mathcal{M}_0$.

Suppose that $E_k \in \mathcal{M}_0\,(j = 1, 2, \dots)$ and $E = \bigcup_{j=1}^{\infty} E_j$. Our next aim is to prove that $E \in \mathcal{M}_0$. For this we can suppose the E_j are mutually disjoint. First assume that $\mu(E) < \infty$. Thus $\sum_j \mu(E_j) < \infty$. Since the E_j are inner regular, for $\varepsilon > 0$ compact sets K_j can be chosen so that $K_j \subset E_j$ and $\mu(E_j) < \mu(K_j) + \varepsilon/2^j$. Put $K = \bigcup_{j=1}^{n} K_j$, where n is chosen so large that $\sum_{j=n}^{\infty} \mu(E_j) < \varepsilon$. Then K is a compact set, and by a simple computation $\mu(K) + 2\varepsilon > \mu(E)$. Thus $E \in \mathcal{M}_0$.

If $\mu(E) = \infty$, then $\mu(K) \to \infty$ as $n \to \infty$. Thus $E \in \mathcal{M}_0$ in this case, too. □

PROOF OF THEOREM 6.18. We shall establish equation (6.22) with μ defined in the above lemmas. Let u be a real valued function in C_c with support K. For a given $\varepsilon > 0$ put

$$E_j = \{x \in K : \varepsilon j < u(x) \le \varepsilon(j+1)\}, \quad j = 0, \pm 1, \pm 2, \dots .$$

Each E_j is a Borel set, and is empty for $|j| > k$ if k is large enough. Let O_j be open sets such that

$$E_j \subset O_j, \quad \mu(O_j - E_j) < \varepsilon 2^{-|j|}, \ \text{ and } \ u(x) < \varepsilon(j+1) + \varepsilon \quad \text{for } x \in O_j.$$

Since $K \subset \bigcup_{j=-k}^{k} O_j$, there exist $u_j \in C_c\,(j = 0, \pm 1, \cdots, \pm k)$ such that

$$u_j \prec O_j \ \text{ and } \ \sum_{|j| \le k} u_j(x) = 1\,(x \in K).$$

Thus $u = \sum_{|j| \le k} uu_j$. Since $uu_j \le \varepsilon(j+2)u_j$,

$$\begin{aligned} Tu &= \sum_{|j| \le k} T(uu_j) \le \sum_{|j| \le k} \varepsilon(j+2) T(u_j) \\ &\le \sum_{|j| \le k} \varepsilon(j+2)\mu(O_j) \le \sum_{|j| \le k} \varepsilon(j+2)[\mu(E_j) + \varepsilon 2^{-|j|}] \\ &= \sum_{|j| \le k} \varepsilon j \mu(E_j) + 2\varepsilon \sum_{|j| \le k} \mu(E_j) + \sum_{|j| \le k} 2\varepsilon^2 2^{-|j|} \\ &\le \int u\,d\mu + 2\varepsilon\mu(K) + 8\varepsilon^2. \end{aligned}$$

Since $\varepsilon > 0$ is an arbitrarily given number,

$$Tu \leq \int u d\mu.$$

If $-u$ is substituted for u, the last inequality induces $T(-u) \leq \int (-u) d\mu$. Thus $Tu = \int u d\mu$.

To prove the uniqueness of the measure, suppose that ν is a regular Borel measure such that $\int u d\mu = \int u d\nu$ for all $u \in C_c$. Let O be a given open set. If $u \prec O$, then

$$\int u d\mu = \int u d\nu \leq \nu(O).$$

Taking the supremum over u, we get $\mu(O) \leq \nu(O)$.

On the other hand, let K be a compact set and suppose $u \prec K$. Then $\int u d\mu = \int u d\nu \geq \nu(K)$, which reduces to $\mu(K) \geq \nu(K)$. Thus by the regularity of measures we get $\mu(E) = \nu(E)$ for $E \in \mathcal{M}$. □

6.7. Bounded linear functionals on C_0

For a complex valued Borel measure μ the norm of μ is defined by

$$||\mu|| = \inf\{\sum_{j=1}^{k} |\mu(E_j)| \ : E_j \text{ are mutually disjoint Borel sets}\}.$$

Let $M(X)$ be the set of all complex valued regular Borel measures with bounded norms.

For $\mu \in M(X)$ define the linear functional T_μ by

$$T_\mu u = \int u d\mu, \quad u \in C_c(X).$$

LEMMA 6.10. *For $\mu \in M(X)$ we have*

$$||T_\mu|| = ||\mu||. \tag{6.26}$$

Since $C_c(X)$ is dense in the space $C_0(X)$ with the supremum norm, T_μ can be extended to a bounded functional on $C_0(X)$ with the same norm.

PROOF. First Step. $||T_\mu|| \leq ||\mu||$.

Let $u \in C_c$. For a given $\varepsilon > 0$ there exists a simple function $s = \sum a_j \chi_{E_j}$ such that the E_j are mutually disjoint Borel sets,

$$|s(x) - u(x)| < \varepsilon \quad \text{and} \quad \left|\int (s-u) d\mu\right| < \varepsilon.$$

Then, since $|a_j| \leq ||u||_\infty + \varepsilon$, we have

$$\begin{aligned}\left|\int u d\mu\right| &\leq \left|\int s d\mu\right| + \left|\int (u-s) d\mu\right| < \sum |a_j \mu(E_j)| + \varepsilon \\ &\leq \sum ||u||_\infty |\mu(E_j)| + \sum \varepsilon |\mu(E_j)| + \varepsilon \\ &\leq ||\mu|| ||u||_\infty + \varepsilon ||\mu|| + \varepsilon.\end{aligned}$$

Thus $|T_\mu u| \leq ||\mu|| ||u||_\infty$, which implies the assertion of the first step.

Second Step. *Suppose that μ is real valued. Then for any Borel set E and $\varepsilon > 0$ there exists a compact set $K \subset E$ such that $|\mu(E)| < \mu(K) + \varepsilon$.*

In fact, let $E = P \cup N$ be a Jordan decomposition with respect to μ, where P and N are positive and negative sets respectively. We can assume that $\mu(P) \geq -\mu(N)$. By the regurality of μ there exists a compact set $K \subset P$ such that $\mu(P) < \mu(K) + \varepsilon$. Thus $|\mu(E)| \leq \mu(P) < \mu(K) + \varepsilon$.

Furthermore, for any open set O including K, there exists $u \in C_c(X)$ such that $K \prec u \prec O$ and $\mu(K) < \int u d\mu + \varepsilon$. Thus

$$|\mu(E)| < \int u d\mu + 2\varepsilon.$$

Third Step. Let μ be a complex valued measure and $\varepsilon > 0$ be an arbitrarily given number. By the definition of the norm there exist mutually disjoint Borel sets E_j $(j = 1, 2, \ldots, k)$ such that

$$||\mu|| < \sum_{j=1}^{k} |\mu(E_j)| + \varepsilon/2.$$

Choose real numbers a_j and b_j so that $a_j^2 + b_j^2 = 1$ and

$$|\mu(E_j)| = a_j \Re\mu(E_j) + b_j \Im\mu(E_j).$$

Applying the second step to the sets E_j and the measure $\Re\mu$, we get compact sets $K_j \subset E_j$, open sets O_j, and $u_j \in C_c(X)$ such that $K_j \prec u \prec O_j$, and

$$|\Re\mu(E_j)| < \int u_j d\Re\mu + \varepsilon/4k.$$

Here we remark that $\{O_j\}$ are chosen so that the O_j are mutually disjoint, and thus $\operatorname{supp} u_j$ may be assumed to be mutually disjoint.

For the imaginary part $\Im\mu$ we have $v_j \in C_c(X)$ such that $\operatorname{supp} v_j$ are mutually disjoint, $0 \leq v_j \leq 1$ and $|\Im\mu(E_j)| < \int v_j d\Im\mu + \varepsilon/4k$. Therefore we have

$$\begin{aligned} |\mu(E_j)| &< |a_j| \int u_j d\Re\mu + |b_j| \int v_j d\Im\mu + \varepsilon/2k \\ &= \Re \int (|a_j| u_j - i|b_j| v_j) d\mu + \varepsilon/2k. \end{aligned}$$

Thus

$$\begin{aligned} ||\mu|| &< \Re\Big\{ \int \sum_{j=1}^{k} (|a_j| u_j - i|b_j| v_j) d\mu \Big\} + \varepsilon \\ &= \Re T\Big(\sum_{j=1}^{k} (|a_j| u_j - i|b_j| v_j)\Big) + \varepsilon. \end{aligned}$$

Since the $(|a_j| u_j - i|b_j| v_j)$ are, in absolute value, bounded by 1 and have mutually disjoint supports, the last term does not exceed $||T_\mu|| + \varepsilon$. Thus $||\mu|| \leq ||T_\mu||$. □

THEOREM 6.19 (The Riesz representation theorem). *Let T be a continuous linear functional on $C_0(X)$. Then there exists a unique measure μ in $M(X)$ such that*

$$T\mu = \int u d\mu, \quad \textit{for } u \in C_0(X). \tag{6.27}$$

Furthermore,

$$||T|| = ||\mu||.$$

LEMMA 6.11. *Let T be a continuous linear functional on $C_0(X)$. Suppose that T is real valued on the real valued functions. Then there exist positive linear functionals T_+ and T_- such that*

$$T = T_+ - T_-.$$

PROOF. For a nonnegative function u in $C_0(X)$ put

$$T_+u = \sup\{Tv : v \in C_0, 0 \leq v \leq u\}.$$

Positivity of T_+ is obvious. Our first object is to show that T_+ can be extended to a linear functional on $C_0(X)$.

If $u \geq 0$ and $a > 0$, then it is obvious that $T_+(au) = aT_+u$. If $0 \leq v_j \leq u_j$ $(j = 1,2)$, then, since $0 \leq v_1 + v_2 \leq u_1 + u_2$, $Tv_1 + Tv_2 \leq T_+(u_1 + u_2)$. Thus $T_+u_1 + T_+u_2 \leq T_+(u_1 + u_2)$. Suppose that $0 \leq v \leq u_1 + u_2$. If $v_1(x) = \min\{u_1(x), v(x)\}$ and $v_2 = v - v_1$, then $0 \leq v_j \leq u_j$ $(j = 1,2)$ and $v = v_1 + v_2$. Thus $Tv = Tv_1 + Tv_2 \leq T_+u_1 + T_+u_2$, which implies that $T_+(u_1 + u_2) \leq T_+u_1 + T_+u_2$. Therefore we have $T_+(u_1 + u_2) = T_+u_1 + T_+u_2$ for $u_1, u_2 \geq 0$.

For a real valued function u, put $T_+u = T_+(u_+) - T_+(u_-)$. Now, it is easily shown that $T_+(au) = aT_+u$ for $a \in \mathbb{R}$.

For real valued functions u, v we have $(u+v)_+ + u_- + v_- = (u+v)_- + u_+ + v_+$. Thus $T_+(u+v)_+ + T_+u_- + T_+v_- = T_+(u+v)_- + T_+u_+ + T_+v_+$, from which we get $T_+(u+v) = T_+u + T_+v$. Therefore T_+ is linear on real valued functions.

For a complex valued function u, define T_+ by

$$T_+u = T_+(\Re u) + iT_+(\Im u).$$

By an elementary computation T_+ is a linear fuctional on $C_0(X)$. Furthermore, if $u \geq 0$, then $|T_+u| \leq \|T\|\|u\|_\infty$. Thus $\|T_+\| \leq \|T\|$.

Put $T_- = T_+ - T$. Then T_- is also a positive linear functional on $C_0(X)$. □

PROOF OF THEOREM 6.19. T is a linear combination of positive linear functionals. In fact, if we define T_R and T_I by $T_R(u + iv) = \Re Tu + i\Re Tv$ and $T_I(u + iv) = \Im Tu + i\Im Tv$ for real valued functions u and v, respectively, then T_R and T_I are real valued on the real valued functions and $T = T_R + iT_I$ on $C_0(X)$.

By Lemma 6.11 and Theorem 6.18 there exists a Radon measure μ such that $Tu = \int u d\mu$ for $u \in C_c(X)$.

The uniqueness follows from $\|T\| = \|\mu\|$. □

6.8. Problems

6.1. Let $(X, \mathcal{M}, \mu)$ be a positive measure space. If $\mu(X) < \infty$ and if $0 < p < q \leq \infty$, then prove that $L^q(X) \subset L^p(X)$. This inclusion relation is not true in general. Give a counterexample when $X = \mathbb{R}$ and μ is Lebesgue measure.

6.2. If $p \neq 2$, prove that the parallelogram law does not hold in the $L^p(\mathbb{R})$-norm. Therefore $L^p(\mathbb{R})$ is not a Hilbert space.

6.3. Let $1 < p < \infty$ and let $f, f_n \in L^p(X)$. If $\|f - f_n\|_p \to 0$ as $n \to \infty$, then prove that

$$\int_X f_n g d\mu \to \int_X f g d\mu$$

for any $g \in L^{p'}(X)$, where $1/p + 1/p' = 1$.

6.4. Let $1 \le p < \infty$ and let $f, f_n \in L^p(X)$. If $\|f - f_n\|_p \to 0$ as $n \to \infty$, then prove that $\|f_n\|_p \to \|f\|_p$.

Conversely, if $\mu(X) < \infty$ and if $\|f_n\|_p \to \|f\|_p$ and $f_n \to f$ a.e., then prove that $\|f - f_n\|_p \to 0$.

6.5. Suppose that $\|f\|_q < \infty$ for some $q > 0$. Prove that

$$\lim_{p\to\infty} \|f\|_p = \|f\|_\infty.$$

6.6. Let $1 \le p < \infty$. Prove that

$$\left(\int \left| \int f(x,y)d\mu(x) \right|^p d\nu(y)\right)^{1/p} \le \int \left(\int |f(x,y)|^p d\nu(y)\right)^{1/p} d\mu(x)$$

(Minkowski's inequality for integrals).

6.7. Let $1 \le p_1, p_2, \dots, p_n \le \infty$. If $1/p_1 + 1/p_2 + \cdots + 1/p_n = 1$, prove that

$$\int |f_1 f_2 \cdots f_n| d\mu \le \|f_1\|_{p_1} \|f_2\|_{p_2} \cdots \|f_n\|_{p_n} \qquad \text{(Hölder's inequality)}.$$

6.8. Suppose that $1 \le p, q \le \infty$, $1/p + 1/q \ge 1$ and $1/r = 1/p + 1/q - 1$. Prove that

$$\|f * g\|_r \le \|f\|_p \|g\|_q \qquad \text{(Young's inequality)}.$$

6.9. A topological space is said to be *separable* if it has a countable dense subset. Prove the following.

(i) If $1 \le p < \infty$, then $L^p(0,1)$ is a separable metric space.
(ii) The space $L^\infty(0,1)$ is not separable.

CHAPTER 7

Schwartz Space and Distributions

To any function in $L^p(\mathbb{R}^d)$ $(1 < p \leq \infty)$, there corresponds a continuous linear functional on $L^{p'}(\mathbb{R}^d)$, where $1/p + 1/p' = 1$, and the converse is also true by the representation theorem of Riesz. Similarly, a bounded Radon measure on $\mathbb{R}^d$ which is in a wider class than the integrable functions can be represented as a continuous linear functional on the space $C_0(\mathbb{R}^d)$.

From such a perspective, L. Schwartz[1] has defined distributions as continuous linear functionals on spaces of infinitely differentiable functions.

The space of infinitely differentiable functions possesses profitable properties for differential calculus. Let f be a continuously differentiable function on the real line and u be an infinitely differentaible function with compact support. Then integration by parts yields

$$\int_{-\infty}^{\infty} \frac{df}{dx} u dx = -\int_{-\infty}^{\infty} f \frac{du}{dx} dx.$$

This shows that the operator $D : f \mapsto \frac{df}{dx}$ satisfies the equation

$$\langle Df, u \rangle = -\langle f, \frac{du}{dx} \rangle,$$

where $\langle f, u \rangle = \int f u dx$. This implies that the derivation of f is shifted to the derivative du/dx multiplied by -1. In this way the notion of differentation is extended to distributions.

The theory of distributions is one of the basic tools of analysis, and can be applied effectively, in particular, to the theory of partial differential equations.

In this chapter, the reader will find some theorems which are stated without proof. These theorems will help you to form a better understanding of the structure of distributions. The omission of the proof will not interfere with the following arguments in this book.

7.1. Differentiable functions and distributions

Let $\mathbb{Z}^d$ be the set of points $\alpha = (\alpha_1, \ldots, \alpha_d)$ in $\mathbb{R}^d$ whose components are all integers, and $\mathbb{Z}_+^d$ be the set of α in $\mathbb{Z}^d$ whose components are all nonnegative. For $\alpha \in \mathbb{Z}_+^d$ and $x \in \mathbb{R}^d$, write $x^\alpha = x_1^{\alpha_1} \cdots x_d^{\alpha_d}$ and

$$D^\alpha = \left(\frac{\partial}{\partial x_1}\right)^{\alpha_1} \cdots \left(\frac{\partial}{\partial x_d}\right)^{\alpha_d}.$$

The length of α is denoted by $\|\alpha\| = |\alpha_1| + \cdots + |\alpha_d|$.

---

[1] L. Schwartz, 1915-. French mathematician. Not to be confused with the inequality man, H. A. Schwarz (German, 1843-1921).

For a compact set K of $\mathbb{R}^d$ put

$$C^\infty(K) = \{u \in C^\infty(\mathbb{R}^d) : \operatorname{supp} u \subset K\}.$$

Thus $C_c^\infty(\mathbb{R}^d) = \bigcup\{C^\infty(K) : K \text{ is compact}\}$.

DEFINITION 7.1. A sequence $\{u_j\}$ in $C_c^\infty(\mathbb{R}^d)$ is said to converge in $C_c^\infty(\mathbb{R}^d)$ to u_0 if

(i) $\operatorname{supp} u_j \subset K$ $(j = 1, 2, \dots)$ for a compact set K,
(ii) for every $\alpha \in \mathbb{Z}_+^d$, $D^\alpha u_j \to D^\alpha u_0$ uniformly.

DEFINITION 7.2. A linear functional T on $C_c^\infty(\mathbb{R}^d)$ is said to be a *distribution* if $Tu_j \to 0$ when $u_j \to 0$ in $C_c^\infty(\mathbb{R}^d)$.

The collection of all distributions is denoted by $\mathcal{D}'(\mathbb{R}^d)$.

For a distribution T, we often denote Tu by

$$T(u) \quad \text{or} \quad \langle T, u\rangle.$$

For a compact subset K of $\mathbb{R}^d$ and $m \geq 0$ we define a seminorm $p_{m,K}$ by

$$p_{m,K}(u) = \sup_{\|\alpha\| \leq m} \sup_{x \in K} |D^\alpha u(x)|, \quad u \in C^\infty(\mathbb{R}^d).$$

THEOREM 7.1. *A linear functional T on $C_c^\infty(\mathbb{R}^d)$ belongs to $\mathcal{D}'(\mathbb{R}^d)$ if and only if for any compact set K there exist a constant C and $m \geq 0$ such that*

$$|\langle T, u\rangle| \leq C p_{m,K}(u) \quad \text{for } u \in C^\infty(K). \tag{7.1}$$

The constant C and m may depend on K.

PROOF. If inequality (7.1) holds, the functional T is obviously continuous on $C_c^\infty(\mathbb{R}^d)$.

Conversely, suppose that $T \in \mathcal{D}'(\mathbb{R}^d)$ and (7.1) does not hold. Then there exist a compact set K and $u_m \in C^\infty(K)$ $(m = 1, 2, \dots)$ such that

$$|\langle T, u_m\rangle| > m p_{m,K}(u_m) = c_m.$$

Put $v_m = c_m^{-1} u_m$. Then $\operatorname{supp} v_m \subset K$ and

$$p_{m,K}(v_m) = c_m^{-1} p_{m,K}(u_m) = \frac{1}{m} \to 0 \quad \text{as } m \to \infty.$$

Thus $v_m \to 0$ in $C_c^\infty(\mathbb{R}^d)$. On the other hand $|\langle T, v_m\rangle| > 1$, which contradicts the continuity of T. □

Now, we shall consider a topology on the space $C_c^\infty(\mathbb{R}^d)$ which is compatible with Definition 7.1. We describe it without the proof, but the reader may piece together the theoretical scenery from the discussion.

For a compact set K and real numbers $m \geq 0$ and $\varepsilon > 0$, denote

$$U(0 : m, \varepsilon, K) = \{u \in C^\infty(K) : p_{m,K}(u) < \varepsilon\}$$

and

$$U(v : m, \varepsilon, K) = v + U(0 : m, \varepsilon, K), \quad v \in C^\infty(K).$$

Then $\{U(v : m, \varepsilon, K) : \varepsilon > 0,\ m = 0, 1, \dots\}$ comprises a fundamental system of neighborhoods of v and introduces a topology on $C^\infty(K)$. Therefore $C^\infty(K)$ is a Hausdorff locally convex topological vector space (see Proposition 6.2).

Now, a fundamental system of neighborhoods of the space $C_c^\infty(\mathbb{R}^d)$ is defined from the neighborhood system of $C^\infty(K)$ as follows: Put $K_n = \{x \in \mathbb{R}^d : |x| \leq n\}$.

Then the space $C_c^\infty(\mathbb{R}^d)$ is a countable union of $C^\infty(K_n)$. We shall extend the topology of $C^\infty(K_n)$ to $C^\infty(\mathbb{R}^d)$. Let U be a subset of $C_c^\infty(\mathbb{R}^d)$ which satisfies

(a) $aU \subset U$ $(a \in \mathbb{C},\ |a| \leq 1)$,

(b) for every n the set $U \cap C^\infty(K_n)$ is a neighborhood of 0 in $C^\infty(K_n)$.

The family of U satisfying conditions (a) and (b) is a fundamental system of neighborhoods of 0 of the space $C_c^\infty(\mathbb{R}^d)$ and makes $C_c^\infty(\mathbb{R}^d)$ a Hausdorff locally convex topological vector space. The detailed proof is found, for example, in [Tr].

We denote the space $C_c^\infty(\mathbb{R}^d)$ with this topology by $\mathcal{D}(\mathbb{R}^d)$.

A sequence u_j in $\mathcal{D}(\mathbb{R}^d)$ converges in the topology of $\mathcal{D}(\mathbb{R}^d)$ to u_0 if and only if it converges in $C_c^\infty(\mathbb{R}^d)$ in the sense of Definition 7.1.

Therefore a distributon in $\mathcal{D}'(\mathbb{R}^d)$ is defined as a continuous linear functional on $\mathcal{D}(\mathbb{R}^d)$.

EXAMPLE 7.1 (Dirac measure). Let T be a linear functional defined by

$$\langle T, u\rangle = u(0) \quad \text{for } u \in \mathcal{D}(\mathbb{R}^d).$$

Let K be any compact set. By the definition of the seminorms $p_{m,K}$ we have $|\langle T, u\rangle| \leq p_{0,K}(u)$, $u \in C^\infty(K)$. Thus $T \in \mathcal{D}'(\mathbb{R}^d)$. The distribution T is called the *Dirac distribution* or Dirac measure supported at the origin, and is denoted by δ; that is,

$$\langle \delta, u\rangle = u(0).$$

The linear functional T defined by

$$\langle T, u\rangle = (-1)^{\|\alpha\|} D^\alpha u(0)$$

is also a distribution.

EXAMPLE 7.2 (Heaviside function). Let $L^1_{loc}(\mathbb{R}^d)$ be the space of locally integrable functions. For $f \in L^1_{loc}(\mathbb{R}^d)$ the linear functional T_f defined by

$$\langle T_f, u\rangle = \int f u dx \tag{7.2}$$

is a distribution. In fact, $|\langle T_f, u\rangle| \leq \int_K |f| dx \cdot p_{0,K}(u)$ for $u \in C^\infty(K)$. The distribution T_f is often denoted simply by f.

In particular, let $Y(t)$ be the Heaviside function, that is, $Y(t) = 1\,(t \geq 0), = 0\,(t < 0)$. Then the integral

$$\langle Y, u\rangle = \int_0^\infty u(t) dt$$

defines a distribution.

EXAMPLE 7.3 (Singular integral). Let $d = 1$. The principal value

$$\left\langle \text{P.V.}\frac{1}{t}, u\right\rangle = \lim_{\varepsilon\to 0}\int_{|t|>\varepsilon} \frac{u(t)}{t} dt$$

defines a distribution. In fact, if $\operatorname{supp} u \subset K = [-k, k]$, the right hand side equals

$$\lim_{\varepsilon\to 0}\int_{|t|>\varepsilon} \frac{u(t)}{t} dt = u(0) \lim_{\varepsilon\to 0}\int_{\varepsilon<|t|<k} \frac{dt}{t} + \lim_{\varepsilon\to 0}\int_{\varepsilon<|t|<k} \frac{u(t)-u(0)}{t} dt.$$

Since $|[u(t) - u(0)]/t| \leq \sup |u'(t)|$, we have

$$\left|\left\langle \text{P.V.}\frac{1}{t}, u\right\rangle\right| \leq 2k p_{1,K}(u) \quad \text{for } u \in C^\infty(K).$$

7.2. Operations on distributions

We shall define several fundamental operations on distributions.

(I) *Linear combination.* For $S, T \in \mathcal{D}'(\mathbb{R}^d)$ and $a, b \in \mathbb{C}$, the linear functional defined by

$$\langle aS + bT, u\rangle = a\langle S, u\rangle + b\langle T, u\rangle, \quad u \in \mathcal{D}(\mathbb{R}^d),$$

is continuous on $\mathcal{D}(\mathbb{R}^d)$. Thus $aS + bT$ makes sense as an element of $\mathcal{D}'(\mathbb{R}^d)$, and $\mathcal{D}'(\mathbb{R}^d)$ is a vector space over $\mathbb{C}$.

(II) *Multiplication by functions.* Let $\varphi \in C^\infty(\mathbb{R}^d)$. The mapping

$$\mathcal{D}(\mathbb{R}^d) \ni u \mapsto \varphi u \in \mathcal{D}(\mathbb{R}^d)$$

is continuous. In fact, we have, by Leibniz's formula,

$$D^\alpha(\varphi u) = \sum_{\beta \leq \alpha} \binom{\alpha}{\beta} (D^\beta \varphi)(D^{\alpha-\beta} u),$$

where $\alpha = (\alpha_1, \ldots, \alpha_d)$ and $\sum$ runs over all $\beta = (\beta_1, \ldots, \beta_d)$ such that $0 \leq \beta_j \leq \alpha_j$ and

$$\binom{\alpha}{\beta} = \frac{\alpha_1! \cdots \alpha_d!}{\beta_1! \cdots \beta_d!(\alpha_1 - \beta_1)! \cdots (\alpha_d - \beta_d)!}.$$

Therefore, if $T \in \mathcal{D}'(\mathbb{R}^d)$, the mapping $\mathcal{D}(\mathbb{R}^d) \ni u \mapsto \langle T, \varphi u\rangle \in \mathbb{C}$ is continuous. The distribution defined by this mapping is denoted by φT, that is,

$$\langle T, \varphi u\rangle = \langle \varphi T, u\rangle.$$

(III) *Translation.* Let $T \in \mathcal{D}'(\mathbb{R}^d)$ and $a \in \mathbb{R}^d$. The translation $\tau_a T$ of T is defined by

$$\langle \tau_a T, u\rangle = \langle T, \tau_{-a} u\rangle, \quad u \in \mathcal{D}(\mathbb{R}^d).$$

For example, $\langle \tau_a \delta, u\rangle = \langle \delta, u(\cdot + a)\rangle = u(a)$. The distribution $\tau_a \delta = \delta_a$ is called Dirac measure with support at $\{a\}$.

(IV) *Reflection.* Let $\check{u}(x) = u(-x)$. For a distribution T, $\check{T}$ is defined by $\langle \check{T}, u\rangle = \langle T, \check{u}\rangle$.

(V) *Dilation.* For $t > 0$ put $\sigma_t u(x) = u(tx)$. The dilation of a distribution T is defined by the equation

$$\langle \sigma_t T, u\rangle = t^{-d}\langle T, \sigma_{1/t} u\rangle.$$

If a distribution T_f is defined by a locally integrable function f, then the dilation $\sigma_t T_f$ is associated with the function $f(tx)$.

(VI) *Differentiation.* For $T \in \mathcal{D}'(\mathbb{R}^d)$, the distribution $D^\alpha T$ is defined by the equation

$$\langle D^\alpha T, u\rangle = (-1)^{\|\alpha\|}\langle T, D^\alpha u\rangle, \quad u \in \mathcal{D}(\mathbb{R}^d).$$

Note that the mapping $\mathcal{D}(\mathbb{R}^d) \ni u \mapsto (-1)^{\|\alpha\|}\langle T, D^\alpha u\rangle$ is continuous, since the mapping $\mathcal{D}(\mathbb{R}^d) \ni u \mapsto D^\alpha u \in \mathcal{D}(\mathbb{R}^d)$ is continuous.

For example, let $f \in C^{(1)}(\mathbb{R})$, and let T_f be the distribution corresponding to f by (7.2). Applying integration by parts, we get

$$\langle DT_f, u\rangle = -\int_{-\infty}^{\infty} f\frac{du}{dx}dx = \int_{-\infty}^{\infty} \frac{df}{dx} u dx.$$

Thus the differentiation DT_f in the distribution sense coincides with the ordinary differentiation of f.

Denote by DT_f or $\frac{df}{dx}$ the differentiation in the distribution sense and by $\{\frac{df}{dx}\}$ the differentiation in the ordnary sense. From the above argument we have

$$DT_f = T_{\{\frac{df}{dx}\}} \quad \text{or} \quad \frac{df}{dx} = \{\frac{df}{dx}\},$$

if $f \in C^{(1)}(\mathbb{R})$.

EXAMPLE 7.4. By definition we have

$$\langle D^\alpha \delta, u\rangle = (-1)^{\|\alpha\|}\langle \delta, D^\alpha u\rangle = (-1)^{\|\alpha\|} D^\alpha u(0).$$

For the Heaviside function Y we have

$$\langle Y', u\rangle = (-1)\langle Y, u'\rangle = -\int_0^\infty u'(x)dx = u(0) = \langle \delta, u\rangle.$$

Thus

$$Y' = \delta.$$

Remark that $\{Y'\} \neq Y'$. In fact, since $\{Y'\} = 0$ except at 0, it follows that $T_{\{Y'\}} = 0$. On the other hand, $Y' = \delta$. Therefore the differentiation in the ordinary sense and in the distribution sense do not always coinside.

EXAMPLE 7.5. The function $\log|t|$ $(t \in \mathbb{R})$ is locally integrable on the real axis. By definition

$$\langle (\log|t|)', u\rangle = -\langle \log|t|, u'\rangle = -\int_{-\infty}^\infty \log|t|\, u'(t)dt, \quad u \in \mathcal{D}(\mathbb{R}).$$

The right-hand side equals, by integration by parts,

$$\begin{aligned} &-\lim_{\varepsilon\to 0}\int_\varepsilon^\infty \log t[u'(t) + u'(-t)]dt \\ &= -\lim_{\varepsilon\to 0}\Big[\log t[u(t) - u(-t)]\Big]_\varepsilon^\infty + \lim_{\varepsilon\to 0}\int_\varepsilon^\infty \frac{u(t) - u(-t)}{t}dt. \end{aligned}$$

Since $|\log\varepsilon[u(\varepsilon) - u(-\varepsilon)]| \leq 2\varepsilon \log\varepsilon \sup_{0<\theta<1}|u'(\theta\varepsilon)| \to 0$ $(\varepsilon \to 0)$, we have

$$\langle (\log|t|)', u\rangle = \lim_{\varepsilon\to 0}\int_{|t|>\varepsilon}\frac{u(t)}{t}dt,$$

that is,

$$(\log|t|)' = \text{P.V.}\frac{1}{t}.$$

EXAMPLE 7.6. Let f be a function in $\mathbb{R}$, continuous except at point a, and continuous on the right. Let $d = f(a+0) - f(a-0)$. Furthermore, assume that $\{f'\} \in L^1_{loc}(\mathbb{R})$.

Then $g(x) = f(x) - dY(x-a)$ is continuous, and $\{g'\} = \{f'\}$ except at a. Thus integration by parts yields

$$\langle Dg, u\rangle = -\left\langle g, \frac{du}{dx}\right\rangle = \langle \{g'\}, u\rangle,$$

and the last term equals $\langle \{f'\}, u\rangle$. On the other hand, we have

$$-\left\langle g, \frac{du}{dx}\right\rangle = -\left\langle f, \frac{du}{dx}\right\rangle + d\left\langle Y(\cdot - a), \frac{du}{dx}\right\rangle = \langle f', u\rangle - d\langle \delta_a, u\rangle.$$

Thus

$$f' = \{f'\} + d\delta_a. \tag{7.3}$$

EXAMPLE 7.7. We have

$$\left(\frac{d^2}{dx^2} + \omega^2\right) Y(x) \frac{\sin \omega x}{\omega} = \delta. \tag{7.4}$$

In fact, applying formula (7.3) repeatedly, we have

$$\frac{d}{dx}\left(Y(x)\frac{\sin\omega x}{\omega}\right) = Y(x)\cos\omega x,$$

$$\frac{d^2}{dx^2}\left(Y(x)\frac{\sin\omega x}{\omega}\right) = -\omega Y(x)\sin\omega x + \delta,$$

from which (7.4) follows.

Let $L = \sum a_\alpha D^\alpha$ be a finite sum of differential operators with constant coefficients. When $Lf = \delta$ in the distribution sense, $f \in \mathcal{D}'$ is said to be a *fundamental solution*.

As an application of formula (7.3) we shall prove Green's formula, following [Sc].

Let D be a domain of $\mathbb{R}^d$ with a smooth boundary. Let f be a function in $C^{(2)}(D)$ such that f vanishes outsede $\overline{D}$ and the derivatives of f up to order 2 have boundary values at ∂D. Let g be the jump of $f(x)$ when $x \in D$ crosses ∂D.

By the definition and (7.3) we have

$$\begin{aligned}\left\langle \frac{\partial f}{\partial x_1}, u\right\rangle &= \int_{\mathbb{R}^{d-1}} \left\langle \frac{\partial}{\partial x_1} f(\cdot, x_2, \dots, x_d), u(\cdot, x_2, \dots, x_d)\right\rangle dx_2 \cdots dx_d \\ &= \int_{\mathbb{R}^{d-1}} \left(\int_{-\infty}^{\infty} \left\{\frac{\partial f}{\partial x_1}\right\} u dx_1 + gu\right) dx_2 \cdots dx_d \\ &= \int_{\mathbb{R}^d} \left\{\frac{\partial f}{\partial x_1}\right\} u dx + \int_{\partial D} g u dx_2 \cdots dx_d.\end{aligned}$$

Let θ_1 be the angle the x_1-axis makes with the outward normal at a point x of ∂D when x_1 increases. Let $d\sigma$ be the area element of ∂D (see Figure 7.1). Then from the last equation we get

$$\left\langle \frac{\partial f}{\partial x_1}, u\right\rangle = \left\langle T_{\{\frac{\partial f}{\partial x_1}\}}, u\right\rangle + \int_{\partial D} g u \cos\theta_1 d\sigma. \tag{7.5}$$

Let $\delta_{\partial D}$ be the distribution defined by the mapping: $v \mapsto \int_{\partial D} v d\sigma$. Then by (7.5) we have

$$\frac{\partial f}{\partial x_j} = \left\{\frac{\partial f}{\partial x_j}\right\} + (g\cos\theta_j)\delta_{\partial D} \quad (j = 1, \dots, d).$$

Iterating the above argument, we get

$$\frac{\partial^2 f}{\partial x_j^2} = \left\{\frac{\partial^2 f}{\partial x_j^2}\right\} + \frac{\partial}{\partial x_j}[(g\cos\theta_j)\delta_{\partial D}] + (g_j\cos\theta_j)\delta_{\partial D}, \tag{7.6}$$

where g_j is the jump of $\{\partial f/\partial x_j\}$ at ∂D.

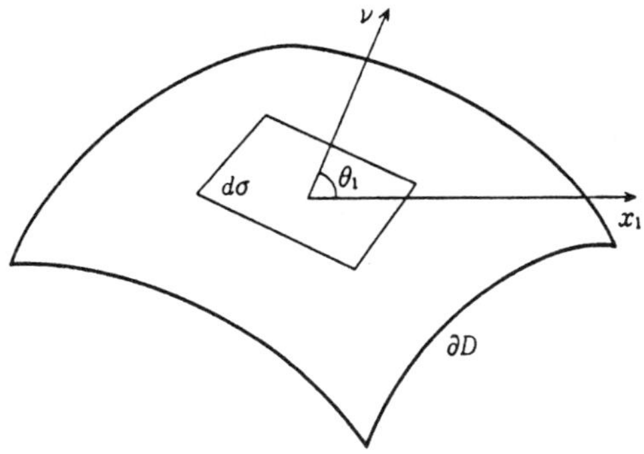

FIGURE 7.1. Normal vector and area element

By the definition of differentition

$$\left\langle \sum_{j=1}^{d} \frac{\partial}{\partial x_j}(g\cos\theta_j)\delta_{\partial D}, u \right\rangle = -\int_{\partial D} \sum_{j=1}^{d} g\cos\theta_j \frac{\partial}{\partial x_j} u d\sigma.$$

Let ν be the outward normal to ∂D. Since $\partial u/\partial\nu = \sum_{j=1}^{d} g\cos\theta_j \partial u/\partial x_j$, the right-hand side equals

$$-\int_{\partial D} g\frac{\partial u}{\partial \nu} d\sigma = -\left\langle g\delta_{\partial D}, \frac{du}{d\nu}\right\rangle = \left\langle \frac{\partial}{\partial\nu}(g\delta_{\partial D}), u\right\rangle.$$

Thus

$$\sum_{j=1}^{d} \frac{\partial}{\partial x_j}[(g\cos\theta_j)\delta_{\partial D}] = \frac{\partial}{\partial \nu}(g\delta_{\partial D}).$$

Note that $\sum_{j=1}^{d} g_j\cos\theta_j$ is equal to the jump of $\partial f/\partial\nu = \sum_{j=1}^{d}\cos\theta_j \partial f/\partial x_j$. Thus we have

$$\sum_{j=1}^{d}(g_j\cos\theta_j)\delta_{\partial D} = \frac{\partial f}{\partial \nu}.$$

Therefore it follows from (7.6) that

$$\Delta f = \{\Delta f\} + \frac{\partial}{\partial\nu}(g\delta_{\partial D}) + \frac{\partial f}{\partial \nu}\delta_{\partial D}, \tag{7.7}$$

where $\Delta = \sum_{j=1}^{d} \partial^2/\partial x_j^2$.

THEOREM 7.2 (Green's theorem). *Let D be a domain of $\mathbb{R}^d$ with smooth boundary ∂D, and let $d\sigma$ be the area element on it. Let ν be the outward nomal to ∂D. Then for $f \in C^{(2)}(\mathbb{R}^d)$ and $u \in \mathcal{D}(\mathbb{R}^d)$ we have*

$$\int_{\partial D}\left(f\frac{\partial u}{\partial\nu} - u\frac{\partial f}{\partial \nu}\right) d\sigma = -\int_D (f\Delta u - u\Delta f)dx. \tag{7.8}$$

PROOF. We have, by (7.7),

$$\langle \Delta f, u\rangle = \langle\{\Delta f\}, u\rangle + \left\langle \frac{\partial}{\partial\nu}(g\delta_{\partial D}), u\right\rangle + \left\langle \frac{\partial f}{\partial\nu}\delta_{\partial D}, u\right\rangle.$$

Since $\langle \partial(g\delta_{\partial D})/\partial\nu, u\rangle = -\int_{\partial D} g\partial u/\partial\nu d\sigma$, and since $f = g$ on ∂D, the right-hand side of the last equation equals

$$\int_D \Delta f u dx + \int_{\partial D} \frac{\partial f}{\partial \nu} u d\sigma - \int_{\partial D} f \frac{\partial u}{\partial \nu} d\sigma.$$

On the other hand, we have

$$\langle \Delta f, u\rangle = \langle f, \Delta u\rangle = \int_D f \Delta u dx.$$

Therefore formula (7.8) is obtained. □

EXAMPLE 7.8. We have

$$\Delta \log|x| = -2\pi\delta$$

on the plane $\mathbb{R}^2$, and

$$\Delta|x|^{-d+2} = -(d-2)\omega_{d-1}\delta$$

on $\mathbb{R}^d$ $(d > 2)$, where ω_{d-1} is the area of the unit sphere.

PROOF. Assume $d > 2$. The case $d = 2$ can be proved similarly.

Note that $|x|^{-d+2}$ is locally integrable and harmonic in $x \neq 0$; that is, $\Delta|x|^{-d+2} = 0$. For $u \in \mathcal{D}(\mathbb{R}^d)$ we have

$$\begin{aligned}\langle \Delta|x|^{-d+2}, u\rangle = \langle |x|^{-d+2}, \Delta u\rangle &= \int_{\mathbb{R}^d} \frac{1}{|x|^{d-2}} \Delta u(x) dx \\ &= \lim_{\varepsilon\to 0} \int_{|x|>\varepsilon} \frac{1}{|x|^{d-2}} \Delta u(x) dx.\end{aligned}$$

Fix any $\varepsilon > 0$ and put $D_\varepsilon = \{x : |x| > \varepsilon\}$. If $f_\varepsilon(x) = |x|^{-d+2}$ $(x \in D_\varepsilon)$, $= 0$ $(x \notin D_\varepsilon)$, then, since $\{\Delta f_\varepsilon\} = 0$, by Green's theorem

$$\int_{D_\varepsilon} \frac{1}{|x|^{d-2}} \Delta u dx = \int_{\partial D_\varepsilon} \frac{1}{\varepsilon^{d-2}} \frac{-\partial u}{\partial r} d\sigma(x) - \int_{\partial D_\varepsilon} \frac{-d+2}{\varepsilon^{d-1}} u d\sigma(x),$$

where $r = |x|$. The first term on the right-hand side is bounded by

$$\sup\left|\frac{\partial u}{\partial r}\right| \int_{\partial D_\varepsilon} \frac{d\sigma(x)}{\varepsilon^{d-2}} = \sup\left|\frac{\partial u}{\partial r}\right| \omega_{d-1}\varepsilon \to 0 \quad (\varepsilon \to 0).$$

On the other hand, since $|u(x) - u(0)| \le \varepsilon \sum_{j=1}^d \sup|\partial u/\partial x_j|$ $(|x| = \varepsilon)$, we have

$$\int_{\partial D_\varepsilon} \frac{u(x)}{\varepsilon^{d-1}} u d\sigma(x) = \int_{\partial D_\varepsilon} \frac{u(x) - u(0)}{\varepsilon^{d-1}} d\sigma(x) + \omega_{d-1} \frac{u(0)}{\varepsilon^{d-1}} \varepsilon^{d-1} \to \omega_{d-1} u(0).$$

Thus

$$\langle \Delta|x|^{-d+2}, u\rangle = -(d-2)\omega_{d-1} u(0).$$

□

7.3. Local property of distributions and convolution

A locally integrable function f defines a distribution by the formula $\langle T_f, u\rangle = \int fudx$. Our first object in this section is to prove that a distribution in $\mathcal{D}'(\mathbb{R}^d)$ is locally expressed as a derivative of an integrable function. This fact is induced from inequality (7.1), which shows that if $T \in \mathcal{D}'(\mathbb{R}^d)$, then for every compact set K there exist a constant C and $m \geq 0$ such that

$$|\langle T, u\rangle| \leq Cp_{m,K}(u), \quad f \in C^\infty(K).$$

THEOREM 7.3. *Let $T \in \mathcal{D}'(\mathbb{R}^d)$, and let K be a compact set. Then there exist $f \in L^2(K)$ and $\alpha \in \mathbb{Z}_+^d$ such that*

$$\langle T, u\rangle = (-1)^{\|\alpha\|} \int_K fD^\alpha udx, \quad u \in C^\infty(K).$$

Note that f and α may depend on the set K.

PROOF. We can assume that K is a cube with center at the origin and side length $2r > 1$. Let m be a number given by (7.1). We have

$$\left|\frac{\partial u(x)}{\partial x_1}\right| = \left|\int_{-r}^{x_1} \frac{\partial^2 u}{\partial x_1^2} dx_1\right| \leq 2r \sup_x \left|\frac{\partial^2 u(x)}{\partial x_1^2}\right|$$

for $u \in C^\infty(K)$. Applying the above inequality repeatedly, we have

$$\sup_x |D^\alpha u(x)| \leq (2r)^{\|\beta-\alpha\|} \sup_x |D^\beta u(x)| \quad (\beta \geq \alpha).$$

By Schwarz's inequality

$$|u(x)| = \left|\int_{-r}^{x_1} \cdots \int_{-r}^{x_d} \frac{\partial^d u}{\partial x_1 \cdots \partial x_d} dx_1 \cdots dx_d\right| \leq (2r)^{d/2} \left\|\frac{\partial^d u}{\partial x_1 \cdots \partial x_d}\right\|_{L^2(K)},$$

for $u \in C^\infty(K)$. Thus if $\alpha = (m+d, \cdots, m+d)$, then we have, by a simple computation,

$$|\langle T, u\rangle| \leq C'\|D^\alpha u\|_{L^2(K)}, \quad u \in C^\infty(K),$$

with some constant C'.

Let $V = \{D^\alpha u : u \in C^\infty(K)\}$. Since the mapping $D^\alpha : C^\infty(K) \ni u \mapsto D^\alpha u \in V$ is surjective, we have

$$|\langle T, D^{-\alpha} v\rangle| \leq C'\|v\|_{L^2(K)}, \quad v \in V,$$

where $D^{-\alpha}$ on V is the inverse of D^α. Let $H = \overline{V}^{L^2}$ be the closure in $L^2(K)$. Since H is a Hilbert space with the inner product $\int_K f\overline{g}dx$, and since the last inequality can be extended to H, there exists, by the Riesz representation theorem, $g \in H \subset L^2(K)$ such that

$$\langle T, D^{-\alpha} v\rangle = \int_K gvdx, \quad v \in V.$$

Therefore we have

$$\langle T, u\rangle = (-1)^{\|\alpha\|} \int_{\mathbb{R}^d} fD^\alpha udx, \quad u \in C^\infty(K),$$

with $f = (-1)^{\|\alpha\|} g$. □

If $T \in \mathcal{D}'(\mathbb{R}^d)$ and $u \in \mathcal{D}(\mathbb{R}^d)$, then $\langle \tau_x T, u\rangle$ is a C^∞-function of x. In fact, let $x \in \mathbb{R}^d$ and fix a positive number $r > 0$. Choose a compact set K_0 so that $K_0 \supset \operatorname{supp} u - B(x,r) = \{y - z : y \in \operatorname{supp} u,\ z \in B(x,r)\}$. By Theorem 7.3 there exist a function $f \in L^2(K_0)$ and an α such that

$$\langle T, v\rangle = (-1)^{\|\alpha\|}\int f(y)D^\alpha v(y)dy \quad \text{for all } v \in C^\infty(K_0).$$

Since $\operatorname{supp} \tau_{-z} u \subset K_0\,(z \in B(x,r))$,

$$\langle T, \tau_{-z}u\rangle = (-1)^{\|\alpha\|}\int f(y)D^\alpha u(y+z)dy \quad z \in B(x,r),$$

which is infinitely differentiable.

By definition, the *support of a distribution* T of $\mathcal{D}(\mathbb{R}^d)$ is the smallest closed set K which satisfies the condition

$$(7.9) \qquad u \in \mathcal{D}(\mathbb{R}^d) \text{ and } \operatorname{supp} u \subset K^c \quad \Rightarrow \quad \langle T, u\rangle = 0.$$

To prove the existence of such a smallest closed set, it suffices to show that if $\{K_\iota : \iota \in I\}$ is a family of closed sets satisfying condition (7.9), then the intersection $K = \bigcap_{\iota\in I} K_\iota$ also satisfies (7.9). In fact, if $u \in \mathcal{D}(\mathbb{R}^d)$ and $\operatorname{supp} u \subset K^c = \bigcup_{\iota \in I} K_\iota^c$, then there exist, by the compactness of $\operatorname{supp} u$, a finite number of indices $\{\iota_1, \cdots, \iota_n\}$ such that $\operatorname{supp} u \subset \bigcup_{j=1}^n K_{\iota_j}^c$. Then, by a partition of unity, we can take $u_j \in \mathcal{D}(\mathbb{R}^d)$ such that $u_j \prec K_{\iota_j}^c$ and $u_1 + \cdots + u_n = 1$ on $\operatorname{supp} u$ (see Theorem 6.17 and Remark 6.2).

Therefore we have $\langle T, u\rangle = \sum_{j=1}^n \langle Tuu_j\rangle = 0$. Thus $\operatorname{supp} T \subset K$.

If the support of T is compact, then $\langle \tau_x T, u\rangle \in \mathcal{D}(\mathbb{R}^d)$ for $u \in \mathcal{D}(\mathbb{R}^d)$. In fact, let $K = \operatorname{supp} T$. Then the set K satisfies (7.9). If $u \in \mathcal{D}(\mathbb{R}^d)$, then the set $-K + \operatorname{supp} u$ is compact, and $\operatorname{supp} \tau_{-x} u = \operatorname{supp} u - x \subset K^c$ for $x \notin -K + \operatorname{supp} u$. Thus

$$\operatorname{supp} \langle T, \tau_{(-\cdot)} u\rangle \subset -K + \operatorname{supp} u.$$

Therefore the supposrt of $\langle \tau_x T, u\rangle$ is compact.

For any $S \in \mathcal{D}'(\mathbb{R}^d)$ and a distribution T with compact support the convolution $S * T$ is defined by

$$\langle S * T, v\rangle = \langle S, \langle T, \tau_{(-\cdot)}u\rangle\rangle, \quad u \in \mathcal{D}(\mathbb{R}^d).$$

Now, we shall give an example of application of the convolution operation which enables us to construct a solution of a differential equation from a fundamental solution.

Recall Example 7.7. We let S and T_f be the distributions associated with $Y(x)(\sin \omega x)/\omega$ and $f \in C_c^{(2)}(\mathbb{R})$ respectively. Then we have

$$\langle (D^2 + \omega^2)S * T_f, u\rangle = \langle S * T_f, (D^2+\omega^2)u\rangle = \langle S, \langle T_f, \tau_{(-\cdot)}(D^2+\omega^2)u\rangle\rangle.$$

Furthermore, we have

$$\begin{aligned}\langle T_f, \tau_{(-\cdot)}(D^2+\omega^2)u\rangle &= \int f(y)\left(\frac{d^2}{dy^2} + \omega^2\right)u(y+x)dy \\ &= \left(\frac{d^2}{dx^2}+\omega^2\right)\int f(y)u(y+x)dy \\ &= (D^2+\omega^2)\langle T_f, \tau_{(-\cdot)}u\rangle.\end{aligned}$$

Since $(D^2+\omega^2)S=\delta$, it follows that

$$\begin{aligned}\langle (D^2+\omega^2)S * T_f, u\rangle &= \langle S, (D^2+\omega^2)\langle T_f, \tau_{(-\cdot)}u\rangle\rangle \\ &= \langle (D^2+\omega^2)S, \langle T_f, \tau_{(-\cdot)}u\rangle\rangle = \langle T_f, u\rangle.\end{aligned}$$

Therefore we get $(D^2+\omega^2)S * T_f = T_f$, or equivalently,

$$\left(\frac{d^2}{dx^2}+\omega^2\right)\int_{-\infty}^{\infty}\frac{\sin\omega y}{\omega}f(x-y)dy = f(x),$$

which implies that

$$g(x)=\int_{-\infty}^{\infty}\frac{\sin\omega y}{\omega}f(x-y)dy$$

is a solution of the equation $(D^2+\omega^2)g=f$.

Finally, we mention a theorem without proving it.

THEOREM 7.4. *Let $\{T_j\}$ be a sequence of distributions in $\mathcal{D}'(\mathbb{R}^d)$. If $\{\langle T_j, u\rangle\}$ converges for every $u\in\mathcal{D}(\mathbb{R}^d)$, then there exists $T\in\mathcal{D}'(\mathbb{R}^d)$ such that*

$$\lim_{j\to\infty}\langle T_j, u\rangle = \langle T, u\rangle \quad \textit{for all } u\in\mathcal{D}(\mathbb{R}^d).$$

7.4. The Schwartz space and the tempered distributions

Let $\mathcal{S}(\mathbb{R}^d)$ be the collection of all functions $u\in C^\infty(\mathbb{R}^d)$ such that

$$\sup_x |x^\alpha D^\beta u(x)| < \infty \quad \text{for every } \alpha \text{ and } \beta \text{ in } \mathbb{Z}_+^d.$$

The set $\mathcal{S}(\mathbb{R}^d)$ is called the *Schwartz space*. A function u in $\mathcal{S}(\mathbb{R}^d)$ is rapidly decreasing in the sense that u and all its deivatives decrease more quickly than the inverse of any polynomial.

If $u\in\mathcal{S}(\mathbb{R}^d)$, then $x^\alpha u(x)$, $D^\beta u(x)\in\mathcal{S}(\mathbb{R}^d)$ for all α, $\beta\in\mathbb{Z}^d$. The space $\mathcal{S}(\mathbb{R}^d)$ is a vector spacc over $\mathbb{C}$ and, as is easily seen,

$$C_c^\infty(\mathbb{R}^d)\subset\mathcal{S}(\mathbb{R}^d)\subset C_0^\infty(\mathbb{R}^d)\subset C^\infty(\mathbb{R}^d).$$

For α, $\beta\in\mathbb{Z}_+^d$ denote

$$p_{\alpha,\beta}(u)=\sup_x |x^\alpha D^\beta u(x)|.$$

We have $p_{\alpha,\beta}(u)<\infty$ for every $u\in\mathcal{S}(\mathbb{R}^d)$ and α, $\beta\in\mathbb{Z}_+^d$. Furthermore, $p_{\alpha,\beta}$ is a seminorm on $\mathcal{S}(\mathbb{R}^d)$. We define a topology on $\mathcal{S}(\mathbb{R}^d)$ by the family of seminorms $\{p_{\alpha,\beta} : \alpha,\ \beta\in\mathbb{Z}_+^d\}$. The fundamental system of neighborhoods of u_0 consists of the family $\{u+u_0 : u\in\mathcal{S}(\mathbb{R}^d),\ p_{\alpha_j,\beta_j}(u)<\varepsilon_j, j=1,2,\dots n\}$, where $\varepsilon_1,\dots,\varepsilon_n$ are any finite positive numbers and $\alpha_j,\beta_j\in\mathbb{Z}_+^d$ (see Proposition 6.2).

EXAMPLE 7.9. The function $u(x)=e^{-|x|^2}$ belongs to $\mathcal{S}(\mathbb{R}^d)$. In fact, for any $\alpha\in\mathbb{Z}_+^d$ we have

$$D^\alpha u(x)=e^{-|x|^2}\sum_{\|\gamma\|\le 2\|\alpha\|}c_{\alpha,\gamma}x^\gamma$$

with some constants $c_{\alpha,\gamma}$. By simple considerations $|x^\gamma|e^{-|x|^2}$ is bounded on $\mathbb{R}^d$ for each $\gamma\in\mathbb{Z}_+^d$. Therefore we have $p_{\alpha,\beta}(u)<\infty$.

THEOREM 7.5. *If $1 \le p \le \infty$, then we have*

$$\mathcal{S}(\mathbb{R}^d) \subset L^p(\mathbb{R}^d).$$

PROOF. Let $u \in \mathcal{S}(\mathbb{R}^d)$. From the inequalities $|u(x)| \le p_{0,0}(u)$ and

$$|(x_1^2 + \cdots + x_d^2)^d u(x)| \le \sum\nolimits_{||\alpha||=d} p_{2\alpha,0}(u),$$

it follows that

$$|u(x)| \le \min(1, |x|^{-2d}) \sum_{||\alpha|| \le 2d} p_{\alpha,0}(u).$$

Thus if we take into account that $\int_{\mathbb{R}^d} \min(1, |x|^{-2d})^p dx = C^p < \infty$, then we obtain

$$\|u\|_p \le C \sum_{||\alpha|| \le 2d} p_{\alpha,0}(u), \tag{7.10}$$

and the proof is complete. □

REMARK 7.1. By (7.10) the embedding $\mathcal{S}(\mathbb{R}^d) \mapsto L^p(\mathbb{R}^d)$ is continuous for every $1 \le p \le \infty$.

For u, $v \in \mathcal{S}(\mathbb{R}^d)$ put

$$\rho(u,v) = \sum_{\alpha,\beta \in \mathbb{Z}_+^d} \frac{1}{2^{||\alpha+\beta||}} \frac{p_{\alpha,\beta}(u-v)}{p_{\alpha,\beta}(u-v)+1}.$$

The the mapping $\rho(\cdot,\cdot)$ is a metric on $\mathcal{S}(\mathbb{R}^d)$. In fact, since $a/(a+1) < 1\,(a \ge 0)$, we have a bound on $\rho(u,v)$ for $u, v \in \mathcal{S}(\mathbb{R}^d)$:

$$\begin{aligned}&\sum_{\alpha,\beta \in \mathbb{Z}_+^d} \frac{1}{2^{||\alpha+\beta||}} \frac{p_{\alpha,\beta}(u-v)}{p_{\alpha,\beta}(u-v)+1}\\ &\le \sum_{\alpha,\beta \in \mathbb{Z}_+^d} \frac{1}{2^{||\alpha+\beta||}} \le \sum_{\alpha,\beta \in \mathbb{Z}_+^d} \frac{1}{2^{\alpha_1+\cdots+\alpha_d+\beta_1+\cdots+\beta_d}} = 2^{2d} < \infty.\end{aligned}$$

Observe that

$$\frac{|a-b|}{|a-b|+1} \le \frac{|a|+|b|}{|a|+|b|+1} \le \frac{|a|}{|a|+1} + \frac{|b|}{|b|+1} \quad (a, b \in \mathbb{R}). \tag{7.11}$$

Then we have $\rho(u,v) \le \rho(u,w) + \rho(w,v)$ for $u, v, w \in \mathcal{S}(\mathbb{R}^d)$. Finally, if $p_{0,0}(u) = 0$, then $u = 0$.

PROPOSITION 7.1. *The topology of $\mathcal{S}(\mathbb{R}^d)$ induced by a family of seminorms $\{p_{\alpha.\beta}\}$ coincides with the topology defined by the metric ρ.*

Thus $\rho(u_n, u) \to 0$ if and only if $p_{\alpha,\beta}(u_n - u) \to 0$. for every α, β.

PROOF. For given $1 > \varepsilon > 0$ and $m \ge 0$ choose $\varepsilon' > 0$ so small that $2^{2m+1}\varepsilon' < \varepsilon$. Then $\rho(u,0) < \varepsilon'$ implies that $p_{\alpha,\beta}(u) < \varepsilon$ for $||\alpha||$, $||\beta|| \le m$.

On the other hand, if m and $\varepsilon'' > 0$ are chosen so that $\sum_{||\alpha||,||\beta|| \ge m} 1/2^{||\alpha||+||\beta||} < \varepsilon/2$ and $\sum_{||\alpha||,||\beta|| \le m} 1/2^{||\alpha||+||\beta||}\varepsilon'' < \varepsilon/2$, then by an elementary computation we have $\rho(u,0) < \varepsilon$ for u such that $p_{\alpha,\beta}(u) < \varepsilon''$ for $||\alpha||$, $||\beta|| \le m$. □

THEOREM 7.6. *$\mathcal{S}(\mathbb{R}^d)$ is a complete metric space.*

PROOF. Let $\{u_j\}$ be a Cauchy sequence in $\mathcal{S}(\mathbb{R}^d)$. Since $\rho(u_i, u_j) \to 0\,(i, j \to \infty)$, we see that

$$p_{\alpha,\beta}(u_i - u_j) \to 0 \quad (i, j \to \infty) \tag{7.12}$$

for every α, β. Thus there exists a constant $C_{\alpha,\beta}$, independent of j, such that $p_{\alpha,\beta}(u_j) \le C_{\alpha,\beta}$. Thus

$$|x^\alpha D^\beta u_j(x)| \le C_{\alpha,\beta}. \tag{7.13}$$

Since

$$|x^\alpha D^\beta (u_i(x) - u_j(x))| \to 0 \text{ (uniformly)}, \tag{7.14}$$

the limit $\lim_{j\to\infty} D^\beta u_j(x) = v^\beta(x)$ exists and v^β is continuous for every $\beta \in \mathbb{Z}_+^d$. Let $u = v^0$.

Now we shall prove that u is infinitely differentiable and $D^\beta u = v^\beta$. Then we have $u \in \mathcal{S}(\mathbb{R}^d)$ by (7.13) and $p_{\alpha,\beta}(u_j - u) \to 0\,(j \to \infty)$ by (7.14). In fact, we have

$$u_j(x_1, x_2, \dots, x_d) = \int_{-\infty}^{x_1} \frac{\partial}{\partial t} u_j(t, x_2, \dots, x_d)dt.$$

By virtue of (7.13) the integrand is by dominated by $C\min(1, |x|^{-2d})$, which is integrable and independent of j. Thus by Lebesgue's convergence theorem

$$u(x_1, x_2, \dots, x_d) = \int_{-\infty}^{x_1} v^{(1,0,\dots,0)}(t, x_2, \dots, x_d)dt.$$

Therefore u is differentiable with respect to x_1 and $\partial u/\partial x_1 = v^{(1,0,\dots,0)}$. Iterating the argument, we get $D^\beta u = v^\beta$. □

DEFINITION 7.3. A continuous linear functional on $\mathcal{S}(\mathbb{R}^d)$ is called a *tempered distribution*, and the collection of all tempered distributions is denoted by $\mathcal{S}'(\mathbb{R}^d)$.

THEOREM 7.7. *A linear functional on $T \in \mathcal{S}(\mathbb{R}^d)$ is a tempered distribution if and only if there exist constants $m > 0$ and $C > 0$ such that*

$$|\langle T, u\rangle| \le C \sum_{\|\alpha\|,\|\beta\|\le m} p_{\alpha,\beta}(u), \quad u \in \mathcal{S}(\mathbb{R}^d). \tag{7.15}$$

PROOF. The "if" part is obvious. To prove the "only if" part, put $p_m(u) = \sum_{\|\alpha\|,\|\beta\|\le m} p_{\alpha,\beta}(u)$.

If (7.15) were not true, for every $m = 1, 2, \dots$ there would exist $u_m \in \mathcal{S}(\mathbb{R}^d)$ such that $|\langle T, u_m\rangle| > m p_m(u_m) > 0$. Put $v_m = [m p_m(u_m)]^{-1} u_m$. Then for $\|\alpha\|$, $\|\beta\| \le m$ we have $p_{\alpha,\beta}(v_m) \le p_m(v_m) = 1/m$. Thus $p_{\alpha,\beta}(v_m) \to 0$ for any $\alpha, \beta \in \mathbb{Z}_+^d$. On the other hand, $|\langle T, v_m\rangle| > 1$, which means that T is not continuous at 0. □

The function $f(x) = e^{|x|^2}$ is locally integrable. Therefore it defines a distribution in $\mathcal{D}'(\mathbb{R}^d)$, but it is not identified with any tempered distribution, because the functional $\langle T_f, u\rangle = \int f u dx$ does not satisfy (7.15).

Any function f in $L^p(\mathbb{R}^d)$ $(1 \le p \le \infty)$ is identified with a tempered distribution T_f by the integral $\langle T_f, u\rangle = \int f u dx$. In fact, as in the proof of Theorem 7.5 we have $\int |fu| dx \le \int \min(1, |x|^{-2d})|f| dx \sum_{\|\alpha\|\le 2d} p_{\alpha,0}(u)$. Thus by Hölder's inequality, we have

$$|\langle T_f, u\rangle| \le C\|f\|_p \sum_{\|\alpha\|\le 2d} p_{\alpha,0}(u), \quad u \in \mathcal{S}(\mathbb{R}^d),$$

where C is a constant independent of u.

7.5. Operations on tempered distributions

The functions in Schwartz space and tempered distributions are closely related with the theory of Fourier analysis, and we will study their properties more closely in the next chapter, where the Fourier transform is discussed.

(I) *Differentiation.* The mapping $\mathcal{S}(\mathbb{R}^d) \ni u \mapsto D^\alpha u \in \mathcal{S}(\mathbb{R}^d)$ is continuous. Thus for every $T \in \mathcal{S}'(\mathbb{R}^d)$ the mapping $\mathcal{S}(\mathbb{R}^d) \ni u \mapsto \langle T, D^\alpha u\rangle$ is continuous. Therefore the derivative $D^\alpha T$ of T is defined by the equation

$$\langle D^\alpha T, u\rangle = (-1)^{\|\alpha\|}\langle T, D^\alpha u\rangle.$$

(II) *Multiplication.* Let $v \in \mathcal{S}(\mathbb{R}^d)$ and $T \in \mathcal{S}'(\mathbb{R}^d)$. Since the mapping $\mathcal{S}(\mathbb{R}^d) \ni u \mapsto vu \in \mathcal{S}(\mathbb{R}^d)$ is continuous, the multiplication vT is defined by $\langle vT, u\rangle = \langle T, vu\rangle$.

Let $P(x)$ be a polynomial. Since $\mathcal{S}(\mathbb{R}^d) \ni u \mapsto Pu \in \mathcal{S}(\mathbb{R}^d)$ is continuous, PT is also defined by $\langle PT, u\rangle = \langle T, Pu\rangle$.

(III) *Convolution.* For $u, v \in \mathcal{S}(\mathbb{R}^d)$ we have

$$x^\alpha D^\beta u * v(x) = x^\alpha D^\beta \int u(x-y)v(y)dy = \int x^\alpha D^\beta u(x-y)v(y)dy.$$

Since $x^\alpha = \sum_{\gamma\le\alpha}\binom{\alpha}{\gamma}(x-y)^\gamma y^{\alpha-\gamma}$, we have

$$|x^\alpha D^\beta u * v(x)| \le \int \sum_{\gamma\le\alpha}\binom{\alpha}{\gamma} p_{\gamma,\beta}(u)|y^{\alpha-\gamma}v(y)|dy.$$

Since each function $y^{\alpha-\gamma}v(y)$ belongs to $\mathcal{S}(\mathbb{R}^d)$, we obtain by (7.10)

$$|x^\alpha D^\beta u * v(x)| \le C\left(\sum_{\gamma\le\alpha}\binom{\alpha}{\gamma} p_{\gamma,\beta}(u)\right)\sum_{\|\delta\|\le 2d+\|\alpha\|} p_{\delta,0}(v).$$

Therefore for any fixed $v \in \mathcal{S}(\mathbb{R}^d)$ the mapping $\mathcal{S}(\mathbb{R}^d) \ni u \mapsto v * u \in \mathcal{S}(\mathbb{R}^d)$ is continuous.

Let $T \in \mathcal{S}'(\mathbb{R}^d)$ and $v \in \mathcal{S}(\mathbb{R}^d)$. Then the above argument shows that the mapping $\mathcal{S}(\mathbb{R}^d) \ni u \mapsto \langle T, v * u\rangle \in \mathbb{C}$ is continuous. Thus the convolution $T * v \in \mathcal{S}'(\mathbb{R}^d)$ is defined by the equation

$$\langle T * v, u\rangle = \langle T, \check{v} * u\rangle,$$

where $\check{v}(x) = v(-x)$.

Finally we mention without proof a remarkable property of the tempered distributions which is an analog to Theorem 7.1.

THEOREM 7.8. *Let $\{T_j\}$ be a sequence in $\mathcal{S}'(\mathbb{R}^d)$. If $\langle T_j, u\rangle$ converges for every $u \in \mathcal{S}(\mathbb{R}^d)$, then there exists $T \in \mathcal{S}'(\mathbb{R}^d)$ such that*

$$\lim_{j\to\infty}\langle T_j, u\rangle = \langle T, u\rangle, \quad u \in \mathcal{S}(\mathbb{R}^d).$$

7.6. Problems

7.1. Let μ be a Borel measure on $\mathbb{R}^d$. Define $\langle T_\mu, u\rangle = \int u d\mu$. Then prove that $T_\mu \in \mathcal{D}'(\mathbb{R}^d)$.

7.2. Let P_t be the Poisson integral. Prove that $\langle P_t, u\rangle \to \langle \delta, u\rangle$ $(t \to 0)$ for every $u \in \mathcal{D}(\mathbb{R}^d)$.

7.3. Let T be a distribution in $T \in \mathcal{D}'(\mathbb{R}^3)$ with compact support. Let $U(x) = (-(4\pi)^{-1}|x|^{-1}) * T$. U is called the *Newton potential.* Prove that $\Delta U = T$.

7.4. Compute the derivatives of

$$\frac{\partial^2}{\partial s \partial t} Y(s)Y(t) \quad \text{and} \quad \frac{d^2 |\sin t|}{dt^2}$$

in the sense of distributions.

7.5. Let $T \in \mathcal{D}'(\mathbb{R}^d)$ and $\varphi \in C^\infty(\mathbb{R}^d)$. Show that

$$D^\alpha(\varphi T) = \sum_{\beta \le \alpha} \binom{\alpha}{\beta} (D^{\alpha-\beta}\varphi)(D^\beta T).$$

7.6. Prove that

$$\left(\frac{d}{dx} - \lambda\right) Y(x)e^{\lambda x} = \delta \quad \text{and} \quad \frac{d^n}{dx^n} \frac{Y(x)x^{n-1}}{(n-1)!} = \delta.$$

CHAPTER 8

Fourier Analysis

The basic idea of Fourier[1] analysisis to decompose a periodic function into simple harmonic motions which are expressed by trigonometric functions, and, conversely, to synthesize a function from trigonometric functions.

The theory has been found to be deeply connected with many fields of mathematics: real analysis, complex analysis, integration theory, theory of partial differential equations, group theory, number theory, wavelet theory, and so on. Special attention should be paid to its contribution to the theory of sets, which was developed by Cantor in connection with the uniqueness problem of trigonometric series.

The problem of expressing a function as a trigonometric series

$$\frac{a_o}{2} + \sum_{n=1}^{\infty}(a_n \cos 2\pi nx + b_n \sin 2\pi nx)$$

was solved rigorously by Riemann when he introduced his integral theory. He defined the coefficients for a function f by the integrals

$$a_n = 2\int_0^1 f(x)\cos 2\pi nx dx,$$

$$b_n = 2\int_0^1 f(x)\sin 2\pi nx dx.$$

When a trigonometric series is given, it naturally raises the following questions: In what sense does the series represent a function f? This problem is intimately related to the theory of the Lebesgue integral, and nowadays Fourier analysis is lucidly described based on the theory of the Lebesgue integral. I mention two examples which show how Fourier analysis is essentially connected with the Lebesgue integral. One of them is the Plancherel theorem (see § 8.3) and the others are the Carleson and Kahane-Katznelson theorems (see § 8.5.3). The former shows that the Lebesgue space L^2 is one of the most reasonable space on which Fourier analysis can be developed. The later implies that the notion of Lebesgue measure is "perfect" for the pointwise (a.e.) convergence problem of Fourier series.

The study of N. Wiener[2] in the 1930s represented important progress in Fourier analysis. Wiener observed that the Lebesgue space $L^1(\mathbb{R})$ is a ring with multiplication defined by convolution, and he shed light on the algebraic aspect of Fourier analysis. In § 8.2 we describe a first step to this theory.

---

[1] J. B. J. Fourier, 1768-1830.

[2] N. Wiener, 1894-1964

8.1. Fourier transform

8.1.1. Fourier transform of a variable. For $f \in L^1(\mathbb{R})$ the *Fourier transform* is defined by

$$\hat{f}(\xi) = \int_{\mathbb{R}} f(x) e^{-2\pi i \xi x} dx, \quad \xi \in \mathbb{R}.$$

$\hat{f}(\xi)$ is often written as $\mathcal{F}f(\xi)$ or $f^{\wedge}(\xi)$.

THEOREM 8.1 (The Riemann-Lebesgue theorem). *If $f \in L^1(\mathbb{R})$, then $\hat{f}(\xi) \to 0$ as $\xi \to \infty$.*

PROOF. For $\varepsilon > 0$ choose a simple function $s(x) = \sum_{j=1}^{k} a_j \chi_{I_j}(x)$ such that $\int |f - s| dx < \varepsilon$, and let the I_j be intervals. For any finite interval $I = (a, b)$ we have

$$\hat{\chi}_I(\xi) = \int_a^b e^{-2\pi i \xi x} dx = \frac{e^{-2\pi i \xi b} - e^{-2\pi i \xi a}}{-2\pi i \xi},$$

from which it follows that $\hat{\chi}_I(\xi) \to 0$ as $\xi \to \infty$, and thus $\hat{s}(\xi) \to 0$ as $\xi \to \infty$.

Since $|\hat{f}(\xi) - \hat{s}(\xi)| \leq \int |f(x) - s(x)| dx < \varepsilon$ for all ξ, it follows that

$$\limsup_{\xi \to \infty} |\hat{f}(\xi)| \leq \varepsilon.$$

Thus $\hat{f}(\xi) \to 0$ as $\xi \to \infty$. □

THEOREM 8.2. *If $f \in L^1(\mathbb{R})$, then $\hat{f} \in C_0(\mathbb{R})$.*

PROOF. We shall prove that $\hat{f}$ is continuous. Then the theorem follows from the Riemann-Lebesgue theorem.

We have

$$(8.1) \qquad \begin{aligned} |\hat{f}(\xi + \eta) - \hat{f}(\xi)| &= \left| \int_{\mathbb{R}} (e^{-2\pi i \eta x} - 1) f(x) e^{-2\pi i \xi x} dx \right| \\ &\leq \int_{\mathbb{R}} |e^{-2\pi i \eta x} - 1| |f(x)| dx. \end{aligned}$$

Since $|e^{-2\pi i \eta x} - 1||f(x)| \leq 2|f(x)| \in L^1(\mathbb{R})$, and since $e^{-2\pi i \eta x} - 1 \to 0$ as $\eta \to 0$ for each x, the right-hand side of (8.1) converges to 0 by the dominated convergence theorem. □

EXAMPLE 8.1 (Dirichlet kernel). Let $\chi_{[-N,N]}$ be the characteristic function of the interval $[-N, N]$. By a simple calculation, we have

$$\hat{\chi}_{[-N,N]}(\xi) = \int_{-N}^{N} e^{-2\pi i \xi x} dx = 2 \int_0^N \cos 2\pi \xi x dx = \frac{\sin 2\pi N \xi}{\pi \xi}.$$

The function

$$(8.2) \qquad D_N(x) = \frac{\sin 2\pi N x}{\pi x}$$

is called the *Dirichret*[3] *kernel* (see Figure 8.1).

[3]P. G. L. Dirichlet, 1805–1859

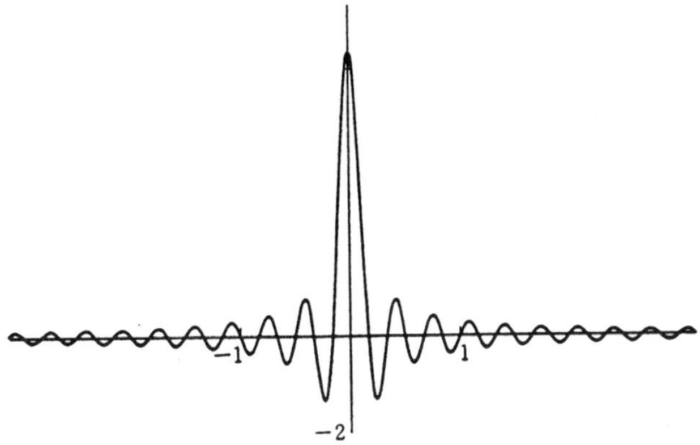

FIGURE 8.1. Dirichlet kernel $D_N(x)$

EXAMPLE 8.2 (Fejér kernel). Let $\Delta_N(\xi) = (1 - |\xi|/N)_+$, $N > 0$, and let F_N be the Fourier transform of Δ_N. F_N is called the *Fejér*[4] *kernel* (see Figure 8.2).

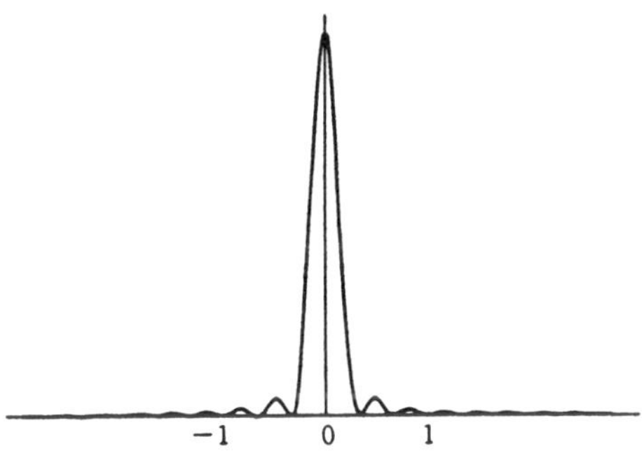

FIGURE 8.2. Fejér kernel $F_N(x)$

The Fejér kernel is explicitly given by

$$F_N(x) = \frac{1}{N}\left(\frac{\sin \pi N x}{\pi x}\right)^2. \tag{8.3}$$

In fact, since the integrand is an even function,

$$F_N(\xi) = \int_{-N}^{N}\left(1 - \frac{|x|}{N}\right)e^{-2\pi i \xi x}dx = 2N\int_0^1 (1-x)\cos 2\pi N\xi x dx.$$

Applying integration by parts, we have

$$\begin{aligned} F_N(\xi) &= 2N\left[(1-x)\frac{\sin 2\pi N\xi x}{2\pi N\xi}\right]_0^1 + \frac{1}{\pi\xi}\int_0^1 \sin 2\pi N\xi x dx \\ &= \frac{2(1-\cos 2\pi N\xi)}{N(2\pi\xi)^2}. \end{aligned}$$

Using the formula $\cos 2\theta = 1 - 2\sin^2\theta$, we get (8.3).

[4]L. Fejér, 1880–1959

EXAMPLE 8.3 (Poisson kernel). The Poisson kernel is given by

$$P_t(x) = \frac{1}{\pi}\frac{t}{x^2+t^2}, \quad t>0$$

(see also Example 6.5). We have

$$P_t(x) = \int_{\mathbb{R}} e^{-2\pi|\xi|t} e^{2\pi i x\xi} d\xi. \tag{8.4}$$

In fact, denote the integral on the right-hand side by I, that is,

$$I(x) = 2\int_0^\infty e^{-2\pi t\xi}\cos 2\pi x\xi d\xi.$$

Integrating by parts twice, we get

$$\begin{aligned} I(x) &= \frac{1}{\pi x}\left[e^{-2\pi t\xi}\sin 2\pi x\xi\right]_0^\infty + \frac{2t}{x}\int_0^\infty e^{-2\pi t\xi}\sin 2\pi x\xi d\xi \\ &= \frac{-t}{\pi x^2}\left[e^{-2\pi t\xi}\cos 2\pi x\xi\right]_0^\infty + \frac{-t^2}{x^2}I(x). \end{aligned}$$

Since the bracketed term is -1, we have $I(x) = t/\pi(x^2+t^2)$.

Observe that

$$\frac{i}{\pi z} = \frac{1}{\pi}\frac{t}{x^2+t^2} + \frac{i}{\pi}\frac{x}{x^2+t^2} \quad (z = x+it),$$

which shows that $P_t(x)$ is the real part of an analytic function in $\Im z > 0$. Its imaginary part

$$Q_t(x) = \frac{1}{\pi}\frac{x}{x^2+t^2}$$

is called the *conjugate Poisson kernel.*

EXAMPLE 8.4 (Gauss-Weierstrass kernel). The Gauss-Weierstrass kernel is given by

$$W_t(x) = \frac{1}{\sqrt{4\pi t}}e^{-x^2/4t}, \quad t>0$$

(see slso Example 6.6). We show that

$$W_t(x) = \int_{\mathbb{R}} e^{-4\pi^2\xi^2 t} e^{2\pi i x\xi} d\xi. \tag{8.5}$$

In fact, expanding $e^{2\pi i\xi x}$ in power series, the right-hand side of (8.5) can be formally expressed as

$$\sum_{n=0}^\infty \frac{1}{n!}\int_{-\infty}^\infty (2\pi i x\xi)^n e^{-4\pi^2\xi^2 t} d\xi = \sum_{n=0}^\infty \frac{1}{(2n)!}\int_{-\infty}^\infty (-4\pi^2 x^2\xi^2)^n e^{-4\pi^2\xi^2 t} d\xi.$$

By formula (B.3) in Appendix we have $\int_{-\infty}^\infty e^{-4\pi^2\xi^2 t} d\xi = (4\pi t)^{-1/2}$. Differentiating both sides n times with respect to t, we have

$$\int_{-\infty}^\infty (-4\pi^2\xi^2)^n e^{-4\pi^2\xi^2 t} d\xi = \frac{1}{\sqrt{4\pi t}}\frac{(2n)!}{2^{2n}n!}\left(\frac{-1}{t}\right)^n,$$

which implies the above formal calculation is legitimate and

$$W_t(x) = \frac{1}{\sqrt{4\pi t}}\sum_{n=0}^\infty \frac{x^{2n}}{2^{2n}n!}\left(\frac{-1}{t}\right)^n = \frac{1}{\sqrt{4\pi t}}e^{-x^2/4t}.$$

8.1.2. Fourier transform of a variable. The formula

$$f(x) = \int_{\mathbb{R}} \widehat{f}(\xi) e^{2\pi i x \xi} d\xi \tag{8.6}$$

is called the *Fourier inversion formula.* The integral does not always converge, and even if it converges at x, the right-hand side is not necessarily equal to $f(x)$. In this section we shall study interpretations of this formal definition. Since the behavior of the integral of (8.6) depends strongly on the dimension, we begin by the one dimensional case.

We call the integral

$$s_N(f)(x) = \int_{-N}^{N} \widehat{f}(\xi) e^{2\pi i x \xi} d\xi$$

the *Nth partial sum* of the Fourier transform of f. By Fubini's theorem

$$s_N(f)(x) = \int_{-N}^{N} \int_{\mathbb{R}} f(y) e^{-2\pi i \xi y} e^{2\pi i x \xi} d\xi dy = \int_{\mathbb{R}} f(y) \frac{\sin 2\pi N(x-y)}{\pi(x-y)} dy.$$

Thus

$$s_N(f)(x) = \int_{\mathbb{R}} f(y) D_N(x-y) dy = f * D_N(x),$$

where D_N is the Nth Dirichlet kernel.

THEOREM 8.3 (Localization theorem of Riemann). *Let $f \in L^1(\mathbb{R})$. Assume that $f = 0$ in an open interval I. Then $s_N(f)(x)$ converges to zero at each point in I, and the convergence is uniform in any compact subinterval K of I.*

From this theorem it follows that if f, $g \in L^1(\mathbb{R})$, and $f = g$ on an open interval (a, b), then $s_N(f)(x) - s_N(g)(x) \to 0$ uniformly on any compact set contained in (a, b).

PROOF. The Nth partial sum of f is written as

$$\begin{aligned} s_N(f)(x) &= \int_{-\infty}^{\infty} \frac{f(x-y)}{\pi y} \sin 2\pi N y dy \\ &= \int_{-\infty}^{\infty} \varphi_x(y) \sin 2\pi N y dy, \end{aligned}$$

where $\varphi_x(y) = f(x-y)/\pi y$. Let $x \in I$ and $\delta = \operatorname{dist}(x, I^c)$. Then, since $f = 0$ in I, $|\varphi_x(y)| \le |f(x-y)/\pi\delta|$ for all y, which means that $\varphi_x(\cdot) \in L^1(\mathbb{R})$. Since $\sin 2\pi N y = [e^{2\pi i N y} - e^{-2\pi i N y}]/2i$, the partial sum $s_N(f)(x)$ is considered as a linear combination of Fourier transforms of integrable functions. Therefore, by the Riemann-Lebesgue theorem it converges to zero as $N \to \infty$.

To prove the uniform convergence, let K be a compact subset of I. Let $\varepsilon > 0$. For each $x \in K$ there exists an integer N_x such that

$$|s_N(f)(x)| < \varepsilon/2 \quad \text{for } N \ge N_x.$$

Since $\operatorname{dist}(K, I^c) > 0$, the mapping $K \ni x \mapsto \varphi_x(\cdot) \in L^1(\mathbb{R})$ is continuous (see Lemma 6.5). Therefore, we can choose $r_x > 0$ so that

$$|s_N(f)(x) - s_N(f)(x')| \le \|\varphi_x - \varphi_{x'}\|_1 < \varepsilon/2 \quad \text{for } x \in I_x = (x - r_x, x + r_x).$$

By compactness of K, there exists a finite covering $\{I_{x_1}, \ldots, I_{x_n}\}$ of K. Put $N = \max\{N_{x_1}, \ldots, N_{x_n}\}$. Then for any $x \in K$ we have

$$|s_N(f)(x)| \leq |s_N(f)(x_j)| + |s_N(f)(x) - s_N(f)(x_j)| < \varepsilon/2 + \varepsilon/2 = \varepsilon,$$

where x_j is chosen so that $x \in I_j$. □

For $\delta > 0$, decompose the Nth partial sum as follows:

$$\begin{aligned} s_N(f)(x) &= \int_0^\infty [f(x-y) + f(x+y)] \frac{\sin 2\pi N y}{\pi y} dy \\ &= \left(\int_0^\delta + \int_\delta^\infty \right) [f(x-y) + f(x+y)] \frac{\sin 2\pi N y}{\pi y} dy. \end{aligned}$$

Since the second term on the right-hand side can be expressed as a linear combination of Fourier transform of integrable functions, it tends to 0 as $N \to \infty$ by the Riemann-Lebesgue theorem. Therefore we have

$$\lim_{N\to\infty} s_N(f)(x) = \lim_{N\to\infty} \int_0^\delta [f(x-y) + f(x+y)] \frac{\sin 2\pi N y}{\pi y} dy, \tag{8.7}$$

for every $\delta > 0$.

LEMMA 8.1. *We have, for any real number* λ,

$$\lim_{a\to\infty} \int_0^a \frac{\sin \lambda y}{y} dy = \frac{\pi}{2} \operatorname{sign} \lambda. \tag{8.8}$$

PROOF. It suffices to prove the case $\lambda > 0$. Furthermore, by a change of variable, we can assume that $\lambda = 1$.

Put $w(x) = e^{-\pi x^2}$. Then we have $\hat{w}(\xi) = e^{-\pi \xi^2}$ by Example 8.4. Thus

$$\lim_{N\to\infty} s_N(w)(0) = \lim_{N\to\infty} \int_{-N}^N \hat{w}(\xi) d\xi = \int_{\mathbb{R}} e^{-\pi\xi^2} d\xi = 1.$$

On the other hand, for any $\delta > 0$ we have, by (8.7),

$$\begin{aligned} \lim_{N\to\infty} s_N(w)(0) &= \lim_{N\to\infty} \int_0^\delta 2e^{-\pi y^2} \frac{\sin 2\pi N y}{\pi y} dy \\ &= \lim_{N\to\infty} \left(\int_0^\delta 2 \frac{e^{-\pi y^2} - 1}{\pi y} \sin 2\pi N y dy + \int_0^\delta \frac{2 \sin 2\pi N y}{\pi y} dy \right) \\ &= I + J, \quad \text{say}. \end{aligned}$$

Since the function $y^{-1}(e^{-\pi y^2} - 1)\chi_{(0,\delta)}(y)$ is integrable on $\mathbb{R}$, I is supposed to be a linear combination of Fourier transforms of integrable functions. Thus $I = 0$ again by the Riemann-Lebesgue theorem. Therefore we get

$$1 = \lim_{N\to\infty} s_N(w)(0) = \lim_{N\to\infty} \int_0^\delta \frac{2 \sin 2\pi N y}{\pi y} dy.$$

Now the formula (8.8) follows by a change of scale. □

For a function f put

$$\psi_x(y) = \frac{1}{2}[f(x+y) + f(x-y)] - f(x).$$

THEOREM 8.4. *Let $f \in L^1(\mathbb{R})$. If*

$$\int_0^\delta \frac{|\psi_x(y)|}{y} dy < \infty \tag{8.9}$$

for some $\delta > 0$, then we have

$$\lim_{N\to\infty} s_N(f)(x) = f(x).$$

PROOF. By (8.7) and (8.8) we have

$$\begin{aligned}
&\lim_{N\to\infty} s_N(f)(x) - f(x) \\
&= \lim_{N\to\infty} \left\{ \int_0^\delta [f(x-y) + f(x+y)] \frac{\sin 2\pi N y}{\pi y} dy - \int_0^\delta f(x) \frac{2\sin 2\pi N y}{\pi y} dy \right\} \\
&= \lim_{N\to\infty} \int_0^\delta \frac{\psi_x(y)}{\pi y} 2 \sin \pi N y dy.
\end{aligned}$$

Since the function $y^{-1}\psi_x(y)\chi_{(0,\delta)}(y)$ is integrable by assumption, the last term vanishes. □

Condition (8.9) is called the *Dini condition*. The Dini condition at a point x does not necessarily imply that a function is continuous at x.

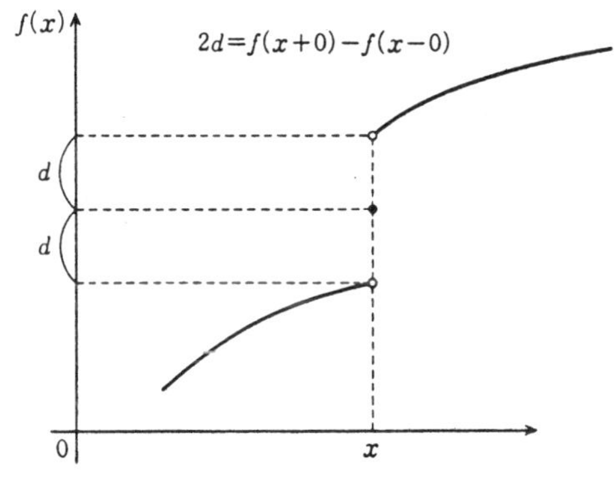

FIGURE 8.3. Convergence at a point x

For example, if

$$f(x) = \frac{1}{2}[f(x+0) + f(x-0)],$$

and $\psi_x(\cdot)$ is smooth, then $s_N(f)(x) \to f(x)$ $(N \to \infty)$ (see Figure 8.3).

If $f(x)$ satisfies the Lipschitz condition of order α, $0 < \alpha \leq 1$, then it satisfies the Dini condition. In fact, there exists a constant $C > 0$ such that $|\psi_x(y)| \leq C|y|^\alpha$ $(|y| < \delta)$ with a constant C independent of y, and thus $\int_0^\delta |\psi_x(y)||y|^{-1} dy < \infty$.

8.1.3. Fourier transform of several variables. Let $f \in L^1(\mathbb{R}^d)$. The Fourier transform of f is, by definition,

$$\hat{f}(\xi) = \int_{\mathbb{R}^d} f(x) e^{-2\pi i \xi x} dx \quad (\xi \in \mathbb{R}^d),$$

where $\xi x = \xi_1 x + \cdots + \xi_d x$.

THEOREM 8.5. *Let $f \in L^1(\mathbb{R}^d)$. Then we have*

(i) $(\widehat{\tau_y f})(\xi) = \widehat{f}(\xi) e^{-2\pi i y\xi} \quad (y, \xi \in \mathbb{R}^d)$.

(ii) $(\widehat{f e^{-2\pi i \eta \cdot}})(\xi) = \tau_{-\eta}\widehat{f}(\xi) \quad (\eta \in \mathbb{R}^d)$.

(iii) *If* $\widetilde{f}(x) = \overline{f(-x)}$, *then* $\widehat{\widetilde{f}}(\xi) = \overline{\widehat{f}(\xi)}$.

(iv) *If* $f_t(x) = t^{-d} f(t^{-1}x)$, *then* $\widehat{f_t}(\xi) = \widehat{f}(t\xi) \quad (t > 0)$.

PROOF. (i) By definition

$$(\widehat{\tau_y f})(\xi) = \int_{\mathbb{R}^d} f(x-y) e^{-2\pi i \xi x} dx.$$

(i) is an easy consequence of a change of variable. To prove (iii) observe that

$$\overline{\widehat{f}(\xi)} = \int_{\mathbb{R}^d} \overline{f(x) e^{-2\pi i \xi x}} dx = \int_{\mathbb{R}^d} \overline{f(x)} e^{2\pi i \xi x} dx.$$

Substituting $-x$ for x, we get (iii). The proof of (ii) and (iv) are similar and left to the reader. □

If a function $f(x)$ is written as a product $f(x) = f_1(x_1) \ldots f_d(x_d)$ of functions of a variable, then the Fourier transform of f is again of product type:

$$\widehat{f}(\xi) = \int_{-\infty}^{\infty} \cdots \int_{-\infty}^{\infty} \prod_{j=1}^{d} f_j(x_j) e^{-2\pi \xi_j x_j} dx_1 \ldots dx_d = \widehat{f_1}(\xi_1) \ldots \widehat{f_d}(\xi_d).$$

A function f is called *radial* if $f(x)$ is independent of the angle of $x \in \mathbb{R}^d$, that is, depends only on the distance $|x|$.

Let $SO(d)$ be the special orthogonal group, that is, the set of all $d \times d$ orthogonal matrices A with det $A = 1$. If $A \in SO(d)$, then $A\xi \cdot x = \xi \cdot A^{-1}x$ for $x, \xi \in \mathbb{R}^d$. In particular, we have $|Ax| = |x|$. Note that for $x, y \in S^{d-1}$ there exists $A \in SO(d)$ such that $y = Ax$.

A function f is radial if and only if $f(x) = f(Ax)$ for all $x \in \mathbb{R}^d$ and $A \in SO(d)$. For $f \in L^1(\mathbb{R}^d)$ and $A \in SO(d)$ we have

$$\widehat{f}(A\xi) = \int_{\mathbb{R}^d} f(x) e^{-2\pi i A\xi \cdot x} dx = \int_{\mathbb{R}^d} f(x) e^{-2\pi i \xi \cdot A^{-1}x} dx.$$

Suppose that the function f is radial. Since det$A = 1$, the last integral equals $\widehat{f}(\xi)$. This implies that the Fourier transform of a radial function is also radial.

THEOREM 8.6. *Let $f \in L^1(\mathbb{R}^d)$.*

(i) *If $f(x)$ is written as a product of functions of a variable, then $\widehat{f}(\xi)$ is the product of Fourier transforms of each function of a variable.*

(ii) *If $f(x)$ is a radial function, then $\widehat{f}(\xi)$ is also radial.*

More detailed discussion on the Fourier transform of radial functions will be given in § 8.3.5 and Appendix B.4.

EXAMPLE 8.5 (Dirichlet kernel). The characteristic function of the cube $[-N, N] \times \cdots \times [-N, N]$ of $\mathbb{R}^d$ can be written as the product $\prod_{j=1}^{d} \chi_{[-N,N]}(x_j)$. Thus the Fourier transform D_N, which we call the Dirichlet kernel for the square sum, is given by

$$D_N(x) = \prod_{j=1}^{d} \frac{\sin 2\pi N x_j}{\pi x_j}.$$

EXAMPLE 8.6 (The Gauss-Weierstrass kernel). The Gauss-Weierstrass kernel is defined by

$$W_t(x) = (\sqrt{4\pi t}\,)^{-d} e^{-|x|^2/4t} = \prod_{j=1}^{d} (\sqrt{4\pi t}\,)^{-1} e^{-x_j^2/4t}$$

for $x = (x_1, \dots, x_d) \in \mathbb{R}^d$ and $t > 0$ (see § 6.5(b)). The kernel W_t is a product of one dimensional kernels and also a radial function. Thus the Fourier transform of the Gauss-Weierstrass kernel is of product type, and it is also radial. By the formula for the one dimensional Gauss-Weierstrass kernel, we have

$$W_t(x) = \prod_{j=1}^{d} \int_{\mathbb{R}} e^{-4\pi^2 \xi_j^2 t} e^{2\pi i x_j \xi_j} d\xi_j = \int_{\mathbb{R}^d} e^{-4\pi^2 |\xi|^2 t} e^{2\pi i x \xi} d\xi. \tag{8.10}$$

By a change of variable we get from (8.10)

$$\widehat{W_t}(\xi) = (\sqrt{4\pi t}\,)^{-d} \int_{\mathbb{R}^d} e^{-|x|^2/4t} e^{-2\pi i \xi x} dx = e^{-4\pi^2 |\xi|^2 t}.$$

EXAMPLE 8.7 (The Poisson kernel). The Poisson kernel $P_t(x)$ on $\mathbb{R}^{d+1}_+$ is defined by

$$P_t(x) = c_d \frac{t}{(|x|^2 + t^2)^{(d+1)/2}},$$

where $c_d = \Gamma\left(\frac{d+1}{2}\right)/\pi^{(d+1)/2}$ (see § 6.5(b)).

We shall prove that

$$P_t(x) = \int_{\mathbb{R}^d} e^{-2\pi |\xi| t} e^{2\pi i x \xi} d\xi. \tag{8.11}$$

First we observe that

$$\sqrt{\frac{\pi}{4}} = \int_0^\infty e^{-(s-\alpha/s)^2} ds \quad (\alpha \in \mathbb{R}). \tag{8.12}$$

In fact, differentiating the right-hand side with respect to α, we get

$$-2 \int_0^\infty \frac{1}{s} \left(\frac{\alpha}{s} - s \right) e^{-(s-\alpha/s)^2} ds,$$

which vanishes, because by a change of variable $t = \alpha/s$ we have

$$\int_0^\infty \frac{\alpha}{s^2} e^{-(s-\alpha/s)^2} ds = \int_0^\infty e^{-(\alpha/t - t)^2} dt.$$

Thus the integral on the right-hand side of (8.12) is independent of α. In particular, when $\alpha = 0$, it equals $\int_0^\infty e^{-s^2} ds = \sqrt{\pi/4}$, which implies (8.12).

It follows from (8.12) that

$$e^{-2\alpha} = \sqrt{\frac{4}{\pi}} \int_0^\infty e^{-s^2 - \alpha^2/s^2} ds. \tag{8.13}$$

Applying (8.13) with $\alpha = \pi|\xi|t$, we get

$$\begin{aligned} \int_{\mathbb{R}^d} e^{-2\pi|\xi|t} e^{2\pi i x \xi} d\xi &= \int_{\mathbb{R}^d} e^{2\pi i x \xi} d\xi \sqrt{\frac{4}{\pi}} \int_0^\infty e^{-s^2 - \pi^2 |\xi|^2 t^2 / s^2} ds \\ &= \sqrt{\frac{4}{\pi}} \int_0^\infty e^{-s^2} ds \int_{\mathbb{R}^d} e^{2\pi i x \xi} e^{-\pi^2 |\xi|^2 t^2 / s^2} d\xi. \end{aligned}$$

By a formula for the Fourier transform of the Gauss-Weierstrass kernel, the last term becomes

$$\sqrt{\frac{4}{\pi}}\int_0^\infty e^{-s^2}\left(\frac{1}{\sqrt{\pi t^2/s^2}}\right)^d e^{-|x|^2 s^2/t^2}ds,$$

which is, by a change of variable $u = s^2(1+|x|^2/t^2)$ and by a simple calculation, equal to

$$\frac{1}{\sqrt{\pi}^{d+1}}\frac{1}{t^d}\frac{1}{(1+|x|^2/t^2)^{(d+1)/2}}\int_0^\infty u^{(d-1)/2}e^{-u}du = \frac{c_d t}{(|x|^2+t^2)^{(d+1)/2}}.$$

Thus the formula (8.11) is obtained.

8.1.4. Summability of the Fourier transform. Let k_t be a summability kernel, which implies that k_t satisfies the conditions (S1), (S2) and (S3) in § 6.6. As we have shown,

$$\|f * k_t - f\|_p \to 0 \quad (t \to 0)$$

for $f \in L^p(\mathbb{R}^d)$ $(1 \le p < \infty)$.

Now we consider the following conditions on k_t:

(S1) $\int_{\mathbb{R}^d} k_t(x)dx = 1 \quad (x \in \mathbb{R}^d,\ t > 0)$,

(S2)′ $|k_t(x)| \le ct^{-d}$,

(S3)′ $|k_t(x)| \le ct^\eta/|x|^{d+\eta}$, where c and η are positive constants not depending on t and x.

PROPOSITION 8.1. *If the kernel k_t $(t > 0)$ satisies the conditions* (S1), (S2)′, *and* (S3)′, *then it is a summability kernel.*

PROOF. To estimate the integral $\int_{\mathbb{R}^d}|k_t(x)|dx$ we divide the integral domin $\mathbb{R}^d$ into $|x| < t$ and $|x| \ge t$, and then apply condition (S2)′ for the integral on the former domain and (S3)′ for the rest. Then the integral is bounded by

$$\int_{|x|<t} ct^{-d}dx + \int_{|x|\ge t}\frac{ct^\eta}{|x|^{d+\eta}}dx.$$

The first term equals $c\Omega_d$, where Ω_d is the volume of the unit sphere and the second term equals $c\,\omega_{d-1}\int_t^\infty t^\eta/r^{(\eta+1)}dr = c\,\omega_{d-1}/\eta$. Thus $\|k_t\|_1 (t > 0)$ is bounded by a constant, which implies condition (S2).

Condition (S3), that $\int_{|x|>\delta}|k_t(x)|dx \to 0\ (t \to 0)$, follows easily from (S3)′. □

For a locally integrable function f put

$$\Phi(r) = \int_{|x|<r}|f(x-y) - f(x)|dy.$$

When $\Phi(r)r^{-d} \to 0\ (r \to 0)$, we denote

$$\Phi(r) = o(r^d) \quad (r \to 0). \tag{8.14}$$

Condition (8.14) holds, by definition, at Lebesgue points of f, and almost every point is a Lebesgue point (see §4.2).

THEOREM 8.7. *Suppose that k_t satisfies conditions* (S1), (S2)′, *and* (S3)′. *If $f(x)/(|x|+1)^{d+\eta} \in L^1(\mathbb{R}^d)$, then we have*

$$k_t * f(x) \to f(x) \quad (t \to 0)$$

at Lebesgue points of f. In particular, it holds almost everywhere.

PROOF. By condition (S1) we have

$$k_t * f(x) - f(x) = \int_{\mathbb{R}^d} k_t(y)[f(x-y) - f(x)]dy.$$

Let x be a Lebesgue point of f. For a given $\varepsilon > 0$ choose $\delta > 0$ so that $\Phi(r) < \varepsilon r^d$ for $r < \delta/2$. Divide the last integral into three parts as follows:

$$\left(\int_{|y|<t} + \int_{t<|y|<\delta} + \int_{\delta<|y|}\right) k_t(y)[f(x-y) - f(x)]dy. \tag{8.15}$$

By condition (S2)′ the first term on the right-hand side of (8.15) is bounded by $\left|\int_{|y|<t}\right| \le ct^{-d}\Phi(t) < c\varepsilon\ (t < \delta)$. By condition (S3)′ we have

$$\left|\int_{\delta<|y|}\right| \le ct^\eta \int_{\delta<|y|} \frac{|f(x-y)| + |f(x)|}{|y|^{d+\eta}} dy.$$

Since $1/|y|^{d+\eta} \le \delta^{-d-\eta}$ for $|y| > \delta$, and since $1/|y|^{d+\eta}$ is integrable in $|y| > \delta$, the integral on the right-hand side is finite by the assumption. Thus $\left|\int_{\delta<|y|}\right| \to 0$ as $t \to 0$.

Again applying (S3)′ we have

$$\left|\int_{t<|y|<\delta}\right| \le ct^\eta \int_{t<|y|<\delta} \frac{|f(x-y) - f(y)|}{|y|^{d+\eta}} dy. \tag{8.16}$$

Notice that $|y|^{-d-\eta} < (t2^n)^{-\eta-d}$ when $t2^n < |y| \le t2^{n+1}$. Then we have

$$t^\eta \int_{t2^n<|y|\le t2^{n+1}} \frac{|f(x-y) - f(y)|}{|y|^{d+\eta}} dy \le t^\eta (t2^n)^{-d-\eta}\Phi(t2^{n+1}) < \varepsilon 2^{-\eta n}.$$

Let N be an integer such that $t2^N < \delta \le t2^{N+1}$. Then the integral domain $t < |y| < \delta$ is covered by $\{t2^n < |y| \le t2^{n+1}\}$, $n = 0, 1, \ldots, N$. Therefore the left-hand side of (8.16) is bounded by

$$c\varepsilon \sum_{n=0}^{N} 2^{-\eta n} < \left(\frac{c}{1 - 2^{-\eta}}\right)\varepsilon.$$

Since $\varepsilon > 0$ is arbitrarily chosen, the proof is complete. □

THEOREM 8.8. *Let $f \in L^p(\mathbb{R}^d)$, $1 \le p \le \infty$.*

(i) *If $p \ne \infty$, then $P_t * f$ and $W_t * f$ converge in L^p to f.*

(ii) *$P_t * f(x)$ and $W_t * f(x)$ converge to $f(x)$ at every Lebesgue point.*

PROOF. (i) is given in Theorem 6.13. We shall show that the Poisson and Gauss-Weierstrass kernels satisfy conditions (S1), (S2)′ and (S3)′. Then (ii) follows from Theorem 8.7.

The Poisson kernel $P_t(x)$ satisfies condition (S1) (see (6.19)). Obviously, we have that $P_t(x) \le c_d t^{-d}$ and $P_t(x) \le c_d t|x|^{-d-1}$. Thus condition (S2)′ holds and (S3)′ with $\eta = 1$.

For the Gauss-Weierstrass kernel $W_t(x) = (2\sqrt{\pi t}\,)^{-d} e^{-|x|^2/4t}$ we have

$$W_t(x) \le (2\sqrt{\pi})^{-d} t^{-d/2} \quad \text{for all } t > 0, \tag{8.17}$$

which implies that (S2)′ is satisfied when t is replaced by $t^{1/2}$.

In particular, if $|x|^2 \le 4dt$, then we have $W_t(x) \le (2\sqrt{\pi})^{-d}t^{-d/2}(4dt|x|^{-2})^d = (2/\sqrt{\pi})^d t^{d/2}|x|^{-2d}$. On the other hand, since $e^{-a} \le a^{-d}$ $(a > d)$, we have, for $|x|^2 > 4dt$, $W_t(x) \le (2\sqrt{\pi t})^{-d}(|x|^2/4t)^{-d} = (2^{-1}\sqrt{\pi})^{-d}t^{d/2}|x|^{-2d}$. Therefore

$$W_t(x) \le (2/\sqrt{\pi})^d t^{d/2}|x|^{-2d} \tag{8.18}$$

for all $x \in \mathbb{R}^d$ and $t > 0$, which shows that (S3)$'$ is satisfied with $\eta = d$. □

8.2. Convolution and Fourier transform

THEOREM 8.9. *For $f, g \in L^1(\mathbb{R}^d)$ we have*

$$\widehat{(f * g)}(\xi) = \widehat{f}(\xi)\widehat{g}(\xi).$$

PROOF. By definition

$$\widehat{(f * g)}(\xi) = \int_{\mathbb{R}^d}\left(\int_{\mathbb{R}^d} \tau_y f(x)g(y)dy\right)e^{-2\pi i\xi x}dx.$$

If we change the order of integration and apply Theorem 8.5(i), the right-hand side becomes

$$\int_{\mathbb{R}^d} \widehat{f}(\xi)g(y)e^{-2\pi i\xi y}dy = \widehat{f}(\xi)\widehat{g}(\xi).$$

□

As we have shown in § 6.5, the Lebesgue space $L^1(\mathbb{R}^d)$ is a commutative Banach algebra with the product defined by convolution, and the space $C_0(\mathbb{R}^d)$ is also a Banach algebra with pointwise multiplication.

THEOREM 8.10. *The Fourier transform $\mathcal{F}$ is a homomorphism of the Banach algebra $L^1(\mathbb{R}^d)$ into the algebra $C_0(\mathbb{R}^d)$ with norm one.*

PROOF. The mapping $\mathcal{F}$ of $L^1(\mathbb{R}^d)$ to $C_0(\mathbb{R}^d)$ is linear, and $\mathcal{F}(f*g) = \mathcal{F}(f)\mathcal{F}(g)$ by Theorem 8.9. By the inequality

$$|\widehat{f}(\xi)| \le \int_{\mathbb{R}^d} |f(x)e^{-2\pi i\xi x}|dx = \|f\|_1$$

the norm of $\mathcal{F}$ does not exceed 1. On the other hand, for the Gauss-Weierstrass kernel we have

$$1 = \|W_t\|_1 = \int_{\mathbb{R}^d} W_t(x)dx = \mathcal{F}W_t(0) \le \|\mathcal{F}W_t\|_\infty.$$

Thus the norm of $\mathcal{F}$ is not less than 1. Thus the norm of $\mathcal{F}$ equals 1. □

The Fourier transform $\mathcal{F}$ is a one-to-one mapping from L^1 to C_0, which will be shown in Corollary 8.1 below (see also Remark 8.1), but is not surjective. In fact, let

$$\varphi(\xi) = \begin{cases} \xi & (|\xi| < 1), \\ \frac{\operatorname{sign}\xi}{1+\log|\xi|} & (|\xi| \ge 1). \end{cases}$$

Then clearly $\varphi \in C_0(\mathbb{R})$. Suppose that $\varphi = \widehat{f}$ for some fuction f in $L^1(\mathbb{R})$. Then

$$\begin{aligned}\int_{1<|\xi|<a} \frac{\varphi(\xi)}{\xi}d\xi &= \int_{1<|\xi|<a} \frac{d\xi}{\xi}\int_{-\infty}^{\infty} f(x)e^{-2\pi i\xi x}dx \\ &= \int_{-\infty}^{\infty} f(x)dx\int_{1<|\xi|<a} e^{-2\pi i\xi x}\frac{d\xi}{\xi}.\end{aligned}$$

The inner integral, which equals $-2i\int_1^a \frac{\sin 2\pi\xi x}{\xi}d\xi$, is uniformly bounded in x. Thus the right-hand side is bounded in a. On the other hand, the left-hand side equals $2\int_1^a[\xi(1+\log\xi)]^{-1}d\xi$, which tends to infinity as $a \to \infty$.

The converse of Theorem 8.10 is also true.

THEOREM 8.11. *Let Φ be a non-zero bounded linear functional on $L^1(\mathbb{R}^d)$. Suppose that*

$$\Phi(f*g) = \Phi(f)\Phi(g) \quad \textit{for } f,\, g \textit{ in } L^1(\mathbb{R}^d),$$

that is, Φ is a homomorphism. Then there exists a unique $\xi \in \mathbb{R}^d$ such that

$$\Phi(f) = \mathcal{F}f(\xi), \quad f \in L^1(\mathbb{R}^d).$$

PROOF. First Step. Since Φ is a bounded linear functional, by the Riesz representation theorem there exists a unique nonzero function φ in $L^\infty(\mathbb{R}^d)$ such that

$$\Phi(f) = \int_{\mathbb{R}^d} f(x)\varphi(x)dx, \quad f \in L^1(\mathbb{R}^d).$$

Thus the assumption $\Phi(f*g) = \Phi(f)\Phi(g)$ can be written as

$$\int\left(\int f(x-y)g(y)dy\right)\varphi(x)dx = \int f(x)\varphi(x)dx\int g(y)\varphi(y)dy,$$

and therefore

$$(8.19) \qquad \int_{\mathbb{R}^d}\int_{\mathbb{R}^d}[\varphi(x+y)-\varphi(x)\varphi(y)]f(x)g(y)dxdy = 0$$

for all $f,\, g \in L^1(\mathbb{R}^d)$. In particular, (8.19) holds for any characteristic functions $f(x)$ and $g(y)$. Thus for the characteristic function $F(x,y)$ of any cube in $\mathbb{R}^d\times\mathbb{R}^d = \mathbb{R}^{2d}$ we have

$$(8.20) \qquad \int_{\mathbb{R}^{2d}}[\varphi(x+y)-\varphi(x)\varphi(y)]F(x,y)dxdy = 0.$$

This implies that (8.20) holds for characteristic function F of any ball in $\mathbb{R}^{2d}$. Thus by Lebesgue's differentiation theorem we get

$$(8.21) \qquad \varphi(x+y)-\varphi(x)\varphi(y) = 0 \quad \text{for a.e. } (x,y) \text{ in } \mathbb{R}^{2d}.$$

Second Step. *The function φ is differentiable.* Since φ is a nonzero bounded function, we can choose an interval $I = [a_1,b_1]\times\cdots\times[a_d,b_d]$ so that $\int_I\varphi(y)dy \neq 0$. By (8.21) we have

$$(8.22) \qquad \varphi(x)\int_I\varphi(y)dy = \int_I\varphi(x+y)dy = \int_{x+I}\varphi(y)dy \quad \text{a.e.}$$

The last integral is a continuous function of x, so the function $\varphi(x)$ is identified with a continuous function, too. Again (8.22) implies that the right-hand side is a $C^{(1)}$-function of x. Thus $\varphi \in C^{(1)}(\mathbb{R}^d)$. In particular, since φ is continuous, (8.21) holds for all x, y.

Third Step. *There exists a unique $\xi \in \mathbb{R}^d$ such that $\varphi(x) = e^{-2\pi i\xi x}$.* Put $\varphi_j(t) = \varphi(0,\ldots,0,t^{(j\text{th coord.})},0,\ldots,0)$. Since $x = \sum_{j=1}^d(0,\ldots,0,x_j,0,\ldots,0)$, we have, by (8.21), $\varphi(x) = \varphi_1(x_1)\ldots\varphi_d(x_d)$. Furthermore,we have

$$\varphi_j(s+t) = \varphi_j(s)\varphi_j(t) \quad (s,t\in\mathbb{R}).$$

Thus we get a differntaial equation

$$\frac{d\varphi_j(s)}{ds} = \varphi_j'(0)\varphi_j(s),$$

whose solution is given by $\varphi_j(s) = A_j e^{a_j s}$, where $a_j = \varphi_j'(0)$. Since $\varphi_j(s)$ is bounded, a_j must be a pure imaginary number, which we write as $-2\pi i\xi_j$. From the equation $\varphi_j(s+t) = \varphi_j(s)\varphi_j(t)$, it follows that $A_j = 1$. Thus the solution of (8.21) is given by

$$\varphi(x) = \prod_{j=1}^{d} \varphi_j(x_j) = e^{-2\pi i(\xi_1 x_1 + \cdots + \xi_d x_d)} = e^{-2\pi i \xi x}.$$

Therefore we have

$$\Phi(f) = \int_{\mathbb{R}^d} f(x) e^{-2\pi i \xi x} dx.$$

□

Let $\Delta(L^1)$ be the set of all nonzero bounded[5] homomorphisms of $L^1(\mathbb{R}^d)$. By Theorems 8.10 and 8.11 there exists a one-to-one correspondence between $\Delta(L^1)$ and $\mathbb{R}^d$, and the correspondence is given by the Fourier transform $\Phi(u) = \mathcal{F}u(\xi)$.

8.3. Fourier inversion formula and the Plancherel theorem

8.3.1. Fourier inversion formula. We begin with a simple but useful observation.

LEMMA 8.2. *We have, for f, $G \in L^1(\mathbb{R}^d)$,*

$$\int_{\mathbb{R}^d} f(y)\widehat{G}(y)dy = \int_{\mathbb{R}^d} \widehat{f}(\xi)G(\xi)d\xi.$$

PROOF. By Fubini's theorem,

$$\int_{\mathbb{R}^d} f(y)\left(\int_{\mathbb{R}^d} G(\xi)e^{-2\pi i y\xi}d\xi\right)dy = \int_{\mathbb{R}^d} G(\xi)\left(\int_{\mathbb{R}^d} f(y)e^{-2\pi i y\xi}dy\right)d\xi,$$

from which the lemma follows. □

We apply Lemma 8.2 to the functions $f(x-y)$ and $G(\xi) = e^{-4\pi^2|\xi|^2 t}$ $(t > 0)$. Snce $\widehat{f}(x - \cdot)(\xi) = \widehat{f}(-\xi)e^{-2\pi i \xi x}$,

$$\int_{\mathbb{R}^d} f(x-y)\widehat{G}(y)dy = \int_{\mathbb{R}^d} \widehat{f}(\xi)e^{-4\pi^2|\xi|^2 t}e^{2\pi i \xi x}d\xi.$$

By formula (8.10) we have $\widehat{G}(y) = (2\sqrt{\pi t})^{-d}e^{-|y|^2/4t} = W_t(y)$. Therefore we get the formula

$$f * W_t(x) = \int_{\mathbb{R}^d} \widehat{f}(\xi)e^{-4\pi^2|\xi|^2 t}e^{2\pi i \xi x}d\xi. \tag{8.23}$$

[5] The condition "bounded" is superfluous. In fact every homomorphism on $L^1(R^d)$ is bounded (cf. [Ka]).

THEOREM 8.12. *If both a function f and its Fourier transform $\widehat{f}$ are integrable on $\mathbb{R}^d$, then the Fourier inversion formula*

$$f(x) = \int_{\mathbb{R}^d} \widehat{f}(\xi) e^{2\pi i x\xi} d\xi \tag{8.24}$$

holds almost everywhere. In particular, (8.24) *holds at any point of continuity of f.*

PROOF. Let $t \to 0$ in the equation (8.23). The right-hand side converges, by the dominated convergence theorem, to

$$\int_{\mathbb{R}^d} \widehat{f}(\xi) e^{2\pi i x\xi} d\xi.$$

On the other hand, $f * W_t(x)$ converges to $f(x)$ at any Lebesgue point. Thus (8.24) is obtained. □

EXAMPLE 8.8. Let $t > 0$. The Poisson kernel is an integrable function and is the Fourier transform of $e^{-2\pi|\xi|t}$, which is integrable, too. Therefore by Theorem 8.12 we have

$$\widehat{P}_t(\xi) = e^{-2\pi|\xi|t}.$$

DEFINITION 8.1. The operator $\mathcal{F}^{-1}$ defined by

$$\mathcal{F}^{-1}F(x) = \int_{\mathbb{R}^d} F(\xi) e^{2\pi i x\xi} d\xi$$

is called the *inverse Fourier transform*.

We have $\mathcal{F}^{-1}F = \mathcal{F}\check{F}$, where $\check{F}(\xi) = F(-\xi)$.

Theorem 8.12 means that, if both functions f and $\widehat{f}$ are integrable, then

$$\mathcal{F}^{-1}\mathcal{F}f = f \text{ a.e.}$$

COROLLARY 8.1. *Let $f, g \in L^1(\mathbb{R}^d)$. If $\widehat{f} = \widehat{g}$, then $f = g$ a.e.*

PROOF. Put $h = f - g$. Then the Fourier transform of h is identically zero, which is of course integrable. Thus, by Theorem 8.12, we have $h = 0$ a.e. □

REMARK 8.1. The homomorphism $\mathcal{F} : L^1(\mathbb{R}^d) \mapsto C_0(\mathbb{R}^d)$ is a one-to-one mapping by Corollary 8.1.

8.3.2. Fourier transform in the Schwartz space.

LEMMA 8.3. *Let u be a function on the real line.*

(i) *If u and the derivative u' are both integrable, then*

$$\left(\frac{du}{dx}\right)^{\wedge}(\xi) = 2\pi i\xi\widehat{u}(\xi).$$

(ii) *If $u(x)$ and $xu(x)$ are integrable, then the Fourier transform $\widehat{u}$ is differentiable, and*

$$\frac{d}{d\xi}\widehat{u}(\xi) = [-2\pi i x u(x)]^{\wedge}(\xi).$$

PROOF. (i) First remark that $u(x) \to 0\,(x \to \pm\infty)$, and apply integration by parts. Then we get

$$\left(\frac{du}{dx}\right)^{\wedge}(\xi) = \int_{\mathbb{R}} u'(x)e^{-2\pi i\xi x}dx = 2\pi i\xi \int_{\mathbb{R}} u(x)e^{-2\pi i\xi x}dx = 2\pi i\xi\widehat{u}(\xi),$$

which proves (i).

(ii) We have

$$\frac{\widehat{u}(\xi+\eta)-\widehat{u}(\xi)}{\eta} = \int_{\mathbb{R}} u(x)\left[\frac{e^{-2\pi i\eta x}-1}{\eta}\right]e^{-2\pi i\xi x}dx. \tag{8.25}$$

The integrand on the right-hand side equals $-2\pi ix[\sin\pi x\eta/\pi\eta x]e^{-\pi i\eta x}u(x)e^{-2\pi i\xi x}$, which converges to $-2\pi ixu(x)e^{-2\pi i\xi x}$as $\eta \to 0$ for $x \neq 0$, and which is, in absolute value, bounded by

$$\left|2\pi x\frac{\sin\pi x\eta}{\pi\eta x}u(x)\right| \leq |2\pi xu(x)|.$$

Thus, by the dominated convergence theorem, (ii) follows from (8.25). □

For a polynomial $p(x) = \sum_\alpha a_\alpha x^\alpha$ of $x \in \mathbb{R}^d$, let $p(D) = \sum_\alpha a_\alpha D^\alpha$ be a differential operator. If $u \in \mathcal{S}(\mathbb{R}^d)$, then for any polynomial p the functions $p(x)u(x)$ and $p(D)u(x)$ belong to $\mathcal{S}(\mathbb{R}^d)$ and thus to $L^1(\mathbb{R}^d)$.

THEOREM 8.13. *Let $u \in \mathcal{S}(\mathbb{R}^d)$, and let p be a polynomial. Then we have*

(i) $[p(D)u]^\wedge(\xi) = p(2\pi i\xi)\widehat{u}(\xi)$,
(ii) $p(D)\widehat{u}(\xi) = [p(-2\pi ix)u(x)]^\wedge(\xi)$.

PROOF. It is enough to prove the theorem for a monomial $p(x) = x^\alpha$, in which case the theorem follows by repeated application of Lemma 8.3. □

THEOREM 8.14. *The mapping $\mathcal{F} : \mathcal{S}(\mathbb{R}^d) \mapsto \mathcal{S}(\mathbb{R}^d)$ is a continuous bijection.*

PROOF. Let $u \in \mathcal{S}(\mathbb{R}^d)$ and $\alpha, \beta \in \mathbb{Z}^d_+$. By Theorem 8.13 we have

$$(2\pi i\xi)^\alpha D^\beta\widehat{u}(\xi) = [D^\alpha((-2\pi ix)^\beta u(x))]^\wedge(\xi).$$

Since $D^\alpha(x^\beta u(x)) \in \mathcal{S}(\mathbb{R}^d)$,

$$p_{\alpha,\beta}(\widehat{u}) = \sup_\xi |\xi^\alpha D^\beta\widehat{u}(\xi)| \leq (2\pi)^{\|\beta\|-\|\alpha\|}\|D^\alpha(x^\beta u(x))\|_1 < \infty.$$

Thus $\widehat{u} \in \mathcal{S}(\mathbb{R}^d)$.

The continuity of the mapping $\mathcal{F}$ follows from the fact that the mapping $\mathcal{S}(\mathbb{R}^d) \ni u \mapsto D^\alpha(x^\beta u(x)) \in \mathcal{S}(\mathbb{R}^d)$ and the embedding $\mathcal{S}(\mathbb{R}^d) \mapsto L^1(\mathbb{R}^d)$ are continuous (see Remark 7.1).

Since the Fourier transform of a function in $\mathcal{S}(\mathbb{R}^d)$ is integrable, we have $\mathcal{F}\mathcal{F}^{-1}u = u$. Thus the mapping $\mathcal{F}$ is bijective on $\mathcal{S}(\mathbb{R}^d)$. □

For a distribution $T \in \mathcal{S}'(\mathbb{R}^d)$ the Fourier transform $\widehat{T}$ is defined by

$$\left\langle\widehat{T}, u\right\rangle = \langle T, \widehat{u}\rangle, \quad u \in \mathcal{S}(\mathbb{R}^d).$$

We observe that, since the mapping $u \mapsto \langle T, \widehat{u}\rangle$ is continuous by Theorem 8.14, $\widehat{T}$ is a distribution in $\mathcal{S}'(\mathbb{R}^d)$.

Let $\alpha \in \mathbb{Z}^d$. By Theorem 8.13 we have

$$\left\langle \widehat{D^\alpha T}, u \right\rangle = (-1)^{\|\alpha\|} \langle T, D^\alpha \widehat{u} \rangle = (-1)^{\|\alpha\|} \langle T, ((-2\pi i x)^\alpha u(x))^\wedge \rangle$$
$$= \left\langle (2\pi i x)^\alpha \widehat{T}, u \right\rangle$$

for $u \in \mathcal{S}(\mathbb{R}^d)$. In the same way we have

$$\langle ((-2\pi i x)^\alpha T)^\wedge, u \rangle = \langle T, (-2\pi i x)^\alpha \widehat{u}(x) \rangle = (-1)^{\|\alpha\|} \left\langle T, \widehat{D^\alpha u} \right\rangle = \left\langle D^\alpha \widehat{T}, u \right\rangle.$$

Therefore we have, for $T \in \mathcal{S}'(\mathbb{R}^d)$,

$$(D^\alpha T)^\wedge = (2\pi i \xi)^\alpha \widehat{T},$$

and

$$((-2\pi i x)^\alpha T)^\wedge = D^\alpha \widehat{T}.$$

These formulas are extentions of Theorem 8.13 to a distribution in $\mathcal{S}'(\mathbb{R}^d)$.

EXAMPLE 8.9. Since $\left\langle \widehat{\delta_a}, u \right\rangle = \langle \delta, \tau_{-a} \widehat{u} \rangle = \int e^{-2\pi i a \xi} u(x) dx$, we have

$$\widehat{\delta_a} = e^{-2\pi i a \xi}.$$

In particular, $\widehat{\delta} = 1$. Thus

$$\widehat{(D^\alpha \delta)} = (2\pi i \xi)^\alpha.$$

EXAMPLE 8.10. The Fourier transform of the distribution P.V.$\frac{1}{x}$ is given by the formula $\left\langle (\text{P.V.}\frac{1}{x})^\wedge, u \right\rangle = \langle \text{P.V.}\frac{1}{x}, \widehat{u} \rangle$. By definition and a change of order of integration we have

$$\left\langle \text{P.V.}\frac{1}{x}, \widehat{u} \right\rangle = \lim_{M \to \infty,\, a \to 0} \int_{a<|x|<M} \frac{1}{x} dx \int_{-\infty}^{\infty} u(\xi) e^{-2\pi i x \xi} d\xi$$
$$= \lim_{M \to \infty,\, a \to 0} \int_{-\infty}^{\infty} u(\xi) \left(-2i \int_a^M \frac{\sin 2\pi i \xi x}{x} dx \right) d\xi.$$

The inner integral of the last term is uniformly bounded in a and M, and converges to $-2i \int_0^\infty (\sin 2\pi \xi x)/x dx = -\pi i \operatorname{sign} \xi$ as $a \to 0$ and $M \to \infty$ by Lemma 8.1. Therefore we get

$$\left\langle \left(\text{P.V.}\frac{1}{x} \right)^\wedge, u \right\rangle = \int_{-\infty}^{\infty} (-\pi i \operatorname{sign} \xi) u(\xi) d\xi,$$

or equivalently

$$\left(\text{P.V.}\frac{1}{x} \right)^\wedge = -\pi i \operatorname{sign} \xi.$$

8.3.3. Fourier transform in L^2-space. Let $f \in L^1 \cap L^2(\mathbb{R}^d)$. Then $f * W_t$ and its Fourier transform are integrable. In fact, $\widehat{(W_t * f)}(\xi) = e^{-4\pi^2|\xi|^2 t}\widehat{f}(\xi)$, and $\widehat{f}$ is bounded. Thus by the Fourier inversion formula we have $\overline{(W_t * f)}(x) = \overline{\mathcal{F}^{-1}\mathcal{F}(W_t * f)} = \mathcal{F}(e^{-4\pi^2|\cdot|^2 t}\overline{\widehat{f}})$. Applying Lemma 8.2, we get

$$\int f(x)\overline{(W_t * f)}(x)dx = \int f(x)\mathcal{F}(e^{-4\pi^2|\cdot|^2 t}\overline{\widehat{f}})(x)dx \tag{8.26}$$
$$= \int |\widehat{f}(\xi)|^2 e^{-4\pi^2|\xi|^2 t}d\xi.$$

The right-hand side of (8.26) converges to $\int |\widehat{f}(\xi)|^2 d\xi$ by B. Levi's theorem. On the other hand, by Schwarz's inequality

$$|\int f[\overline{f - (W_t * f)}]dx| \leq ||f||_2||f - W_t * f||_2 \to 0$$

as $t \to 0$. Thus the left-hand side converges to $\int |f(x)|^2 dx$. Therefore we obtain

$$\int |f(x)|^2 dx = \int |\widehat{f}(\xi)|^2 d\xi \tag{8.27}$$

for $f \in L^1 \cap L^2(\mathbb{R}^d)$.

For a function $f \in L^2(\mathbb{R}^d)$ choose a sequence $\{f_n\}$ in $L^1 \cap L^2(\mathbb{R}^d)$ such that $||f - f_n||_2 \to 0$. For example, put $f_n(x) = f(x)\,(|x| \leq n)$, $= 0\,(|x| \geq n)$. From the above argument we have $||f_n - f_m||_2 = ||\widehat{f_n} - \widehat{f_m}||_2 \to 0$ as $m, n \to \infty$. Therefore $\{\widehat{f_n}\}$ is a Cauchy sequence in L^2. Denote its L^2-limit by $\widehat{f}$ or $\mathcal{F}f$. The function $\widehat{f}$ belongs to L^2 and is also called the Fourier transform of f. By the equation (8.27) we have the formula

$$||f||_2 = ||\widehat{f}||_2$$

for $f \in L^2(\mathbb{R}^d)$. The function $\widehat{f}$ does not depend on the choice of the sequence $\{f_n\}$. In fact, let $\{g_n\}$ be any sequence in $L^1 \cap L^2$ such that $||g_n - f||_2 \to 0\,(n \to \infty)$. Then we have $||\widehat{g}_n - \widehat{f}||_2 \leq ||\widehat{g}_n - \widehat{f}_n||_2 + ||\widehat{f}_n - \widehat{f}||_2$. The first term on the right-hand side equals, by (8.27), $||g_n - f_n||_2$, which is estimated by $||g_n - f||_2 + ||f - f_n||_2 \to 0$. The second term converges to zero by the assumption. Thus $\widehat{g}_n \to \widehat{f}$ in L^2.

Summarizing the above argument, we get

THEOREM 8.15 (The Plancherel theorem). *The Fourier transform which is defined in $L^1 \cap L^2(\mathbb{R}^d)$ extends uniquely to functions in $L^2(\mathbb{R}^d)$ so that the equation*

$$||f||_2 = ||\widehat{f}||_2$$

holds.

Since the mapping $\mathcal{F} : \mathcal{S} \mapsto \mathcal{S}$ is surjective and since $\mathcal{S}$ is dense in L^2, the extension $\mathcal{F} : L^2 \mapsto L^2$ is surjective.

THEOREM 8.16. *If $f, g \in L^2(\mathbb{R}^d)$, then we have*

$$\int f\overline{g}dx = \int \widehat{f}\overline{\widehat{g}}d\xi.$$

PROOF. We have

$$\int |f + g|^2 dx = \int |f|^2 dx + 2\Re\int f\overline{g}dx + \int |g|^2 dx,$$

and

$$\int |\widehat{f} + \widehat{g}|^2 d\xi = \int |\widehat{f}|^2 d\xi + 2\Re \int \widehat{f}\,\overline{\widehat{g}} d\xi + \int |\widehat{g}|^2 d\xi.$$

By the Plancherel theorem two right-hand sides coincide, and furthermore we have $\int |f|^2 dx = \int |\widehat{f}|^2 d\xi$ and $\int |g|^2 dx = \int |\widehat{g}|^2 d\xi$. Therefore we get

$$\Re \int f \overline{g} dx = \Re \int \widehat{f}\,\overline{\widehat{g}} d\xi.$$

Substituting ig for g, we get

$$\Im \int f \overline{g} dx = \Im \int \widehat{f}\,\overline{\widehat{g}} d\xi.$$

The last two equations imply the theorem. □

Denote the inner product on the space L^2 by $(\cdot,\cdot)_{L^2}$. By Theorem 8.16 we have

$$(f,g)_{L^2} = (\mathcal{F}f, \mathcal{F}g)_{L^2} = (\mathcal{F}^*\mathcal{F}f, g)_{L^2}$$

for $f, g \in L^2$, where $\mathcal{F}^*$ is the adjoint of $\mathcal{F}$. Therefore $\mathcal{F}^*\mathcal{F}f = f$, and thus the Fourier transform is a unitary operator from L^2 to L^2. Furthermore, the extension of the inverse Fourier transform $\mathcal{F}^{-1}$ to L^2 coincides with $\mathcal{F}^*$.

REMARK 8.2. If $f \in L^p(\mathbb{R}^d)$ and $1 < p < 2$, then f can be written as a sum $f = f_1 + f_2$, where $f_1 \in L^1$ and $f_2 \in L^2$. In fact, let $f_1 = f$ if $|f| \geq 1$ and $= 0$ otherwise, and let $f_2 = f - f_1$. Then $f_j \in L^j$ $(j = 1, 2)$. Therefore the Fourier transform of f is defined by $\mathcal{F}f = \mathcal{F}f_1 + \mathcal{F}f_2$. It is an easy exercise to show that the definition is independent of the decomposition of f.

Let $1 \leq p \leq 2$ and $1/p + 1/p' = 1$. Then we have, for $f \in L^p(\mathbb{R}^d)$,

$$\|\widehat{f}\|_{p'} \leq c_p \|f\|_p \qquad \text{(the Hausdorff-Young inequality)},$$

with a constant c_p independent of f[6].

8.3.4. Fourier transform in Sobolev spaces. We shall introduce spaces, called Sobelev spaces, which are intrinsically related with the Fourier transform as well as the Lebesgue space L^2 or the Schwartz space $\mathcal{S}$.

Let k be a nonnegative integer. Let $W^{k,2}(\mathbb{R}^d)$ be the totality of distributions in $\mathcal{S}'(\mathbb{R}^d)$ such that $D^\alpha f$ belongs to $L^2(\mathbb{R}^d)$ for every $\alpha \in \mathbb{Z}_+^d$ with $\|\alpha\| \leq k$.

LEMMA 8.4. *For every $f \in W^{k,2}(\mathbb{R}^d)$ we have*

$$c\|(1 + |\xi|^2)^{k/2} \widehat{f}(\xi)\|_2 \leq \sum_{\|\alpha\| \leq k} \|D^\alpha f\|_2 \leq c'\|(1 + |\xi|^2)^{k/2} \widehat{f}(\xi)\|_2,$$

where c, $c' > 0$ are constants not depending on f.

PROOF. First Step. *We have*

$$c_1(1 + |\xi|^2)^k \leq \sum_{\|\alpha\| \leq k} |\xi^\alpha|^2 \leq c_1'(1 + |\xi|^2)^k, \tag{8.28}$$

with constants $c_1, c_1' > 0$ independent of $\xi \in \mathbb{R}^d$.

[6] The best constant is known to be $c_p = \{p^{1/p}/p'^{1/p'}\}^{d/2} < 1$ (W. Beckner).

In fact, (8.28) follows from the inequalities

$$c_2|\xi|^{2l} \le \sum_{\|\alpha\|=l} |\xi^\alpha|^2 \le c_2'|\xi|^{2l}, \tag{8.29}$$

$l = 0, 1, \dots, k$, where c_2 and c_2' are constants independent of ξ. (8.29) is obvious when $|\xi| = 1$, and the general case follows from the homogenuity.

Second Step. *For $f \in W^{k,2}$ we have*

$$\int |D^\alpha f|^2 dx = (2\pi)^{2\|\alpha\|} \int |\xi^\alpha \widehat{f}(\xi)|^2 d\xi. \tag{8.30}$$

In fact, by the definitions of differentiation and the Fourier transform of distributions we have

$$\begin{aligned}\langle D^\alpha f, u\rangle &= (-1)^{\|\alpha\|}\langle f, D^\alpha u\rangle = (-1)^{\|\alpha\|}\left\langle \widehat{f}, \widehat{D^\alpha u}\right\rangle \\ &= (-1)^{\|\alpha\|}\left\langle \widehat{f}, (2\pi i\xi)^\alpha \widehat{u}\right\rangle = (-2\pi i)^{\|\alpha\|}\left\langle \xi^\alpha \widehat{f}, \widehat{u}\right\rangle\end{aligned}$$

for every $u \in \mathcal{S}$. Thus

$$\int D^\alpha f(x)u(x)dx = (-2\pi i)^{\|\alpha\|}\int \xi^\alpha \widehat{f}(\xi)\widehat{u}(\xi)d\xi.$$

Taking the supremum on $u \in \mathcal{S}$ with $\|u\|_2 = \|\widehat{u}\|_2 = 1$, we get (8.30).

Third Step. Lemma 8.4 is obtained by adding the two sides of equation (8.30) and then applying (8.28). □

For a real number s define the operator $\Lambda^s u$ by the Fourier transform

$$\widehat{\Lambda^s u}(\xi) = (1 + |\xi|^2)^{s/2}\widehat{u}(\xi), \quad u \in \mathcal{S}(\mathbb{R}^d).$$

If s is a nonnegative integer, a function $\Lambda^s u$ is estimated by $D^\alpha u$ with $\|\alpha\| \le s$ in the sense of L^2.

DEFINITION 8.2. For a real number s let $W^{s,2}(\mathbb{R}^d)$ be the set of distributions f in $\mathcal{S}'(\mathbb{R}^d)$ such that $\Lambda^s f \in L^2(\mathbb{R}^d)$. The space $W^{s,2}(\mathbb{R}^d)$ is called the *Sobolev space* of order s.

PROPOSITION 8.2. *We have the following:*

(i) $W^{0,2} = L^2$.
(ii) *If $s > t$, then $W^{s,2} \subset W^{t,2}$.*
(iii) *For every real number α the differential operator D^α is a bounded mapping of $W^{s,2}$ to $W^{s-\|\alpha\|,2}$.*

PROOF. (i) and (ii) are obvious by the definition of Sobolev spaces. (iii) follows from the inequality $|\xi^\alpha| \le (1 + |\xi|^2)^{\|\alpha\|/2}$. □

EXAMPLE 8.11. The space $W^{s,2}(\mathbb{R}^2)$ does not always consist of functions. In fact, since the Fourier transform of the Dirac measure δ is identically 1, we have $\widehat{\delta}(\xi)(1 + |\xi|^2)^s \in L^2(\mathbb{R}^d)$ if $s < -d/2$. Thus $\delta \in W^{s,2}(\mathbb{R}^d)$.

We define an inner product in $W^{s,2}(\mathbb{R}^d)$ by

$$(f, g)_{W^{s,2}} = \int_{\mathbb{R}^d} (\Lambda^s f)\overline{(\Lambda^s g)}dx.$$

By the Plancherel theorem we have

$$(f,g)_{W^{s,2}} = \int_{\mathbb{R}^d} \widehat{f}(\xi)\overline{\widehat{g}(\xi)}(1+|\xi|^2)^s dx.$$

We remark that the right-hand side of the last equation is the inner product of the Hilbert space $L^2(\mathbb{R}^d, (1+|\xi|^2)^s d\xi)$.

THEOREM 8.17. *The Sobolev space $W^{s,2}(\mathbb{R}^d)$ is a Hilbert space and isometrically isomorphic to the weighted Lebesgue space $L^2(\mathbb{R}^d, (1+|\xi|^2)^s d\xi)$.*

THEOREM 8.18 (The Sobolev imbedding theorem). *Let k be a nonnegative integer and $s > k + \frac{d}{2}$. Then $W^{s,2}(\mathbb{R}^d) \subset C^{(k)}(\mathbb{R}^d)$. More precisely, a function in $W^{s,2}(\mathbb{R}^d)$ belongs to $C^{(k)}(\mathbb{R}^d)$ after a correction on a null set.*

PROOF. Let $f \in W^{s,2}$. If $\widehat{D^\alpha f} \in L^1(\mathbb{R}^d)$, then the inverse Fourier transform $D^\alpha f$ is continuous after a correction on a null set.

Now, observe that $\widehat{(D^\alpha f)}(\xi) = (2\pi i)^{\|\alpha\|}\widehat{f}(\xi)$. Then for $\|\alpha\| \le k$ we have

$$(2\pi)^{-\|\alpha\|}\int |\widehat{(D^\alpha f)}(\xi)|d\xi = \int |\xi^\alpha \widehat{f}(\xi)|d\xi \le \int (1+|\xi|^2)^{k/2}|\widehat{f}(\xi)|d\xi,$$

which does not exceed, by Schwarz's inequality,

$$\left(\int (1+|\xi|^2)^s|\widehat{f}(\xi)|^2 d\xi\right)^{1/2}\left(\int (1+|\xi|^2)^{k-s} d\xi\right)^{1/2}.$$

The last integral is finite when $k - s < -d/2$. Thus we have $\widehat{D^\alpha f} \in L^1(\mathbb{R}^d)$ for every α with $\|\alpha\| \le k$, which proves the theorem. □

COROLLARY 8.2. *If $f \in W^{s,2}(\mathbb{R}^d)$ for every real number s, then f is infinitely differentiable after a correction on a null set.*

8.4. Fourier transform of radial functions

Let $J_\mu(t)$, $\Re\mu > -1$, be the Bessel function of order μ, which is defined by the sum

$$J_\mu(t) = \left(\frac{t}{2}\right)^\mu \sum_{k=0}^\infty \frac{(-1)^k}{2^{2k}\Gamma(k+1)\Gamma(\mu+k+1)} t^{2k}.$$

Put

$$V_\mu(t) = \frac{J_\mu(t)}{t^\mu}$$

(for the Bessel functions, see Appendix B.3).

In this section we assume that $d \ge 2$.

LEMMA 8.5. *We have*

$$\int_{S^{d-1}} e^{-2\pi i \xi w} d\sigma(w) = (2\pi)^{d/2} V_{(d-2)/2}(2\pi|\xi|), \quad \xi \in \mathbb{R}^d. \tag{8.31}$$

PROOF. Since the integral on the left-hand side is independent of the argument of ξ, we can assume $\xi = (0, \ldots, 0, |\xi|)$.

For $w \in S^{d-1}$ put $w = ((1-t^2)^{1/2}\overline{w}, t)$, where $-1 \leq t \leq 1$ and $\overline{w} = (w_1, \dots, w_{d-1}) \in S^{d-2}$. Then the integral in (8.31) is written as

$$\int_{S^{d-1}} e^{-2\pi i|\xi|t} d\sigma(w) = \int_{-1}^{1} e^{-2\pi i|\xi|t}(1-t^2)^{(d-3)/2} dt \int_{S^{d-2}} d\sigma(\overline{w})$$
$$= \omega_{d-2} \sum_{k=0}^{\infty} \frac{(-2\pi i|\xi|)^k}{k!} \int_{-1}^{1} t^k (1-t^2)^{(d-3)/2} dt.$$

The kth integral vanishes when k is odd. If $k = 2n$, it becomes, on putting $u = t^2$,

$$\int_0^1 u^{n-(1/2)}(1-u)^{(d-3)/2} du = \frac{\Gamma(n+\frac{1}{2})\Gamma(\frac{d-1}{2})}{\Gamma(n+\frac{d}{2})}$$

(see Appendix B.1). Therefore we have

$$\int_{S^{d-1}} e^{-2\pi i|\xi|t} d\sigma(w) = \omega_{d-2} \sum_{n=0}^{\infty} \frac{(-1)^n \Gamma(n+\frac{1}{2})\Gamma(\frac{d-1}{2})}{\Gamma(2n+1)\Gamma(n+\frac{d}{2})} (2\pi|\xi|)^{2n},$$

which by formula (B.5) is transformed to

$$\sqrt{\pi}\Gamma(\frac{d-1}{2})\omega_{d-2} \sum_{n=0}^{\infty} \frac{(-1)^n (2\pi|\xi|)^{2n}}{2^{2n}\Gamma(n+1)\Gamma(n+\frac{d}{2})} = (2\pi)^{d/2} V_{(d-2)/2}(2\pi|\xi|).$$

This completes the proof. □

THEOREM 8.19 (Bochner's formula). *Let f^* be a radial function in $L^1(\mathbb{R}^d)$, and put $f(x) = f^*(|x|)$. Then we have*

$$\widehat{f}(\xi) = (2\pi)^{d/2} \int_0^{\infty} f^*(r) V_{(d/2)-1}(2\pi\rho r) r^{d-1} dr,$$

where $\xi = \rho\omega$ with $\rho \geq 0$ and $\omega \in S^{d-1}$.

PROOF. By polar coordinates we have

$$\widehat{f}(\xi) = \int_0^{\infty} f^*(r) r^{d-1} dr \int_{S^{d-1}} e^{-2\pi i \rho r \omega w} d\sigma(w),$$

which implies Theorem 8.19 by Lemma 8.5. □

For a more general form of Bochner's formula see Appendix B.4 and also [SW].

8.5. Fourier transform and analytic functions

8.5.1. The Hardy spaces. Let f be a function in $L^p(\mathbb{R})$ $(1 \leq p \leq \infty)$. Then the Poisson integral $f(x,y) = P_y * f(x)$ is harmonic in the upper half-plane $\mathbb{R}^2_+ = \{(x,y) : x \in \mathbb{R},\ y > 0\}$. Furthermore, by Young's inequality we have $||f(\cdot,y)||_p \leq ||f||_p$ for all $y > 0$ (see Problem 6.8 and Example 6.4).

DEFINITION 8.3. Let $0 < p < \infty$. For a function $f(z)$ analytic in the upper half-plane $\Re z > 0$, define

$$||f||_{H^p} = \sup_{y>0} \left(\int_{-\infty}^{\infty} |f(x+iy)|^p dx \right)^{1/p}.$$

The *Hardy space* $H^p(\mathbb{R}^2_+)$ is the collection of all analytic functions f in the upper half-plane such that $||f||_{H^p} < \infty$.

When $1 \le p$, it is not difficult to show that $\|\cdot\|_{H^p}$ satisfies the conditions for a norm. In the case $0 < p < 1$, we can prove that $\|\cdot\|_{H^p}^p$ defines a metric on $H^p(\mathbb{R}^2_+)$ in the same way as in the case L^p. Therefore, in any case the space $H^p(\mathbb{R}^2_+)$ $(0 < p \le \infty)$ is a metric space.

LEMMA 8.6. *Let $\Im z > 0$. Then*

$$\frac{1}{2\pi i}\int_{-\infty}^{\infty}\frac{1}{s-z}e^{2\pi i\xi s}\,ds = \begin{cases} e^{2\pi i\xi z} & (\xi > 0), \\ 0 & (\xi < 0). \end{cases}$$

PROOF. First suppose $\xi > 0$. Applying the Cauchy integral theorem to the entire function $e^{2\pi i\xi\zeta}$, we have

$$\frac{1}{2\pi i}\int_{C_1\cup C_2}\frac{1}{\zeta - z}e^{2\pi i\xi\zeta}\,d\zeta = e^{2\pi i\xi z},$$

where $C_1 = [-R, R]$ is a line segment and C_2 is the semicircle $\{Re^{i\theta} : 0 < \theta < \pi\}$ in the upper half-plane.

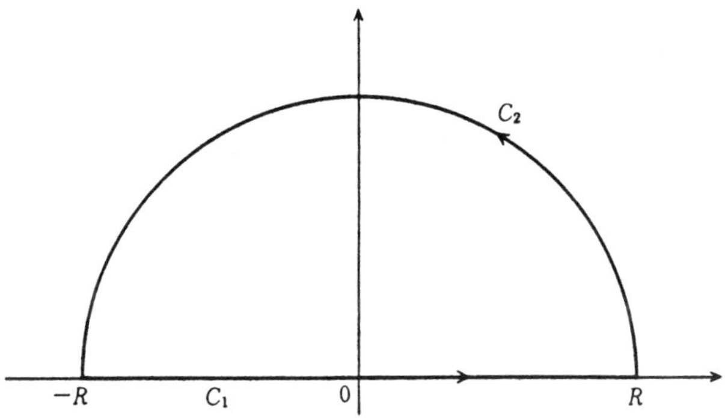

FIGURE 8.4. $C_1 \cup C_2$

We choose $R > 0$ so large that z is in the interior of the curve $C_1 \cup C_2$ (see Figure 8.4). For the integral on the semicircle we have the estimate

$$\left|\int_{C_2}\right| \le \int_0^{\pi}\frac{|e^{2\pi i\xi Re^{i\theta}}|}{|Re^{i\theta} - z|}R\,d\theta \le \int_0^{\pi}\frac{e^{-2\pi R\xi\sin\theta}}{R - |z|}R\,d\theta.$$

Since $\xi > 0$, the integrand of the last integral is uniformly bounded and converges to 0 as $R \to \infty$. Thus, by Lebesgue's dominated convergence theorem, $\int_{C_2} d\zeta \to 0$. On the other hand,

$$\int_{C_1} \to \frac{1}{2\pi i}\int_{-\infty}^{\infty}\frac{e^{2\pi i\xi s}}{s - z}\,ds$$

as $R \to \infty$. Thus the lemma is proved when $\xi > 0$.

Let C_2 be a smicircle in the lower half plane. Since the function $e^{2\pi i\xi\zeta}/(\zeta - z)$ is analytic in the lower half plane,

$$\frac{1}{2\pi i}\int_{C_1\cup C_2}\frac{1}{\zeta - z}e^{2\pi i\xi\zeta}\,d\zeta = 0.$$

If $\xi < 0$, then we can prove that $\int_{C_2} \to 0$ in a way similar to the case $\xi > 0$. Thus the lemma is obtained. □

Recall that

$$\frac{-1}{\pi i z} = P_y(x) + iQ_y(x) \tag{8.32}$$

for a complex number $z = x + iy$, $y > 0$, where P_y is a Poisson kernel and Q_y is a conjugate Poisson kernel (see Example 8.3). By Lemma 8.6 we have

$$\frac{-1}{\pi i}\mathcal{F}\left(\frac{1}{\cdot + iy}\right)(\xi) = 2e^{-2\pi y\xi}\chi_{(0,\infty)}(\xi),$$

and

$$\mathcal{F}P_y(\xi) = e^{-2\pi y|\xi|}.$$

Therefore we get

$$\mathcal{F}Q_y(\xi) = -i \operatorname{sign}\xi e^{-2\pi y|\xi|}.$$

For $f \in L^2(\mathbb{R})$ put $f(x,y) = f * P_y(x)$ and $\tilde{f}(x,y) = f * Q_y(x)$. Then $f(x,y) + \tilde{f}(x,y) = f * [-1/\pi i(\cdot + iy)](x)$. Therefore we obtain

$$\frac{1}{2}[f(x,y) + i\tilde{f}(x,y)] = \frac{1}{2\pi i}\int_{-\infty}^{\infty} \frac{f(s)}{z-s}ds \quad (z = x+iy).$$

By Lemma 8.6 the last equation can be expressed as

$$\frac{1}{2}[f(x,y) + i\tilde{f}(x,y)] = \frac{1}{2\pi i}\int_0^{\infty} \widehat{f}(\xi)e^{2\pi i z\xi}d\xi.$$

In particular, if f is a real valued function, $f(x,y)$ and $\tilde{f}(x,y)$ are the real and imaginary parts of an analytic function of $z = x + y$.

THEOREM 8.20 (The Paley-Wiener theorem). [7] *A function f belongs to the Hardy space $H^2(\mathbb{R}^2_+)$ if and only if there exists a function f^* in $L^2(\mathbb{R})$ such that $\widehat{f^*}(\xi) = 0$ $(\xi < 0)$ and*

$$f(z) = \int_{-\infty}^{\infty} \widehat{f^*}(\xi)e^{2\pi i z\xi}d\xi, \tag{8.33}$$

or, equivalently,

$$f(z) = \frac{1}{2\pi i}\int_{-\infty}^{\infty} \frac{f^*(s)}{z-s}ds. \tag{8.34}$$

PROOF. First Step. *If $f^* \in L^2(\mathbb{R})$ and $f^*(\xi) = 0$ for $\xi < 0$, and f is given by (8.33), then $f \in H^2(\mathbb{R}^2_+)$.*

In fact, $f(z)$ is obviously analytic in the upper half plane. Furthermore, by the Plancherel theorem $\|f(\cdot + iy)\|_2 = \|\widehat{f^*}(\cdot)e^{-2\pi y\cdot}\|_2 \le \|\widehat{f^*}\|_2 = \|f^*\|_2 < \infty$.

Next, assume that $f \in H^2(\mathbb{R}^2_+)$. We divide the proof into two parts.

Second Step. *For any $y_1 > 0$ we have*

$$f(z + iy_1) = \frac{1}{2\pi i}\int_{-\infty}^{\infty} \frac{f(s+iy_1)}{s-z}ds \quad (\Im z > 0). \tag{8.35}$$

To prove (8.35) we introduce the following function:

$$f_\varepsilon(z) = e^{i\varepsilon z}\frac{1}{2\varepsilon}\int_{-\varepsilon}^{\varepsilon} f(z+u)du,$$

[7] R. E. A. C. Paley, 1907–1933.

where ε is any given positive number. The function $f_\varepsilon(z)$ $(\Im z > 0)$ has the following properties:

(a) $f_\varepsilon(z)$ is analytic in $\Im z > 0$.
(b) $f_\varepsilon(z) \to f(z)$ $(\varepsilon \to 0)$.
(c) By Schwarz's inequality

$$|f_\varepsilon(z)| \le \frac{e^{-\varepsilon\Im z}}{\sqrt{2\varepsilon}}\left(\int_{-\varepsilon}^{\varepsilon} |f(z+u)|^2 du\right)^{1/2} \le \frac{e^{-\varepsilon\Im z}}{\sqrt{2\varepsilon}}\|f\|_{H^2}.$$

(d) $|f_\varepsilon(z)| \le Mf(\cdot + iy)(x)$, $(z = x + iy, y > 0)$, where Mf is the Hardy-Littlewood maximal function.

First we show that

$$f_\varepsilon(z + iy_1) = \frac{1}{2\pi i}\int_{-\infty}^{\infty} \frac{f_\varepsilon(s + iy_1)}{s - z} ds \quad (\Im z > 0). \tag{8.36}$$

In fact, since $f_\varepsilon(z + iy_1)$ is analytic in $\Im z > 0$, by the Cauchy integral theorem

$$f_\varepsilon(z + iy_1) = \frac{1}{2\pi i}\int_{C_1 \cup C_2} \frac{f_\varepsilon(\zeta + iy_1)}{\zeta - z} d\zeta, \tag{8.37}$$

where $C_1 = [-R, R]$ and $C_2 = \{Re^{i\theta} : 0 < \theta < \pi\}$ with R sufficiently large. If $\zeta \in C_2$, then $e^{-\varepsilon\Im\zeta} \to 0$ as $R \to \infty$. Thus by (c) $\int_{C_2} \to 0$ as $R \to \infty$. Thus (8.36) follows from (8.37).

Now, let $\varepsilon \to 0$ in (8.36). Then the left-hand side of (8.36) tends to $f(z)$ by (b). Remark that $1/(\cdot - z)$ belongs to $L^2(\mathbb{R})$, and also $Mf(\cdot + iy)(x)$ belongs to $L^2(\mathbb{R})$ by the Hardy-Littlewood maximal theorem. Therefore by Schwarz's inequality, the integrand $f_\varepsilon(\cdot + iy_1)/(\cdot - z)$ of (8.36) is bounded by a function in L^1 which is independent of ε. Therefore (8.35) follows from (8.36) by the convergence theorem as $\varepsilon \to 0$.

Third Step. By (8.35) and Lemma 8.6 the Fourier transform of $f(\cdot + iy + iy_1)$ can be written as $\widehat{f}(\cdot + iy_1)(\xi)e^{-2\pi y\xi}\chi_{(0,\infty)}(\xi)$. Interchanging y and y_1, we get

$$\widehat{f}(\cdot + iy_1)(\xi)e^{-2\pi y\xi}\chi_{(0,\infty)}(\xi) = \widehat{f}(\cdot + iy)(\xi)e^{-2\pi y_1\xi}\chi_{(0,\infty)}(\xi),$$

which implies that the function $\widehat{f^*}$ defined by

$$\widehat{f^*}(\xi) = \widehat{f}(\cdot + iy)(\xi)e^{2\pi y\xi}\chi_{(0,\infty)}(\xi)$$

is independent of $y > 0$.

By the Plancherel theorem

$$\int_0^{\infty} |\widehat{f^*}(\xi)e^{-2\pi y\xi}|^2 d\xi = \int_{-\infty}^{\infty} |f(x + iy)|^2 dx,$$

which is uniformly bounded in y. Thus we get $\widehat{f^*} \in L^2(\mathbb{R})$. Furthernore, by the Fourier inversion formula,

$$\int_0^{\infty} \widehat{f^*}(\xi)e^{2\pi iz\xi} d\xi = \int_{-\infty}^{\infty} \widehat{f}(\cdot + iy)(\xi)\chi_{(0,\infty}(\xi)e^{2\pi y\xi}e^{2\pi ix\xi - 2\pi\xi y} d\xi = f(x + iy).$$

Thus the proof is complete. □

REMARK 8.3. Theorem 8.20 holds for functions in the space H^p $(p \ge 1)$. The proof is similar, but a slight modification will be needed.

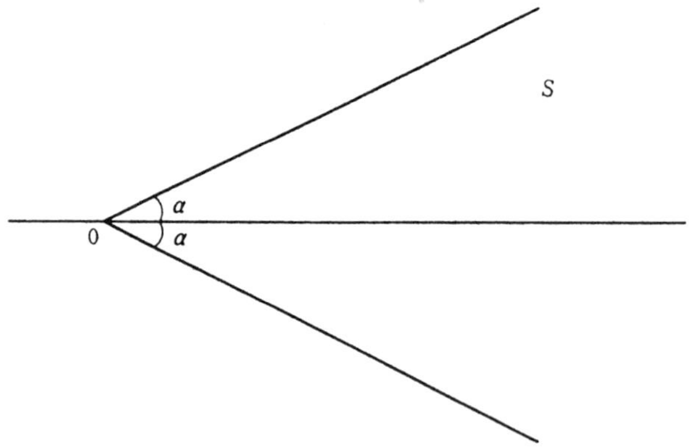

FIGURE 8.5. Angle S

8.5.2. Fourier transform and entire functions of exponential type. Let f be a function in $L^2(\mathbb{R})$. If the support of $\widehat{f}$ is contained in a bounded interval $[-a/2\pi, a/2\pi]$, then the function $f(z)$ defiend by

$$f(z) = \int_{-\infty}^{\infty} \widehat{f}(\xi) e^{2\pi i z \xi} d\xi \tag{8.38}$$

is analytic in the whole complex space $\mathbb{C}$, that is, $f(z)$ is an entre function. Furthermore, we have

$$|f(z)| \leq \int_{-a/2\pi}^{a/2\pi} |\widehat{f}(\xi)| e^{2\pi|\xi y|} d\xi \leq (a/\pi)^{1/2} \|f\|_2 e^{a|y|}, \quad z = x + iy.$$

In general, if a function $f(z)$ is entire and if there exist positive constants C and a such that

$$|f(z)| \leq C e^{a|z|}, \tag{8.39}$$

then f is called an *entire function of exponential type*. The lower bound of such a is called the degree of f.

THEOREM 8.21 (The Paley-Wiener theorem). *Let $f \in L^2(\mathbb{R})$. A necessary and sufficient condition that a function f be extendable to an entire function of degree at most a is that* $\operatorname{supp} \widehat{f} \subset [-a/2\pi, a/2\pi]$.[8]

LEMMA 8.7 (Phragmén-Lindelöf maximum principle). *Let $0 < \alpha < \pi/2$ and let S be the angle with the directed half lines with directions $\pm\alpha$ (Figure 8.5). Assume that*

(a) *$F(z)$ is a function analytic in S and satisfies* (8.39) *in S for some positive constants C and a,*
(b) *F is continuous on the closure $\overline{S}$ and $|F(z)| \leq B$ $(z \in \partial S)$.*

Then, we have

$$|F(z)| \leq B \quad \textit{for } z \in S.$$

[8] If $\operatorname{supp} \widehat{f} \subset [-\alpha, \alpha]$, then f is said sometimes to be band limited to $[-\alpha, \alpha]$. On the other hand a function with compact support is said to be time limited.

PROOF OF THE LEMMA. Choose $\gamma > 1$ so that $\gamma\alpha < \pi/2$. Put $F_A(z) = F(z)e^{-Az^\gamma}$ where A is any given positive number.

For $z = re^{\pm i\alpha}$, we have $|e^{-Az^\gamma}| = e^{-Ar^\gamma \cos \alpha\gamma} \leq 1$. Thus $|F_A(z)| \leq B\,(z \in \partial S)$. On the other hand, since F satisfies (8.39) in S,

$$|F_A(Re^{i\theta})| \leq Ce^{aR}e^{-AR^\gamma \cos\gamma\alpha} \leq B \quad (|\theta| \leq \alpha)$$

for a large number R. Thus by the maximum principle we have $|F_A(z)| \leq B\,(z \in S)$. Letting $A \to 0$, we get the desired estimate. □

PROOF OF THEOREM 8.21. The sufficiency is already shown. To prove the neccesity, let $f(z)$ be an entire function such that $|f(z)| \leq Ce^{b|z|}$ for any $b > a$ with some constant $C > 0$ which may depend on b. Put

$$F(z) = e^{ibz} \int_{-1/2}^{1/2} f(z+s)ds.$$

Then $F(z)$ is also an entire function, and we have

$$\widehat{F}(\xi) = \int_{-1/2}^{1/2} \widehat{f}\left(\xi - \frac{b}{2\pi}\right) e^{2\pi i\xi s - ibs} ds.$$

In fact, the function $g(x) = f(x+s)e^{ibx}$ belongs to $L^2(R)$ by assumption, and the partial sum $\int_{-N}^{N} g(x)e^{-2\pi i\xi x}dx$ converges in L^2 to $\widehat{g}(\xi) = \widehat{f}(\xi - (b/2\pi))e^{2\pi i\xi s - ibs}$. Thus

$$\widehat{F}(\xi) = \widehat{f}\left(\xi - \frac{b}{2\pi}\right) \frac{\sin(\pi\xi - b/2)}{\pi\xi - b/2}.$$

We shall show that $\widehat{F}(\xi) = 0$ a.e. $(\xi < 0)$, which is equivalent to saying that $\widehat{f}(\xi) = 0$ a.e. in $(-\infty, -b/2\pi)$. By a similar argument with $-b$ in place of b we can show that $\widehat{f}(\xi) = 0$ a.e. in $(b/2\pi, \infty)$.

First we observe that

$$|F(x)| \leq \left(\int_{-1/2}^{1/2} |f(x+s)|^2 ds\right)^{1/2} \leq \|f\|_2,$$

which implies that $F(z)$ is bounded on the real axis. Since

$$|F(iy)| \leq Ce^{-by}e^{b|iy+1/2|} \leq Ce^{b/2}$$

for $y > 0$, $F(z)$ is bounded on the upper half imaginary axis. Therefore $F(z)$ is bounded in the first and second quadrants by the Phragmén-Lindelöf maximum principle.

Let

$$F_1(z) = \frac{F(z)}{2\pi i(z+i)}.$$

Then $F_1(z)$ belongs to $H^2(\mathbb{R}^2_+)$. In fact, $F_1(z)$ is analytic in the upper half plane and

$$\int_{-\infty}^{\infty} |F_1(x+iy)|^2 dx \leq \sup_{\Im z > 0} \left(\frac{|F(z)|}{2\pi}\right)^2 \int_{-\infty}^{\infty} \frac{1}{x^2 + (1+y)^2} dx.$$

The last integral is bounded in $y > 0$. Thus, by Theorem 8.20, $\widehat{F_1}(\xi) = 0$ a.e. $(\xi < 0)$.

By definition,

$$\widehat{F}_1(\xi) = \frac{1}{2\pi i}\int_{-\infty}^{\infty} \frac{F(x)}{x+i} e^{-2\pi i \xi x} dx.$$

Since $\mathcal{F}(\cdot + i)^{-1}(\eta) = -2\pi i e^{-2\pi\eta}\chi_{(0,\infty)}(\eta)$ by Lemma 8.6, we have, by Theorem 8.16,

$$\widehat{F}_1(\xi) = -\int_{-\infty}^{\xi} \widehat{F}(\eta) e^{-2\pi(\xi-\eta)} d\eta.$$

Since the left-hand side vanishes almost everywhere in $\xi < 0$, $\widehat{F}(\eta) = 0$ a.e. $\eta < 0$. Thus the theorem is proved. □

8.5.3. Heisenberg's inequality. If f is a nonzero function in $L^2(\mathbb{R})$ and the support of $\widehat{f}$ is compact, then by the Paley-Wiener theorem, f can be extended to an entire function. Thus the zeros of $f(z)$ are at most countable. If the roles of f and $\widehat{f}$ are exchanged, the situation is similar.

A function $f(x)$ in $L^2(\mathbb{R})$ is used to express a signal, and its Fourier transform $\widehat{f}(\xi)$ reveals information on the frequencies. The Paley-Wiener theorem suggests that if the distribution Δ_f of the signal f is localized, then the distribution $\Delta_{\widehat{f}}$ of the frequency information cannot be localized, and the converse is also true. That is, f and $\widehat{f}$ cannot be simultaneously localized. This phenomenon is called the uncertainty principle and is one of the factors which make various arguments complicated in the Fourier analysis. The theory of wavelets, which is described in the following chapter, gives a fruitful approach to this problem.

Let us think of showing such a phenomenon quantitatively. Let f be a function in $L^2(\mathbb{R})$ such that the finiteness of the following integarls is guaranteed. We define the average and the distribution of f by

$$\tilde{x} = \frac{1}{\|f\|_2^2}\int_{-\infty}^{\infty} x|f(x)|^2 dx$$

and

$$\Delta_f = \frac{1}{\|f\|_2}\left(\int_{-\infty}^{\infty} (x-\tilde{x})^2 |f(x)|^2 dx\right)^{1/2},$$

respectively. In the same way, for the Fourier transform $\widehat{f}$ we define

$$\tilde{\xi} = \frac{1}{\|f\|_2^2}\int_{-\infty}^{\infty} \xi|\widehat{f}(\xi)|^2 d\xi, \quad \text{and} \quad \Delta_{\widehat{f}} = \frac{1}{\|f\|_2}\left(\int_{-\infty}^{\infty} (\xi-\tilde{\xi})^2 |\widehat{f}(\xi)|^2 d\xi\right)^{1/2}.$$

THEOREM 8.22 (Heisenberg's inequality). [9] *Let $f \in L^2(\mathbb{R})$ be a nonzero function. Then we have*

$$4\pi\Delta_f\Delta_{\widehat{f}} \geq 1.$$

The lower bound is attained by the function $f(x) = ce^{-a(x-m)^2}e^{2\pi i b x}$, $a > 0$ (see Problem 8.5), and the equality holds only in this case.

[9] W. K. Heisenberg, 1901–1976. German physicist.

PROOF. First, suppose that $f \in \mathcal{S}(\mathbb{R})$. Since $\Delta_f \Delta_{\hat{f}}$ is invariant under translations of the function as well as the Fourier transform, we can assume $\tilde{x} = 0$ and $\tilde{\xi} = 0$. Furthermore we can assume that $\|f\|_2 = 1$ (multiply by a constant).

We have the following:

$$\begin{aligned}
(4\pi\Delta_f\Delta_{\hat{f}})^2 &= (4\pi)^2 \int_{-\infty}^{\infty} x^2 |f(x)|^2 dx \int_{-\infty}^{\infty} \xi^2 |\hat{f}(\xi)|^2 d\xi \\
&= (4\pi)^2 \int_{-\infty}^{\infty} |xf(x)|^2 dx \int_{-\infty}^{\infty} |(-2\pi i)^{-1} \widehat{f'}(\xi)|^2 d\xi \qquad \text{(Lemma 8.3)} \\
&= 4 \int_{-\infty}^{\infty} |xf(x)|^2 dx \int_{-\infty}^{\infty} |f'(x)|^2 dx \qquad \text{(the Plancherel theorem)} \\
&\geq 4 \left(\int_{-\infty}^{\infty} |xf(x)f'(x)| dx \right)^2 \qquad \text{(Schwarz's inequality)} \\
&\geq 4 \left(\int_{-\infty}^{\infty} \left| \frac{x}{2} [f(x)\overline{f'(x)} + \overline{f(x)} f'(x)] \right| dx \right)^2 \\
&\geq \left(\int_{-\infty}^{\infty} x \frac{d}{dx} |f(x)|^2 dx \right)^2 \\
&= \left(\int_{-\infty}^{\infty} |f(x)|^2 dx \right)^2 = 1 \qquad \text{(integration by parts)},
\end{aligned}$$

which proves the inequality.

Now, we suppose that f is a nonzero function in $L^2(\mathbb{R})$. Since $\|xf\|_2 > 0$ and $\|\xi\hat{f}\|_2 > 0$, it is enough to consider the case where $\|xf\|_2 < \infty$ and $\|\xi\hat{f}\|_2 < \infty$.

Let $\varphi_t(x)$ be a mollifier (see Example 6.3) and put $f_n(x) = f(x)\chi_{(-n,n)}(x)$. Then $f_n * \varphi_t \in C_c^\infty$. Since the inequality holds for $f_n * \varphi_t$, the remainder of the proof is to check the following inequalities:

$$\limsup_{t \to 0} \limsup_{n \to \infty} \int x^2 |f_n * \varphi_t(x)|^2 dx \leq \int x^2 |f(x)|^2 dx, \tag{8.40}$$

and

$$\limsup_{t \to 0} \limsup_{n \to \infty} \int \xi^2 |\hat{f}_n(\xi)\hat{\varphi}(t\xi)|^2 d\xi \leq \int \xi^2 |\hat{f}(\xi)|^2 d\xi. \tag{8.41}$$

For a fixed t the function $\xi\hat{\varphi}(t\xi)$ is bounded. Thus the left-hand side of (8.41) converges to $\int \xi^2 |\hat{f}(\xi)\hat{\varphi}(t\xi)|^2 d\xi$ as $n \to \infty$, which is bounded by $\int \xi^2 |\hat{f}(\xi)|^2 d\xi$, since $|\hat{\varphi}(\xi)| \leq 1$.

To prove (8.40) we remark that $|f_n * \varphi_t(x)| \leq |f| * \varphi_t(x)$. For a given $\varepsilon > 0$ we have

$$\int_{|x|<\varepsilon} x^2 [|f| * \varphi_t(x)]^2 dx \leq \varepsilon^2 \int [|f| * \varphi_t(x)]^2 dx \leq \varepsilon^2 \int |f|^2 dx.$$

On the other hand, since $[|f| * \varphi_t(x)]^2 \leq \int |f(x-y)|^2 \varphi_t(y) dy$ by Schwarz's inequality, Fubini's theorem gives us

$$\int_{|x|>\varepsilon} x^2 [|f| * \varphi_t(x)]^2 dx \leq \int_{|y|<t} dy \int_{|x|>\varepsilon} x^2 |f(x-y)|^2 \varphi_t(y) dx.$$

If $|y| < t < \varepsilon$ and $|x| > \varepsilon$, then $x^2 = (x-y)^2[x/(x-y)]^2 \le (x-y)^2[\varepsilon/(\varepsilon-t)]^2$. Since $\int \varphi_t(z)dz = 1$, we find that the last term does not exceed

$$\left(\frac{\varepsilon}{\varepsilon - t}\right)^2 \int_{-\infty}^{\infty} x^2|f(x)|^2 dx.$$

Therefore, letting $t \to 0$ and then $\varepsilon \to 0$, we get inequality (8.40). □

8.6. Fourier series

When we work on the theory of Fourier series, we assume that a function in $L^p[0,1)$ is always extended to a periodic function with period 1, that is, $f(x) = f(x+1)$ for all $x \in \mathbb{R}$. A function f may be supposed to be defined on the group

$$\mathbb{T} \cong \mathbb{R}/\mathbb{Z} \cong \mathbb{R} \pmod 1.$$

Also, we may assume that it is defined on the circle of circumference 1.

For a function $f \in L^p(\mathbb{T})$ $(1 \le p < \infty)$ we use the notation

$$||f||_p = \left(\int_0^1 |f(x)|^p dx\right)^{1/p}.$$

For $p = \infty$ we denote $||f||_\infty = \operatorname{ess\,sup} |f(x)|$.

8.6.1. Definition of Fourier series. For $f \in L^1(\mathbb{T})$ the *nth Fourier coefficient* is defined by

$$\widehat{f}_n = \int_{\mathbb{T}} f(x) e^{-2\pi i n x} dx \quad (n \in \mathbb{Z}),$$

and the formal series

$$f \sim \sum_{n=-\infty}^{\infty} \widehat{f}_n e^{2\pi i n x} \tag{8.42}$$

is called the *Fourier series* of f.

The coefficients defined by

$$a_n(f) = 2\int_0^1 f(x)\cos(2\pi n x)dx, \quad \text{and} \quad b_n(f) = 2\int_0^1 f(x)\sin(2\pi n x)dx$$

are called the nth Fourier cosine coefficient and the nth sine coefficient, respectively. By the Euler formula

$$e^{i\theta} = \cos\theta + i\sin\theta$$

series (8.42) can be formally rewritten as

$$f \sim \frac{a_0}{2} + \sum_{n=1}^{\infty}(a_n \cos 2\pi n x + b_n \sin 2\pi n x).$$

Let $\mathbb{T}^d$ be the d-product $\mathbb{T} \times \ldots \times \mathbb{T}$ of $\mathbb{T}$.

For a function f in $L^1(\mathbb{T}^d)$ which is periodic in each variable with period 1, the Fourier coefficients are defined by

$$\widehat{f}_n = \int_{\mathbb{T}^d} f(x) e^{-2\pi i n x} dx \quad (n \in \mathbb{Z}^d),$$

where $x = (x_1, \dots, x_d)$, $n = (n_1, \dots, n_d)$ and $nx = n_1x_1 + \cdots + n_dx_d$, and the Fourier series of f is a formal series

$$f \sim \sum_{n \in \mathbb{Z}^d} \widehat{f}_n e^{2\pi inx}. \tag{8.43}$$

THEOREM 8.23 (The Riemann-Lebesgue theorem). *If $f \in L^1(\mathbb{T}^d)$, then $\widehat{f}_n$ converges to zero as $|n| \to \infty$.*

PROOF. The coefficient $\widehat{f}_n$ can be assumed to be a Fourier transform of integrable function which vanishes outside the interval $[0, 1)^d$. Thus the theorem follows from the Riemann-Lebesgue theorem for Fourier integrals. □

It is known that there exists an integrable function whose Fourier coefficients tend to zero more slowly than any given sequence tending to zero. We also remark that a Fourier series does not always converge.

In the rest of this chapter we restrict ourselves to Fourier series in one variable.

8.6.2. Convolution and Fourier series.

For f, $g \in L^1(\mathbb{T})$ let

$$f * g(x) = \int_{\mathbb{T}} f(x - y)g(y)dy,$$

and call $f * g$ the *convolution* of f and g.

As in the Fourier transform case, we have

THEOREM 8.24. *Let f, $g \in L^1(\mathbb{T})$. Then*

(i) *$f * g \in L^1(\mathbb{T})$ and $\|f * g\|_1 \leq \|f\|_1\|g\|_1$,*

(ii) *$|\widehat{f}_n| \leq \|f\|_1 \quad (n \in \mathbb{Z})$,*

(iii) *the mapping $L^1(\mathbb{T}) \ni f \to \widehat{f}_n \in \mathbb{C}$ is a homomorphism, that is,*

$$\widehat{f * g}_n = \widehat{f}_n \widehat{g}_n \quad (n \in \mathbb{Z}).$$

LEMMA 8.8. *Assume that the series $k(x) = \sum \lambda_n e^{2\pi inx}$ converges uniformly. Then for every $f \in L^1(\mathbb{T})$ we have*

$$f * k(x) = \sum_{n=-\infty}^{\infty} \lambda_n \widehat{f}_n e^{2\pi inx},$$

where the sum on the right-hand side converges uniformly.

PROOF. By definition

$$\int_0^1 f(y)e^{2\pi in(x-y)}dy = \widehat{f}_n e^{2\pi inx}.$$

If we multiply both sides by λ_n and sum over n, then, since $k(x)$ is a uniformly convergent sum, from the left-hand side we obtain

$$\int_0^1 f(y)[\sum_{n=-\infty}^{\infty} \lambda_n e^{2\pi in(x-y)}]dy = f * k(x).$$

□

8.6.3. Convergence of Fourier series. A finite sum

$$D_n(x) = \sum_{j=-n}^{n} e^{2\pi i j x}$$

is called the *nth Dirichlet kernel.*

By Lemma 8.8 the nth partial sum of the Fourier series of f is written as

$$s_n(f)(x) = \sum_{j=-n}^{n} \widehat{f}_j e^{2\pi i j x} = \int_0^1 f(x-y) D_n(y) dy.$$

If $e^{2\pi i x} \neq 1$, by an elementary computation

$$D_n(x) = e^{-2\pi i n x} \frac{e^{2\pi i(2n+1)x} - 1}{e^{2\pi i x} - 1} = \frac{\sin \pi(2n+1)x}{\sin \pi x}.$$

Since $\int_0^1 D_n(x)dx = 1$, integrating term by term, we get

$$s_n(f)(x) - f(x) = \int_0^1 [f(x-y) - f(x)] D_n(y) dy.$$

Replacing the integral domain by $(-1/2, 1/2)$ and splitting it into positive and negative parts we get

$$s_n(f)(x) - f(x) = \int_0^{1/2} \psi_x(y) \frac{2 \sin \pi(2n+1)y}{\sin \pi y} dy,$$

where $\psi_x(y) = \frac{1}{2}[f(x+y) + f(x-y)] - f(x)$.

Let $f \in L^1$ and fix a point x. For $1/2 > \delta > 0$ the integral

$$\int_\delta^{1/2} \frac{2\psi_x(y)}{\sin \pi y} \sin \pi(2n+1)y \, dy$$

can be written as a finite linear combination of Fourier coefficients of an integrable function. Thus it converges to zero as $|n| \to \infty$ by the Riemann-Lebesgue theorem, from which we obtain

THEOREM 8.25. *Let $f \in L^1(\mathbb{T})$ and fix a point x. A necessary and sufficient condition that the partial sum $s_n(f)(x)$ converges to $f(x)$ as $n \to \infty$ is that*

$$\int_0^\delta \psi_x(y) \frac{2 \sin \pi(2n+1)y}{\sin \pi y} dy \to 0 \quad (n \to 0) \tag{8.44}$$

for any $1/2 > \delta > 0$

(8.44) holds if the Dini condition $\int_0^\delta (|\psi_x(y)|/y) dy < \infty$ is satisfied. In particular, if f satisfies the Lipschitz condition of order $1 \geq \alpha > 0$, that is,

$$|f(x) - f(y)| \leq C|x-y|^\alpha,$$

with a constant $C > 0$, then (8.44) holds. The reader may want to compare the convergence condition for the partial sums of a Fourier transform (see § 8.1.2).

Solutions to the problem of convergence of partial sums of a Fourier series have been elaborated for a long time. We mention below some of the remarkable results. Since any known proofs of these theorems require advanced arguments of Fourier analysis, the reader should consult the Bibliography at the end of the book or refer to the original papers.

On the divergence of Fourier series of continuous functions, we have the following results.

(i) du Bois Reymond(1876): There exists a continuous function f such that the partial sum $s_n(f)(x)$ diverges at $x = 0$ as $n \to \infty$.

This result was ultimately strengthened as follows:

(ii) J.-P. Kahane and Y. Katznelson(1965): For any null set N there exists a continuous function whose partial sum diverges on N.

On the convergence problem for the functions in L^p:

(iii) A. Kolmogorov(1923): There exists an integrable function f whose partial sum diverges everywhere.

(iv) L. Carleson(1966): If $f \in L^2(\mathbb{T})$, then $s_n(f)(x)$ converges almost everywhere to $f(x)$. R. Hunt(1967) showed that Carleson's proof can be improved to hold for functions in $L^p(\mathbb{T})\,(1 < p)$.

8.6.4. The Poisson summation formula.

THEOREM 8.26. *Let $f \in L^1(\mathbb{R})$.*

(i) *Put $u(x) = \sum_{j=-\infty}^{\infty} f(x+j)$. Then the right-hand side converges almost everywhere and also in the $L^1(\mathbb{T})$ sense.*

(ii) $\widehat{u}_n = \widehat{f}(n)$.

(iii) *If the series in* (i) *converges at $x = 0$ and the Fourier series of u converges at 0 to $u(0)$, then*

$$\sum_{j=-\infty}^{\infty} f(j) = \sum_{n=-\infty}^{\infty} \widehat{f}(n). \tag{8.45}$$

Equation (8.45) is called the *Poisson summation formula.* For example, it holds for $f \in \mathcal{S}(\mathbb{R})$. In fact, if $f \in \mathcal{S}(\mathbb{R})$, then $\sum f(x+j)$ and $\sum f'(x+j)$ converge absolutely, and thus u is differentiable. Therefore the conditions in (iii) are satisfied.

PROOF. We have

$$\int_0^1 \sum_{j=-k}^{k} |f(x+j)|dx = \int_{-k}^{k+1} |f(x)|dx \le \int_{-\infty}^{\infty} |f(x)|dx.$$

Thus the series $\sum f(x+j)$ converges absolutely almost everywhere by B. Levi's theorem. Therefore we obtain

$$\widehat{u}_n = \sum_{j=-\infty}^{\infty} \int_0^1 f(x+j)e^{-2\pi inx}dx = \sum_{j=-\infty}^{\infty} \int_j^{j+1} f(x)e^{-2\pi inx}dx = \widehat{f}(n).$$

The formula (8.45) follows from the fact that $\sum_j f(j) = u(0) = \sum_n \widehat{u}_n$. □

8.7. Summability kernels for Fourier series

8.7.1. Examples of summability kernels. We shall introduce the summability kernels for Fourier series corresponding to the Fourier transform.

EXAMPLE 8.12 (Fejér kernel). The mean

$$F_n(x) = \frac{\sum_{m=0}^{n} D_m(x)}{n+1} = \sum_{j=-n}^{n} \left(1 - \frac{|j|}{n+1}\right) e^{2\pi ijx} \tag{8.46}$$

is called the *Fejér kernel*. We have

$$F_n(x) = \frac{1}{n+1}\left(\frac{\sin \pi(n+1)x}{\sin \pi x}\right)^2. \tag{8.47}$$

In fact, recall the formulas

$$D_m(x) = \sin \pi(2m+1)x / \sin \pi x$$

and

$$\sin \pi(2m+1)x = [\cos 2\pi m x - \cos 2\pi(m+1)x]/2\sin \pi x.$$

Then from (8.46) we get

$$F_n(x) = \frac{1}{n+1}\frac{1-\cos 2\pi(n+1)x}{2\sin \pi x}.$$

Using the formula $1-\cos 2\pi(n+1)x = 2\sin^2 \pi(n+1)x$, we get (8.47).

The mean value of the partial sum

$$\sigma_n(f)(x) = \frac{1}{n+1}\sum_{m=0}^{n} s_m(f)(x)$$

is called the nth *Cesàro mean*. By a simple observation we have

$$\sigma_n(f)(x) = F_n * f(x).$$

EXAMPLE 8.13 (The Poisson kernel). The infinite sum

$$P_r(x) = \sum_{n=-\infty}^{\infty} r^{|n|} e^{2\pi i n x} \quad (0 \le r < 1)$$

is called the *Poisson kernel*. By an elementary computation,

$$P_r(x) = \frac{1}{1-re^{2\pi i x}} + \frac{re^{-2\pi i x}}{1-re^{-2\pi i x}} = \frac{1-r^2}{1-2r\cos 2\pi x + r^2}.$$

On the other hand,

$$Q_r(x) = \sum_{n=-\infty}^{\infty} -i(\operatorname{sign} n) r^{|n|} e^{2\pi i n x} = \frac{2r \sin 2\pi x}{1-2r\cos 2\pi x + r^2}$$

is calle the *conjugate Poisson kernel*.

Since

$$\frac{1}{2}[P_r(x) + iQ_r(x)] = \frac{1}{2} + \sum_{n=1}^{\infty} z^n \quad (z = re^{2\pi i x},\ 0 \le r < 1),$$

the kernels $P_r(x)$ and $Q_r(x)$ are the real and imaginary parts of an analytic function in the unit disk $|z| < 1$, respectively.

For $f \in L^1(\mathbb{R})$ the functions

$$P_r * f(x) = \sum_{n=-\infty}^{\infty} r^{|n|} \widehat{f}_n e^{2\pi i n x}$$

and

$$Q_r * f(x) = \sum_{n=-\infty}^{\infty} -i(\operatorname{sign} n) r^{|n|} \widehat{f}_n e^{2\pi i n x}$$

are called the *Abel-Poisson mean* and the *conjugate Abel-Poisson mean*, respectively. Let $f(z) = \frac{1}{2}[P_r * f(x) + iQ_r * f(x)]$, where $z = re^{2\pi ix}$. Then

$$f(z) = \frac{1}{2}\widehat{f}_0 + \sum_{n=1}^{\infty} \widehat{f}_n z^n.$$

Since $\{\widehat{f}_n\}$ is bounded, $f(z)$ is analytic in the unit disk. Furthermore, if f is real valued, then $P_r * f(x)$ and $Q_r * f(x)$ are its real and imaginary parts respectively.

EXAMPLE 8.14 (The Gauss-Weierstrass kernel). The series

$$w_t(x) = \sum_{n=-\infty}^{\infty} e^{-4\pi^2 n^2 t} e^{2\pi inx} \quad (t > 0)$$

is called the *Gauss-Weierstrass kernel.*

Recall the Gauss-Weierstrass kernel $W_t(x) = (1/\sqrt{4\pi t})e^{-x^2/4t}$ for a Fourier integral and its Fourier transform $\widehat{W_t}(\xi) = e^{-4\pi^2\xi^2 t}$. Then we can write w_t as follows:

$$w_t(x) = \sum_{n=-\infty}^{\infty} \widehat{W_t}(n)e^{2\pi inx} = \sum_{n=-\infty}^{\infty} \widehat{\tau_{-x}W_t}(n).$$

By the Poisson summation formula we have

$$w_t(x) = \sum_{j=-\infty}^{\infty} \tau_{-x}W_t(j) = \sum_{j=-\infty}^{\infty} \frac{1}{\sqrt{4\pi t}} e^{-(x+j)^2/4t}.$$

8.7.2. Summability of Fourier series. A family $\{k_t(x) : t > 0\}$ is said to be a summability kernel in $L^1(\mathbb{T})$ if the following three conditions are satisfied:

(s1) $\int_{\mathbb{T}} k_t(x)dx = 1$,

(s2) $\|k_t\|_1 \le C$,

(s3) for every δ, $1/2 > \delta > 0$, $\lim_{t\to 0} \int_{1/2>|x|>\delta} |k_t(x)|dx = 0$,

where C is a constant not depending on t.

If there exists constants c and $\eta > 0$ independent of $t > 0$ and $|x| < 1/2$ such that

(s2)′ $|k_t(x)| \le ct^{-1}$,

(s3)′ $|k_t(x)| \le ct^\eta/|x|^{1+\eta}$,

then conditions (s2) and (s3) are satisfied. This is proved similarly as in the $\mathbb{R}^d$ case.

We state a theorem analogous to the case $L^p(\mathbb{R}^d)$ whose proof is also similar as in the $L^p(\mathbb{R}^d)$ case and is omitted.

THEOREM 8.27. *Let $\{k_t\}$ be a summability kernel. We have the following:*

(i) *If $f \in L^p(\mathbb{T})$, $1 \le p < \infty$, then $\|f * k_t - f\|_p \to 0$ as $t \to 0$. If $f \in C(\mathbb{T})$, then $\|f * k_t - f\|_\infty \to 0$ as $t \to 0$.*

Furthermore assume that $k_t(x)$ satisfies conditions (s2)′ *and* (s3)′. *Then*

(ii) *for $f \in L^1(\mathbb{T})$ we have $f * k_t(x) \to f(x)$ a.e. as $t \to 0$.*

The three examples given in the preceding subsection satisfy conditions (s1), (s2)′ and (s3)′.

For the Fejér kernel we take a discrete parameter $t = 1/(n+1)$. Due to term by term integration we have $\int_0^1 F_n(x)dx = 1$ (see (8.46)). Condition (s2)′ follows from

$$F_n(x) \le \sum_{j=-n}^{n} \left| \left(1 - \frac{|j|}{n+1}\right) e^{2\pi i j x} \right| \le n+1 = \frac{1}{t}.$$

Finally, applying the inequality $|\sin \pi x| \ge 2|x|$ ($|x| < 1/2$) (see Figure 8.6) to (8.47), we get $F_n(x) \le 1/(n+1)(2x)^2$, which implies (s3)′ with $\eta = 1$.

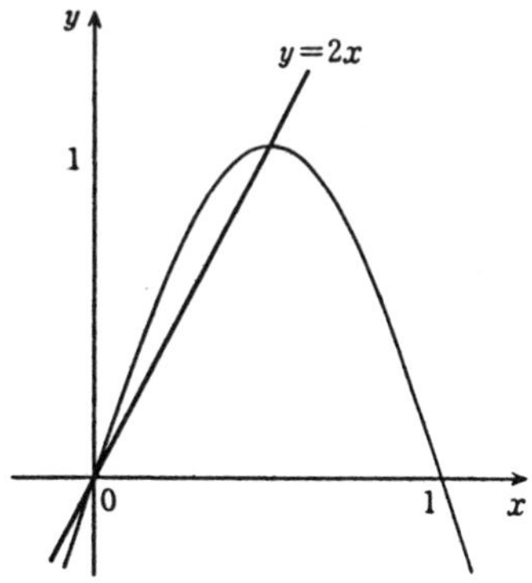

FIGURE 8.6. $\sin \pi x \ge 2x \quad (0 \le x \le 1/2)$

For the Poisson kernel choose a parameter $t = 1-r$. Condition (s1) follows again from term by term integration of $P_t(x) = \sum r^{|n|} e^{2\pi i n x}$. By elementary computation we have $P_r(x) \le \sum_{n=-\infty}^{\infty} r^{|n|} = (1+r)/(1-r) \le 2/t$, which implies (s2)′.

Since $1 - 2r\cos 2\pi x + r^2 = (1-r)^2 + 4r(\sin \pi x)^2 \ge 16rx^2$ for $|x| < 1/2$, we have $P_r(x) \ge (1-r^2)/16rx^2 \ge t/8x^2$ for $0 < t < 1/2$.

For the Gauss-Weierstrass kernel we take the parameter $t^{1/2}$. The proof is left to the reader (see Problem 8.6.).

A finite linear combination of $\{\cos nx, \sin nx : n \in \mathbb{Z}\}$ is called a trigonometric polynomial.

THEOREM 8.28 (The Weierstrass approximation theorem).

(i) *Any function in $C(\mathbb{T})$ can be uniformly approximated by trigonometric polynomials.*

(ii) *Any function in $C[0,1]$ can be uniformly approximated by polynomials.*

Therefore, the set of all trigonometric polynomials is dense in $L^p[0,1]$ $(1 \le p < \infty)$ and in $C[0,1]$, and the same is true for the set of all polynomials.

PROOF. (i) For a function f in $C(\mathbb{T})$ the nth Cesàro mean $\sigma_n(f)(x)$ is a trigonometric polynomial, and we have $\|\sigma_n(f) - f\|_\infty \to 0$ as $n \to \infty$.

(ii) For a given continuous function f in $[0,1/2]$ put $g(x) = f(x)$ for $x \in [0,1/2]$ and $= f(1-x)$ for $x \in (1/2, 1]$. Since g can be extended as a continuous periodic function with period 1, for any $\varepsilon > 0$ there exists a trigonometric polynomial $t(x) = \sum_{n=-N}^{N} a_n e^{2\pi i n x}$ such that $\sup_{x\in[0,1]} |g(x) - t(x)| < \varepsilon/2$.

For each n the quantity

$$\sup_{x\in[0,1]} \left| e^{2\pi i n x} - \sum_{j=0}^{k} \frac{(2\pi i n)^j}{j!} x^j \right|$$

can be as small as desired when k is large. Thus we can choose a poynomial $p(x)$ such that $\sup_{x\in[0,1/2]}|f(x)-p(x)|<\varepsilon$. Thus the set of all polynomials is dense in $C[0,1/2]$. Now the theorem follows by a change of scale. □

8.8. Orthogonal function systems

8.8.1. Orthogonal systems. Let H be a Hilbert space over the complex numbers. A family $\{f_j : j \in J\}$ is said to be *linearly independent* if for any finite number of $j_1, j_2, \ldots, j_n \in J$ and complex numbers $c_1, c_2, \ldots, c_n$ the equality

$$c_1 f_{j_1} + c_2 f_{j_2} + \cdots + c_n f_{j_n} = 0$$

implies $c_1 = c_2 = \cdots = c_n = 0$.

A family $\{f_j : j \in J\}$ is said to be *closed* or *total*[10] if the set of all finite linear combinations of f_j $(j \in J)$, is dense in H. This implies that for every $f \in H$ and $\varepsilon > 0$ there exist complex numbers $c_1, c_2, \ldots, c_n$ and $f_{j_1}, f_{j_2}, \ldots, f_{j_n}$ such that

$$\|f - (c_1 f_{j_1} + c_2 f_{j_2} + \cdots + c_n f_{j_n})\| < \varepsilon.$$

A subset $\{e_j\}$ of H is called an *orthogonal system* if

$$(e_i, e_j) = 0 \quad (i \neq j).$$

Furthermore, if

$$(e_i, e_j) = 1 \quad (i = j),$$

$\{e_j\}$ is called an *orthonormal system*.

The *trigonometric system* $\{e^{2\pi inx} : n \in \mathbb{Z}\}$ is an orthonormal system in $L^2(\mathbb{T})$. In fact,

$$\int_0^1 e^{2\pi imx} e^{-2\pi inx}\,dx = \delta_{m,n},$$

with the *Kronecker δ-symbol* $\delta_{m,n}$ which is defined to be 1 for $m = n$ and 0 for $m \neq n$.

By Theorem 8.28 the trigonometric system is total in $L^p(\mathbb{T})$ $(1 \leq p < \infty)$, and in $C(\mathbb{T})$.

THEOREM 8.29 (The Gram-Schmidt orthogonalization). *Let H be a Hilbert space and $\{f_j : j = 0, 1, 2, \ldots\}$ be a linearly independent family in H. Then there exists an orthonormal system $\{e_k : k = 0, 1, 2, \ldots\}$ such that*

(i) *e_k is a linear combination of $\{f_j\}$,*
(ii) *if $\{f_j\}$ is a total system, then $\{e_k\}$ is total, too.*

PROOF. To prove the existence of an orthonormal system we put $e_0 = f_0/\|f_0\|$, and assume that we have chosen an orthonormal system $e_j, j = 0, 1, \ldots, k$, whose members are linear combinations of $f_0, f_1, \ldots, f_k$. Let

$$g_{k+1} = f_{k+1} - \sum_{j=0}^{k} (f_{k+1}, e_j) e_j.$$

Then, as is easily shown, we have $(g_{k+1}, e_j) = 0$, $j = 0, 1, \ldots, k$. Observe that g_{k+1} is a linear combination of $f_0, f_1, \ldots, f_{k+1}$ and that $g_{k+1} \neq 0$. Therefore if

[10]The terninology total is not in popular use. However, in this book we use it to avoid confusion with the notion "closed" in general topology.

$e_{k+1} = g_{k+1}/||g_{k+1}||$, we have $k+2$ orthonormal elements $e_0, \ldots, e_k, e_{k+1}$. Thus the existence of an orthonormal system is established by induction.

By construction e_k is a linear combination of $\{f_j\}$ and, conversely, f_j is a linear combination of $\{e_k\}$. Thus $\{e_k\}$ is total if $\{f_k\}$ is total. □

The above method of constructing an orthogonal system is called the *Gram-Schmidt ortnogonalization process.*

For $f \in H$ the inner product

$$\widehat{f}_j = (f, e_j)$$

is called the jth Fourier coefficient with respect to the orthonormal system $\{e_j\}$.

LEMMA 8.9. *Let $\{e_k\}$ be an orthonomal system in a Hilbert space H. Fix an integer n. Then the quantity*

$$||f - \sum_{k=1}^{n} c_k e_k||$$

is minimal if and only if $c_k = (f, e_k)$, $k = 1, 2, \ldots, n$.

PROOF. Since $||f - \sum_{k=0}^{n} c_k e_k||^2 = (f - \sum_{k=0}^{n} c_k e_k, f - \sum_{k=0}^{n} c_k e_k)$, expanding the right-hand side we get

$$\begin{aligned} &||f - \sum_{k=0}^{n} c_k e_k||^2 \\ &= (f, f) - \sum_{k=0}^{n} c_k \overline{(f, e_k)} - \sum_{k=0}^{n} \bar{c}_k (f, e_k) + \sum_{j,k=0}^{n} c_j \bar{c}_k (e_j, e_k) \\ &= ||f||^2 + \sum_{k=1}^{n} |(f, e_k) - c_k|^2 - \sum_{k=0}^{n} |(f, e_k)|^2. \end{aligned} \tag{8.48}$$

The right-hand side attains its minimum if and only if its middle term is minimal, which is zero when $c_k = (f, e_k)$. □

Put $c_k = (f, e_k)$ in the equation (8.48). Since the left-hand side of (8.48) is nonnegative, we have

$$||f||^2 \geq \sum_{k=0}^{n} |\widehat{f}_k|^2.$$

Letting $n \to \infty$, we obtain

THEOREM 8.30. *Let $\{e_k : k = 0, 1, 2, \ldots\}$ be an orthonormal system in a Hilbert space H. Then we have*

$$||f||^2 \geq \sum_{k=0}^{\infty} |(f, e_k)|^2 \qquad (\textit{Bessel's inequality})$$

for $f \in H$.

THEOREM 8.31. *For an orthonormal system $\{e_k : k = 0, 1, 2, \ldots\}$ in a Hilbert space H the following are equivalent:*

(i) *If $(f, e_k) = 0$ for all k, then $f = 0$.*
(ii) *The system $\{e_k\}$ is total in H.*

(iii) *For every* $f \in H$ *we have*

$$\|f - \sum_{k=0}^{n}(f, e_k)e_k\| \to 0 \quad as \ n \to \infty.$$

(iv) *For every* $f \in H$ *we have*

$$\|f\| = \left(\sum_{k=0}^{\infty} |(f, e_k)|^2\right)^{1/2} \qquad (Parseval's\ formula).$$

An orthonormal system is said to be *complete* if it satisfies condition (i) of Theorem 8.31. Thus an orthonormal system in a Hilbert space is complete if and only if it is total.

PROOF. The implications (iv)⇒ (i) and (iii) ⇒ (ii) are obvious.
By Lemma 8.9 we have

$$\|f - \sum_{k=0}^{n} \widehat{f}_k e_k\|^2 = \|f\|^2 - \sum_{k=0}^{n} |(f, e_k)|^2.$$

Assume (ii). Then by Lemma 8.9 the left-hand side converges to zero, which implies Parseval's formula. Finally, to prove the implication (i) ⇒ (iii) let M be the closure of the finite linear combinations of $\{e_k\}$. (iii) holds for $f \in M$ by Lemma 8.9 and the definition of M. On the other hand if $g \in M^{\perp}$, then $(g, e_k) = 0$ for all $k = 0, 1, \ldots$. Thus $g = 0$ by the assumption. Thus $M^{\perp} = \{0\}$, from which it follows that $H = M$. □

8.8.2. Examples of orthogonal systems. We shall mention some examples of orthogonal systems. Each of them is closely related to operations of a commutative group, a non-commutative group, or a differential operator, but here we shall pay attention only to the orthogonality.

EXAMPLE 8.15 (Trigonometric system). As we have shown, the system

$$\{e^{2\pi i n x} : n \in \mathbb{Z}\}$$

is a complete orthonormal system of $L^2[0, 1)$. The system

$$\{1, \sqrt{2}\cos 2\pi n x, \sqrt{2}\sin 2\pi n x \ : n = 1, 2, \ldots\}$$

is also called a trigonometric system. In fact, we have

$$2\int_0^1 \cos 2\pi m x \cos 2\pi n x \, dx = \delta_{m,n}, \quad 2\int_0^1 \sin 2\pi m x \sin 2\pi n x \, dx = \delta_{m,n},$$

and

$$2\int_0^1 \cos 2\pi m x \sin 2\pi n x \, dx = 0 \quad (m,\ n = 1, 2, \ldots).$$

Next is a simple example of wavelets, which are treated in the next chapter.

EXAMPLE 8.16 (The Haar system). The step function

$$\psi_H(x) = \begin{cases} 1 & (0 \le x < 1/2), \\ -1 & (1/2 \le x < 1), \\ 0 & (\text{otherwise}) \end{cases}$$

is called the Haar function. The set of functions

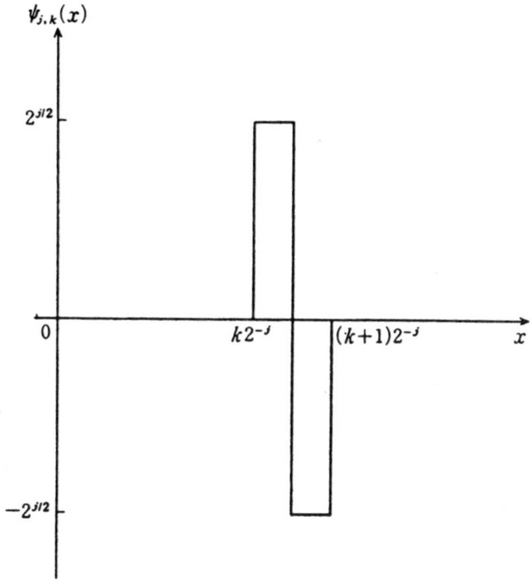

FIGURE 8.7. Haar functions

$$\psi_{j,k}(x) = 2^{j/2}\psi_H(2^j x - k), \quad k = 0, 1, \dots, 2^j - 1; j = 0, 1, \dots,$$

is called the *Haar system* (see Figure 8.7).

The set $\{\psi_{j,k}\}$ is an orthonormal system of $L^2[0,1)$. In fact, if $k \neq l$, then we have $\int \psi_{j,k}\psi_{j,l}dx = 0$, since $\operatorname{supp}\psi_{j,k} \cap \operatorname{supp}\psi_{j,l} = \emptyset$. If $i < j$, then $\psi_{i,k}$ is constant on $\operatorname{supp}\psi_{j,k}$. Thus $\int \psi_{i,k}\psi_{j,k}dx = (\pm 2^{i/2}) \times \int \psi_{j,k}dx = 0$.

In the following we call the system $\{\psi_{j,k}(x) : j, k \in \mathbb{Z}\}$ the Haar system; it is an orthonormal system of $L^2(\mathbb{R})$.

EXAMPLE 8.17 (The Rademacher system). The Rademacher system is a set of functions defined on the product set

$$D(2) = \prod_{j=1}^{\infty}\{x = (x_1, x_2, \dots, x_j, \dots) : x_j = 0 \text{ or } 1\}.$$

For $x = (x_1, x_2, \dots, x_j, \dots) \in D(2)$ define

$$r_0(x) \equiv 1, \quad \text{and} \quad r_j(x) = (-1)^{x_j} \ (j = 1, 2, \dots),$$

so that $r_j(x) = -1$ or 1 for each x.

We remark that the mapping

$$D(2) \ni x \to t = t(x) = \sum_{j=1}^{\infty} x_j 2^{-j} \in [0,1]$$

is surjective and one-to-one except on the dyadic rationals in $[0,1]$, which are the points that can be expressed as finite sums $\sum x_j 2^{-j}$. Therefore we can suppose that the function r_j is defined for almost every t in $[0,1]$ by the formula $r_j(t(x)) = r_j(x)$. The family $\{r_j(t) : j = 0, 1, \dots\}$ is called the *Rademacher system* (see Figure 8.8).

Let $I_{j,k} = (k2^{-j}, (k+1)2^{-j})$, $k = 0, 1, \dots, 2^j - 1$. The function $r_j(t)$ is constant on each interval $I_{j,k}$. More precisely, $r_j(t)$ equals 1 on the left half-interval $I_{j,2k}$ of $I_{j-1,k}$ and equals -1 on the right half-interval $I_{j,2k+1}$ of $I_{j-1,k}$. Thus, if $i < j$, then we have $\int_{I_{j-1,k}} r_i(t)r_j(t)dt = \pm\int_{I_{j-1,k}} r_j(t)dt = 0$ for each k, from which the orthogonality follows:

$$\int_0^1 r_i(t)r_j(t)dt = 0 \quad (i < j).$$

EXAMPLE 8.18 (The Walsh system). The Walsh system is also a family of functions defined on the product set $D(2)$. Let $\Delta(2)$ be a subset of $D(2)$ such that

$$\Delta(2) = \{(\varepsilon_1, \varepsilon_2, \dots, \varepsilon_j, \dots) \in D(2) : \varepsilon_j = 0 \text{ except for a finite number of } j\text{'s}\}.$$

For $\varepsilon = (\varepsilon_1, \varepsilon_2, \dots, \varepsilon_j, \dots) \in \Delta(2)$ the *Walsh function* is defined by

$$w_\varepsilon(x) = \prod_{j=1}^{\infty}(-1)^{x_j\varepsilon_j} \quad (x = (x_1, x_2, \dots) \in D(2)).$$

For the same reason as for the Rademacher functions, $w_\varepsilon(x)$ can be assumed to be a function in $[0,1]$. Furhtermore, since the mapping $\Delta(2) \ni \varepsilon \to n(\varepsilon) = \sum_{j=1}^{\infty} \varepsilon_j 2^{j-1} \in \mathbb{Z}_+$ is bijective, we can express $w_\varepsilon(x)$ as $w_{n(\varepsilon)}(t(x))$. Thus the Walsh system can be taken to be a family of functions in $[0,1]$ enumerated by nonnegative integers.

By definition we have $w_n(t) = \prod_{j=1}^{\infty} r_j^{\varepsilon_j}(t)$, where $n = \varepsilon_0 2^0 + \varepsilon_1 2^1 + \varepsilon_2 2^2 + \cdots$. In particular, we have $w_{2^j}(t) = r_j(t)$.

Now we shall prove the orthogonality of the *Walsh system* $\{w_n(t)\}$. Let $p = n(\varepsilon)$ and $q = n(\eta)$ be distinct integers. Let $\delta = (\delta_0, \delta_1, \dots)$ be an element of $\Delta(2)$ such that $\delta_j \equiv \varepsilon_j + \eta_j \pmod 2$. Since $p \neq q$, $\delta \neq 0$. Let j be the maximal integer such that $\delta_j = 1$. Then we have

$$w_p(t(x))w_q(t(x)) = w_\delta(t(x)) = r_0^{\delta_0}(t)r_1^{\delta_1}(t)\dots r_j^{\delta_j}(t).$$

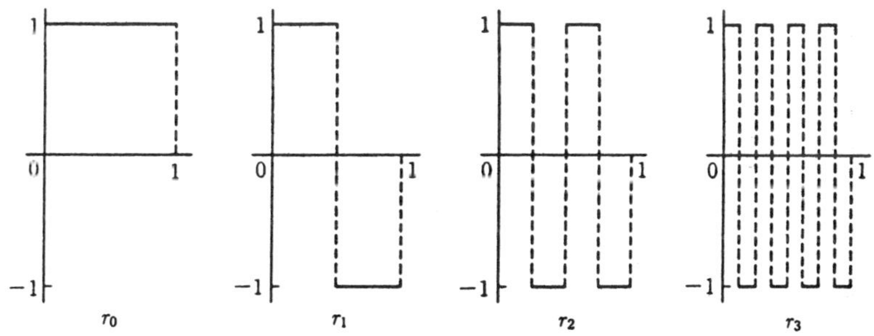

FIGURE 8.8. Rademacher functions

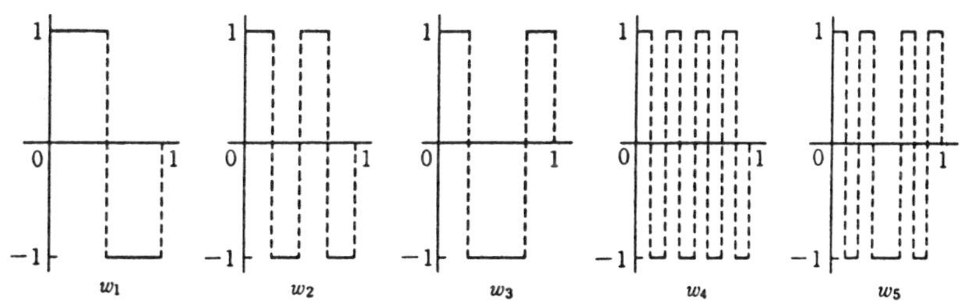

FIGURE 8.9. Walsh functions

Observe that $r_j^{\delta_j}(t)$ has different signs on the left half-interval of the interval $I_{j-1,k}$ and on the right half-interval of $I_{j-1,k}$ for each k. But $r_i^{\delta_i}(t)$ with $i < j$ are constant on each $I_{j-1,k}$. Thus $\int_{I_{j-1,k}} w_\delta(t)dt = 0$ for every k. Summing both sides over k, we get the orthogonal relation $\int_0^1 w_p w_q dt = 0$.

EXAMPLE 8.19 (The Chebyshev polynomial). The nth Chebyshev polynomial $T_n(x)$ is defined by the formula

$$T_n(\cos\theta) = \cos n\theta.$$

Observe that

$$\cos n\theta = \sum_{0\le k\le n/2} \binom{n}{2k} (\cos\theta)^{n-2k}(-\sin^2\theta)^k,$$

which is a consequence of Euler's formula:

$$\cos n\theta = \Re e^{in\theta} = \Re(\cos\theta + i\sin\theta)^n = \Re\left\{\sum_{j=0}^{n} \binom{n}{j} (\cos\theta)^{n-j}(i\sin\theta)^j\right\}.$$

Therefore we have

$$T_n(x) = \sum_{0\le k\le n/2} \binom{n}{2k} x^{n-2k}(x^2-1)^k.$$

The Chebyshev system $\{T_n(x) : n = 0, 1, 2, \dots\}$ is complete and orthonormal in the weighted space $L^2((-1,1); w(x)dx)$ with the weight $w(x) = \pi^{-1}(1-x^2)^{-1/2}$.

In fact, we have

$$\begin{aligned}\int_{-1}^{1} T_m(x)T_n(x)w(x)dx &= \frac{1}{\pi}\int_0^\pi T_m(\cos\theta)T_n(\cos\theta)d\theta\\ &= \frac{1}{\pi}\int_0^\pi \cos m\theta\cos n\theta d\theta = \delta_{m,n}.\end{aligned}$$

To prove the completeness, let f be a function in $L^2((-1,1); w(x)dx)$ such that

$$\int f(x)T_n(x)w(x)dx = 0 \quad \text{for all } n = 0, 1, 2, \dots$$

The left-hand side equals $\int_0^\pi f(\cos\theta)\cos n\theta d\theta$. Therefore, by the completeness in $L^2(0,\pi)$ of the cosine system $\{\cos n\theta\}$ we have $f(x) = 0$ a.e.

EXAMPLE 8.20 (The Legendre, Jacobi, Hermite, and Laguerre polynomials). Let $w(x)$ be a nonnegative function and $-\infty \le a < b \le \infty$. Let $\{p_n(x)\}$ be the Gram-Schmidt orthogonalization of the monomials $\{1, x, x^2, \dots\}$, applied in the weighted Lebesgue space $L^2((a,b); w(x)dx)$.

(i) When $a = -1$, $b = 1$, and $w(x) = 1$, we have $p_n(x) = c_n P_n(x)$, where P_n is a *Legendre polynomial*:

$$P_n(x) = \frac{(-1)^n}{2^n n!}\frac{d^n}{dx^n}(1-x^2)^n,$$

and

$$c_n = \left(\frac{2n+1}{2}\right)^{1/2}.$$

(ii) When $a = -1$, $b = 1$, and $w(x) = (1-x)^\alpha(1+x)^\beta$, we have $p_n(x) = c_n P_n^{(\alpha,\beta)}(x)$, where $P_n^{(\alpha,\beta)}(x)$ is a *Jacobi polynomial* of order (α, β):

$$P_n^{(\alpha,\beta)}(x) = (1-x)^{-\alpha}(1+x)^{-\beta}\frac{(-1)^n}{2^n n!}\frac{d^n}{dx^n}(1-x)^{\alpha+n}(1+x)^{\beta+n},$$

and

$$c_n = \left(\frac{\Gamma(n+1)\Gamma(\alpha+\beta+n+1)(\alpha+\beta+2n+1)}{\Gamma(\alpha+n+1)\Gamma(\beta+n+1)2^{\alpha+\beta+1}}\right)^{1/2}.$$

(iii) When $a = -\infty$, $b = \infty$, and $w(x) = e^{-x^2}$, we have $p_n(x) = c_n H_n(x)$, where

$$H_n(x) = (-1)^n e^{x^2}\frac{d^n}{dx^n}e^{-x^2},$$

and

$$c_n = \left(\frac{1}{2^n n! \sqrt{\pi}}\right)^{1/2}.$$

$H_n(x)$ is called an *Hermite polynomial.*

(iv) When $a = 0$, $b = \infty$ and $w(x) = x^\alpha e^{-x}$, $\alpha > -1$, we have $p_n(x) = c_n L_n^\alpha(x)$, where

$$L_n^\alpha(x) = \frac{1}{n!}e^x x^{-\alpha}\frac{d^n}{dx^n}(e^{-x}x^{\alpha+n}),$$

and

$$c_n = (-1)^n\left(\frac{\Gamma(n+1)}{\Gamma(\alpha+n+1)}\right)^{1/2}.$$

$L_n^\alpha(x)$ is called a *Laguerre polynomial.*

We shall show that in case (i) the orthogonalization $p_n(x)$ is a constant multiple of the Legendre polynomial $P_n(x)$. The proof is similar for the expression by Jacobi, Hermite, or Laguerre polynomials.

First, we remark the following.

(a) The Gram-Schmidt orthogonalization $p_n(x)$ and the Legendre polynomial $P_n(x)$ are polynomials of order n.

(b) Since the system $\{x^n : n = 0, 1, \dots\}$ is total in $L^2(-1, 1)$, the orthogonalization $\{p_n(x)\}$ is also total in $L^2(-1, 1)$.

(c) For $m < n$ we have $\int_{-1}^1 x^m P_n(x)dx = 0$. In fact, the left-hand side becomes, using integration by parts,

$$\left[x^m \frac{d^{n-1}}{dx^{n-1}}(x^2-1)^n\right]_{-1}^1 - m\int_{-1}^1 x^{m-1}\frac{d^{n-1}}{dx^{n-1}}(x^2-1)^n dx.$$

It is obvious that the first term vanishes, and, integrating repeatedly by parts, we see that the second term also vanishes. Therefore $\int_{-1}^1 p_j(x)P_n(x)dx = 0$ for $j < n$.

Let $P_n(x) = \sum_{j=0}^\infty c_j p_j(x)$ be the Fourier series expansion of P_n with respect to $\{p_j\}$. Then $c_j = 0$ for $j < n$. On the other hand, since any x^k can be expressed as a linear combination of $p_0(x), \dots, p_k(x)$, we have $c_j = 0$ for $j > n$. Thus $P_n = c_n p_n$.

Since p_n is normalized, $\int_{-1}^{1} P_n^2 dx = c_n^2 \int_{-1}^{1} p_n^2 dx = c_n^2$. Integrating by parts n times, we get

$$\begin{aligned} c_n^2 &= \int_{-1}^{1} P_n(x)P_n(x)dx \\ &= (-1)^n \left(\frac{1}{2^n n!}\right)^2 \int_{-1}^{1} (1-x^2)^n \frac{d^{2n}}{dx^{2n}}(1-x^2)^n dx. \end{aligned}$$

By the equation $\dfrac{d^{2n}}{dx^{2n}}(1-x^2)^n = (-1)^n(2n)!$ the last term becomes

$$\frac{(2n)!}{(2^n n!)^2} \int_{-1}^{1} (1-x^2)^n dx = \frac{(2n)!\Gamma\left(n+\frac{1}{2}\right)\Gamma\left(\frac{1}{2}\right)}{(2^n n!)^2 \Gamma\left(n+\frac{1}{2}\right)} = \frac{1}{n+\frac{1}{2}}.$$

Thus $c_n = \pm\sqrt{(2n+1)/2}$. Since the coefficients of x^n in $p_n(x)$ and $P_n(x)$ are positive, we obtain $c_n = \sqrt{(2n+1)/2}$.

8.9. Problems

8.1. Using the residue of the function $e^{-az}/(1+z^2)$, compute

$$\frac{1}{\pi}\int_{-\infty}^{\infty} \frac{e^{-2\pi i t|\xi| s}}{1+s^2} ds = e^{-2\pi|\xi| t} \quad \text{for } t > 0.$$

8.2. Compute the Fourier transforms of the following distributions.

(a) x^k $(k = 1, 2, \dots)$,

(b) $Y(x)$ (Heaviside function),

(c) $f(x) = Y(x)e^{-x}$.

8.3. For a bounded Borel measure $\mu \in M(\mathbb{R})$ the integral

$$\widehat{\mu}(\xi) = \int_{\mathbb{R}} e^{-2\pi i \xi x} d\mu(x)$$

is called the *Fourier-Stieltjes transform*. If μ is a positive measure, prove that $\widehat{\mu}$ is *positive definite*, that is, for every $\xi_1, \xi_2, \dots, \xi_n$ in $\mathbb{R}$ and $c_1, c_2, \dots, c_n$ in $\mathbb{C}$ we have

$$\sum_{j,k} c_j \overline{c_k}\, \widehat{\mu}(\xi_j - \xi_k) \geq 0.$$

8.4. The space $L^1(\mathbb{R})$ is a commutative algebra with convolution product (see Theorem 6.11). Prove that $L^1(\mathbb{R})$ does not possess a unit. (Hint: Suppose that it has unit and compute its Fourier transform.)

8.5. If $f(x) = ce^{-a(x-m)^2}e^{2\pi i b x}$, where a, b, and m are real constants, then prove that f attains equality in Heisenberg's inequality.

8.6. Prove that the Gauss-Weierstrass kernel satisfies the conditions for a summability kernel (s1), (s2)′ and (s3)′ with t replaced by $t^{1/2}$.

CHAPTER 9

Wavelet Analysis

In the theory of Fourier analysis a function is analyzed into simple oscillations of which it consists, and is synthesized from all simple oscillations which compose the function. But, roughly speaking, we cannot get all the information on a function in a given neighborhood of the origin from the behavior of its Fourier tranform around the origin, as the Paley-Wiener theorem and Heisenberg's inequality show. The windowed Fourier transform is one attempt to analyse data by time and frequency parameters simultaneously, and the wavelet transform is closely related to that idea.

The theory of wavelets is based on a simple and clear idea, as many distinguished theories are, and is developed in both pure mathematics and engineering. The reader will find a more detailed physical and historical background of wavelet theory in the book of Daubechies [Da].

Nowaday, wavelet theory is mathematically based mainly on Fourier analysis. In this chapter, a short mathematical introduction to wavelet analysis is given. We give basic properties of wavelet transforms and constructions of orthonormal system of wavelets.

9.1. Wavelet transform

Let ψ be a function in $L^2(\mathbb{R})$ such that $\|\psi\|_2 = 1$. Furthermore we assume that

$$C_\psi = \int_{-\infty}^{\infty} |\widehat{\psi}(\xi)|^2 \frac{d\xi}{|\xi|} < \infty. \tag{9.1}$$

If $\psi \subset L^1(\mathbb{R})$, then (9.1) implies that $\widehat{\psi}(0) = 0$, and thus $\int_{-\infty}^{\infty} \psi(x)dx = 0$.

For $a, b \in \mathbb{R}$ put

$$\psi_{a,b}(x) = |a|^{-1/2}\psi(a^{-1}(x-b)).$$

By a change of variable we have $\|\psi_{a,b}\|_2 = \|\psi\|_2 = 1$.

$\psi_{a,b}(x)$ is considered as the wave $\psi(x)$ dilated by a and translated by b (see Figure 9.1).

DEFINITION 9.1. For a function f in the real axis, define

$$(W_\psi f)(a,b) = \int_{-\infty}^{\infty} \overline{\psi_{a,b}(x)} f(x)dx = \int_{-\infty}^{\infty} \frac{1}{\sqrt{|a|}} \overline{\psi\left(\frac{x-b}{a}\right)} f(x)dx.$$

$W_\psi f$ is called the *wavelet transform* of f with *analyzing wavelet* ψ. Here we assume that the integral exists.

The quantity $(W_\psi f)(a,b)$ is interpreted as information on how the function f contains the wave $\psi_{a,b}$ which has frequency a and is concentrated around b, comparing with the original function ψ. For example, if f coincides with $\psi_{a,b}$, then

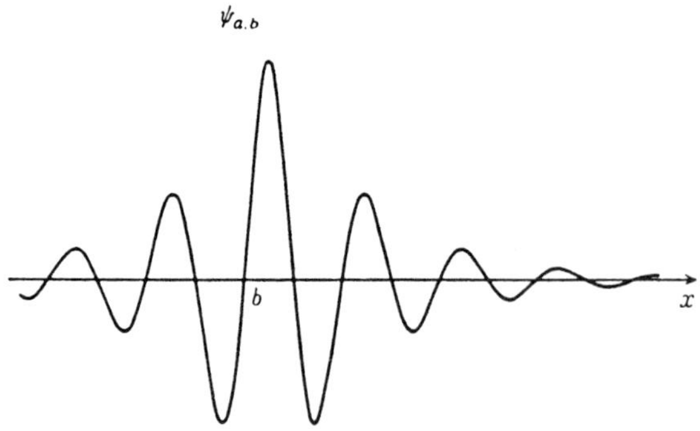

FIGURE 9.1. $\psi_{a,b}$

$(W_\psi f)(a,b) = \int |\psi|^2 dx = 1$. On the other hand, if f is a slowly varying wave, then $(W_\psi f)(a,b)$ equals approximately

$$\frac{1}{\sqrt{|a|}} \int (\text{constant}) \overline{\psi\left(\frac{x-b}{a}\right)} dx = 0.$$

PROPOSITION 9.1. *For $f, g \in L^2(\mathbb{R})$, we have*

$$\int_{-\infty}^{\infty} f\overline{g} dx = \frac{1}{C_\psi} \int_{-\infty}^{\infty} \int_{-\infty}^{\infty} W_\psi f(a,b) \overline{W_\psi g(a,b)} \frac{dadb}{a^2}. \tag{9.2}$$

PROOF. Since $\mathcal{F} W_\psi f(a, \cdot)(\xi) = \widehat{f}(\xi) |a|^{1/2} \overline{\widehat{\psi}(a\xi)}$, by the Plancherel theorem we have

$$\int_{-\infty}^{\infty} \frac{da}{a^2} \int_{-\infty}^{\infty} W_\psi f(a,b) \overline{W_\psi g(a,b)} db = \int_{-\infty}^{\infty} \frac{da}{a^2} \int_{-\infty}^{\infty} \widehat{f}(\xi) \overline{\widehat{g}(\xi)} |a| |\widehat{\psi}(a\xi)|^2 d\xi.$$

Observe that $\int |\widehat{\psi}(a\xi)|^2 da/|a| = C_\psi$. Then, by Fubini's theorem the right-hand side becomes $C_\psi \int \widehat{f}(\xi) \overline{\widehat{g}(\xi)} d\xi = C_\psi \int f(x) \overline{g(x)} dx$. □

Let us denote the inner products on $L^2(\mathbb{R})$ and $L^2(\mathbb{R}^2; a^{-2} dadb)$ by

$$(f, g) = (f, g)_{L^2} = \int f \overline{g} dx$$

and

$$(F, G)' = C_\psi^{-1} \int \int F(a,b) \overline{G(a,b)} a^{-2} dadb$$

respectively. Then the mapping $f \mapsto W_\psi f$ of $L^2(\mathbb{R})$ to $L^2(\mathbb{R}^2; a^{-2} dadb)$ is isometric by (9.2). Let V be the range of the mapping. Then, W_ψ is a unitary mapping of $L^2(\mathbb{R})$ to V. In fact, we have the inversion formula

$$f(x) = \frac{1}{C_\psi} \int_{\mathbb{R}^2} W_\psi f(a,b) \psi_{a,b}(x) \frac{dadb}{a^2}, \tag{9.3}$$

where the integral converges in the L^2 sense for $f \in L^2(\mathbb{R})$.

To pove this formula let $A > \varepsilon > 0$ and $B > 0$, and put

$$s_{\varepsilon,A,B}(f)(x) = \frac{1}{C_\psi} \int_{\varepsilon < |a| < A} \int_{|b| < B} W_\psi f(a,b) \psi_{a,b}(x) \frac{dadb}{a^2}.$$

Since $W_\psi f(a,b) \in L^2(\mathbb{R}^2; a^{-2} dadb)$, the integral converges in the sense of Lebesgue.

THEOREM 9.1. *For* $f \in L^2(\mathbb{R})$

$$\|s_{\varepsilon,A,B}(f) - f\|_2 \to 0 \quad \textit{as } \varepsilon \to 0 \textit{ and } A, B \to \infty.$$

PROOF. First observe that

$$\|f - s_{\varepsilon,A,B}(f)\|_2 = \sup_{\|g\|_2=1} |(f - s_{\varepsilon,A,B}(f), g)|,$$

and by Fubini's theorem

$$(s_{\varepsilon,A,B}(f), g) = \frac{1}{C_\psi}\int_{\varepsilon<|a|<A}\int_{|b|<B} W_\psi f(a,b)\overline{W_\psi g(a,b)}\frac{dadb}{a^2}.$$

Applying formula (9.2) to the inner product (f,g), we have

$$|(f - s_{\varepsilon,A,B}(f), g)| = \sup_{\|g\|_2=1}\left|\frac{1}{C_\psi}\int\int_D W_\psi f(a,b)\overline{W_\psi g(a,b)}\frac{dadb}{a^2}\right|,$$

where $D = \{(a,b) : \varepsilon < |a| < A,\ |b| < B\}^c$. By Schwarz's inequality the last term is bounded by

$$\le \left(\frac{1}{C_\psi}\int\int_D |W_\psi f(a,b)|^2\frac{dadb}{a^2}\right)^{1/2}\left(\frac{1}{C_\psi}\int\int_{\mathbb{R}^2} |W_\psi g(a,b)|^2\frac{dadb}{a^2}\right)^{1/2}$$
$$= \left(\frac{1}{C_\psi}\int\int_D |W_\psi f(a,b)|^2\frac{dadb}{a^2}\right)^{1/2}\|g\|_2.$$

Notice that $W_\psi f \in L^2(\mathbb{R}^2; a^{-2}dadb)$ and that $D \to \emptyset$ as $\varepsilon \to 0$ and $A, B \to \infty$. Then the first factor of the last term converges to 0. □

We can choose as an analyzing wavelet any function in L^2 which satisfies condition (9.1).

EXAMPLE 9.1 (The Haar wavelet). The Haar function ψ_H satisfies condition (9.1), since $\int \psi_H dx = 0$, and is sometimes called the Haar wavelet.

EXAMPLE 9.2 (The French hat wavelet). The French hat wavelet (Figure 9.2) is defined by the function

$$\psi(x) = \begin{cases} 1/\sqrt{3}, & |x| < 1, \\ -1/(2\sqrt{3}), & 1 \le |x| < 3, \\ 0, & 3 \le |x|. \end{cases}$$

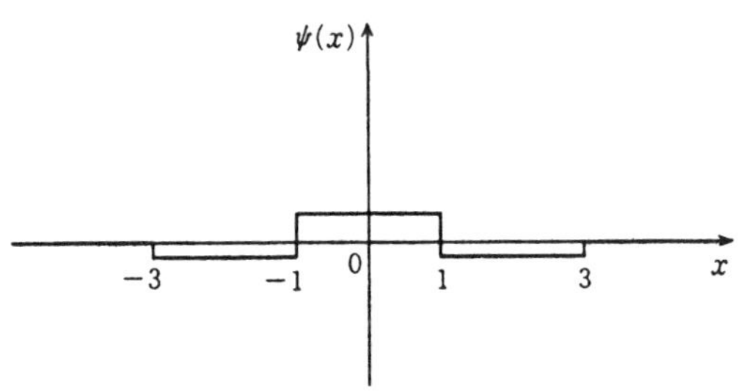

FIGURE 9.2. French hat function

The next example is a smooth wavelet with similar shape.

EXAMPLE 9.3 (The Mexican hat wavelet). The Mexican hat wavelet (Figure 9.3) is defined by the function

$$\psi(x) = \frac{8}{\sqrt{3}}(1 - 2\pi x^2)e^{-\pi x^2}.$$

Since $\psi(x) = -\frac{4}{\sqrt{3}\pi}\frac{d^2}{dx^2}e^{-\pi x^2}$, we have $\widehat{\psi}(\xi) = -\frac{4}{\sqrt{3}\pi}(2\pi i\xi)^2 e^{-\pi\xi^2}$.

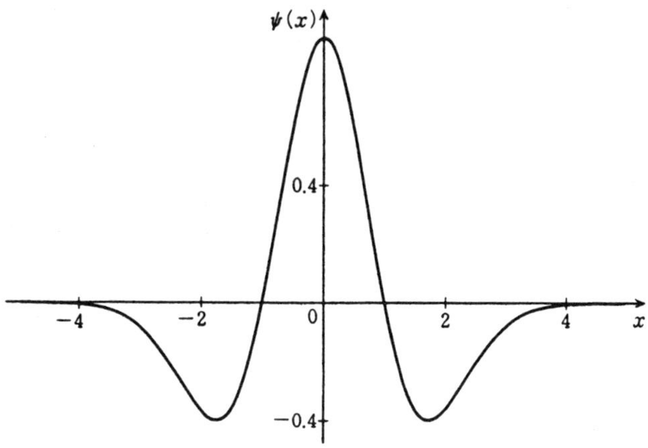

FIGURE 9.3. Mexican hat function

Another smooth wavelet is given by $C(e^{-2\pi i\rho x} - e^{-\pi\rho^2})e^{-\pi\xi^2}$, where C and ρ are nonzero constants.

9.2. Wavelet expansion

9.2.1. Discretization of wavelets. It is not always efficient to compute the wavelet transform $W_\psi f(a,b)$ for all $a > 0$ and b, and then to reconstruct the function f from $W_\psi f(a,b)$ in $(a,b) \in \mathbb{R}^2$. It seems natural to take discrete parameters $(a_0^{-j}, a_0^j b_0 k)$, $k, j \in \mathbb{Z}$, where $a_0 > 1$ and b_0 are positive fixed numbers. This implies choosing data from each box in the time-frequency coordinates (see Figure 9.4).

The most standard discretization of the wavelet transform is given by

$$\psi_{j,k}(x) = 2^{j/2}\psi(2^j x - k), \quad k, j \in \mathbb{Z}.$$

For $f \in L^2(\mathbb{R})$ the wavelet coefficients of f are defined by

$$c_{j,k} = \int f(x)\overline{\psi_{j,k}(x)}dx.$$

Several methods are known to synthesize the function f from the coefficents $\{c_{j,k}\}$. If the system $\{\psi_{j,k}(x)\}$ is complete and orthonormal in $L^2(\mathbb{R})$, then we have already shown that the series

$$\sum_{j,k=-\infty}^{\infty} c_{j,k}2^{j/2}\psi(2^j x - k) \tag{9.4}$$

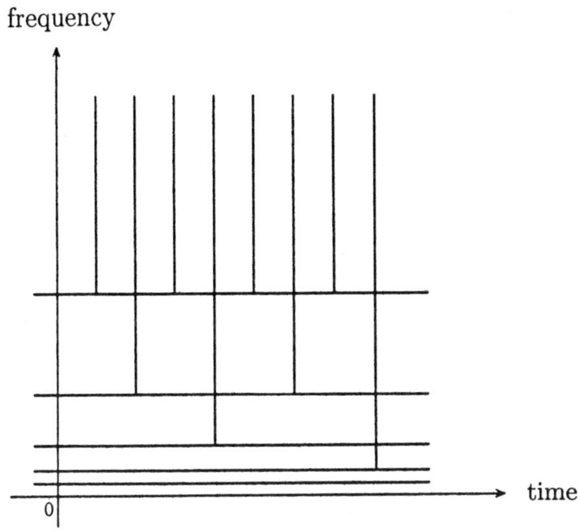

FIGURE 9.4. Time-frequency coordinates

converges to f in L^2. In the following we shall restrict our attension only orthogonal wavelets. Another useful methods are provided by notions of biorthogonal system, frame and so on (see, for example [Da]).

9.2.2. The Haar wavelet. In this section we shall consider the Haar wavelet to make clear the general argument, where wavelets are introduced via multiresolution analysis.

Put $\varphi(x) = \varphi_H(x) = 1\,(0 \leq x < 1),\ = 0\,(\text{otherwise})$.

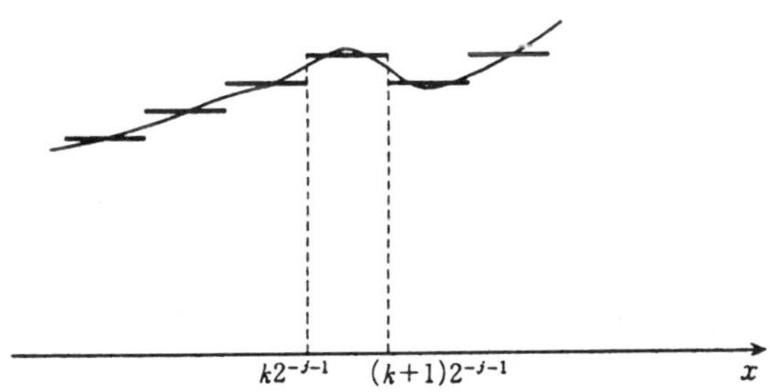

FIGURE 9.5. Approximation of level j

Let $f \in C_c(\mathbb{R})$. For $k = 0, \pm 1, \pm 2, \ldots$ let

$$c_{0,k} = \int_{\mathbb{R}} f(x)\varphi(x-k)dx.$$

Since $c_{0,k}$ is the mean value of f on the interval $[k, k+1)$, the step function

$$f_0(x) = \sum_{k=-\infty}^{\infty} c_{0,k}\varphi(x-k)$$

gives an approximation of $f(x)$. We call this the approximation of f of level 0 (see Figure 9.5).

Next put

$$c_{1,k} = \int_{\mathbb{R}} f(x) 2^{1/2} \varphi(2x-k)dx,$$

which is the mean value of f multiplied by $2^{-1/2}$ on the interval $[k/2, (k+1)/2)$. Thus the funcion defined by

$$f_1(x) = \sum_{k=-\infty}^{\infty} c_{1,k} 2^{1/2} \varphi(2x-k)$$

gives a better approximation, called the approximation of level 1. In the same way the approximation of f of level j is defined by

$$f_j(x) = \sum_{k=-\infty}^{\infty} c_{j,k} 2^{j/2} \varphi(2^j x - k), \quad c_{j,k} = \int_{\mathbb{R}} f(x) 2^{j/2} \varphi(2^j x - k)dx.$$

Let $V_j = \overline{\text{span}}^{L^2}\{\varphi(2^j x - k) : k \in \mathbb{Z}\}$.

The system $\{V_j\}$ of closed subspaces of $L^2(\mathbb{R})$ has the following properties:

(i) By (9.5) we have $V_j \subset V_{j+1}$, and every approximation f_j of level j of a function in $L^2(\mathbb{R})$ belongs to the space V_j.

(ii) If $f \in L^2(\mathbb{R})$, then f_j converges to f in L^2. In fact, for every $\varepsilon > 0$ there exists $u \in C_c(\mathbb{R})$ such that $\|f-u\|_2 < \varepsilon$. We have, by Schwarz's inequality,

$$|f_j(x) - u_j(x)|^2 = |\int [f(t) - u(t)] 2^j \varphi(2^j t - k)dt|^2 \leq 2^j \int_{I_{j,k}} |f-u|^2 dt$$

on the interval $I_{j,k} = [k2^{-j}, (k+1)2^{-j})$. Integrating on $I_{j,k}$ and the summing over k, we get

$$\|f_j - u_j\|_2 \leq \|f-u\|_2 < \varepsilon.$$

Thus

$$\|f_j - f\|_2 \leq \|f_j - u_j\|_2 + \|u_j - u\|_2 + \|u - f\|_2 < 2\varepsilon + \|u_j - u\|_2.$$

The support of u_j is contained in a compact set for $j \geq 0$, and u_j converges uniformly to u. Thus $\|u_j - u\|_2 \to 0$ as $j \to \infty$. Thus $\|f_j - f\|_2 \to 0$ as $j \to \infty$ for every $f \in L^2(\mathbb{R})$.

Therefore the union $\bigcup_{j=-\infty}^{\infty} V_j$ is dense in L^2.

On the other hand, it is an easy exercise to prove that $\bigcap_{j=-\infty}^{\infty} V_j = \{0\}$.

(iii) $f(x) \in V_0 \Leftrightarrow f(2^j x) \in V_j$ for all $j \in \mathbb{Z}$.

(iv) $f(x) \in V_0 \Leftrightarrow f(x-k) \in V_0$ for all $k \in \mathbb{Z}$.

(v) The family $\{\varphi_H(x-k) : k \in \mathbb{Z}\}$ is an orthonomal basis of V_0. We call φ_H the *Haar scaling function*.

The scaling function φ_H belongs to V_1, since $V_0 \subset V_1$. On the other hand, $\{\varphi_{1,k}\}$ is an orthonormal basis in V_1. In particular, we have the Fourier expansion

$$\varphi_H(x) = \varphi_H(2x) + \varphi_H(2x-1). \tag{9.5}$$

Now we recall the Haar function ψ_H, which is defined by $\psi_H(x) = 1\,(0 \leq x < 1/2)$, $= -1\,(1/2 \leq x < 1)$, and $= 0$ (otherwise). The functions

$$\psi_{j,k}(x) = 2^{j/2} \psi_H(2^j x - k) \quad (j, k \in \mathbb{Z})$$

are called Haar wavelets (see Figure 8.7). These functions make up a complete orthonormal system of $L^2(\mathbb{R})$ as we have shown in Chapter 8.

Our next aim is to make clear the relation between the Haar scaling function and the Haar wavelets, and to describe the limiting process of a sequence of approximations of level j's by a Haar wavelet expansion.

As is easily seen,

$$\varphi_H(2x) = \frac{1}{2}[\varphi_H(x) + \psi_H(x)], \quad \varphi_H(2x-1) = \frac{1}{2}[\varphi_H(x) - \psi_H(x)],$$

from which it follows that

$$\varphi_{j,2k}(x) = \frac{1}{\sqrt{2}}[\varphi_{j-1,k}(x) + \psi_{j-1,k}(x)],$$

$$\varphi_{j,2k+1}(x) = \frac{1}{\sqrt{2}}[\varphi_{j-1,k}(x) - \psi_{j-1,k}(x)],$$

and

$$\psi_{j,k}(x) = \sqrt{2}\varphi_{j+1,2k}(x) - \varphi_{j,k}(x).$$

For $f \in L^2(\mathbb{R})$ let f_j be the approximation of level j, that is,

$$f_j(x) = \sum_{k=-\infty}^{\infty} c_{j,k}\varphi_{j,k}(x), \quad c_{j,k} = \int f(x)\varphi_{j,k}(x)dx.$$

Define g_j by

$$g_j(x) = \sum_{k=-\infty}^{\infty} d_{j,k}\psi_{j,k}(x), \quad d_{j,k} = \int f(x)\psi_{j,k}(x)dx.$$

By definition we have

$$c_{j,2k} = 2^{-1/2}[c_{j-1,k} + d_{j-1,k}], \quad c_{j,2k+1} = 2^{-1/2}[c_{j-1,k} - d_{j-1,k}].$$

Thus the approximation of level j is transformed to

$$\begin{aligned} f_j(x) &= \sum_{k=-\infty}^{\infty} [c_{j,2k}\varphi_{j,2k}(x) + c_{j,2k+1}\varphi_{j,2k+1}(x)] \\ &= \sum_{k=-\infty}^{\infty} [c_{j-1,k}\varphi_{j-1,k}(x) + d_{j-1,k}\psi_{j-1,k}(x)] \\ &= f_{j-1}(x) + g_{j-1}(x). \end{aligned}$$

Let

$$W_j = \overline{\text{span}}^{L^2}\{\psi_{j,k} : k \in \mathbb{Z}\}.$$

We have $(\varphi_{j-1,k}, \psi_{j-1,l}) = 0\,(k, l \in \mathbb{Z})$, since $\varphi_{j-1,k}$ is constant on the support of $\psi_{j-1,l}$. Thus $V_{j-1} \perp W_{j-1}$. Furthermore, since $f_{j-1} \in V_{j-1}$ and $g_{j-1} \in W_{j-1}$, it follows from the last equality that

$$V_j = V_{j-1} \oplus W_{j-1}.$$

Iterating the above procedure, we get

$$\begin{aligned} f_j &= g_{j-1} + g_{j-2} + f_{j-2} \\ &= \cdots = g_{j-1} + g_{j-2} + \cdots + g_{j-m} + f_{j-m}. \end{aligned}$$

By § 9.2.2.(ii), $f_j \to f$ in L^2 as $j \to \infty$. On the other hand, $f_j \to 0$ weakly in L^2 as $j \to -\infty$.

In fact, let $g \in L^2(\mathbb{R})$. For a given $\varepsilon > 0$ choose $u \in C_c(\mathbb{R})$ so that $\|g-u\|_2 < \varepsilon$. We have, obviously,

$$(f_j, g) = (f_j, g-u) + (f, u_j).$$

The first term on the right-hand side is bounded by $\|f_j\|_2\|g-u\|_2 < \|f\|_2\varepsilon$. Let K be a finite interval which contains the support of u, and choose j so small that $K \subset [-2^{-j}, 2^{-j})$. Then

$$\begin{aligned}\|u_j\|_2^2 &= 2^j\left|\int_{-2^{-j}}^{0} u dx\right|^2 + 2^j\left|\int_{0}^{2^{-j}} u dx\right|^2 \\ &\le 2^j m(K)\int |u|^2 dx \to 0 \quad \text{as } j \to -\infty.\end{aligned}$$

Thus by Schwarz's inequality we have

$$|(f, u_j)| \le \|f\|_2\|u_j\|_2 \to 0 \quad \text{as } t \to -\infty.$$

Thus $f_j \to 0$ weakly in L^2.

Therefore we have

$$f(x) = \sum_{j=-\infty}^{\infty} g_j = \sum_{j=-\infty}^{\infty}\sum_{k=-\infty}^{\infty} d_{j,k}\psi_{j,k}(x),$$

where the series converge weakly in L^2. Since $\{\psi_{j,k}(x)\}$ is an orthonomal system, the last series converges in the L^2-sense, and thus converges to f in L^2.

In the following sections we shall construct orthonormal wavelets which are associated with a given multiresolution analysis.

9.3. Multiresolution analysis

9.3.1. Definition of multiresolution analysis.

DEFINITION 9.2. A *multiresolution analysis* in $L^2(\mathbb{R})$ is a resolution system $\{V_j\}$ which has the following properties:

(i) Each V_j is a closed subspace of $L^2(\mathbb{R})$ and

$$\cdots V_{-1} \subset V_0 \subset V_1 \subset V_2 \subset \cdots.$$

(ii) The union $\bigcup_{j=-\infty}^{\infty} V_j$ is dense in $L^2(\mathbb{R})$, and $\bigcap_{j=-\infty}^{\infty} V_j = \{0\}$.

(iii) $f(x) \in V_j \Leftrightarrow f(2x) \in V_{j+1}$ for all $j \in \mathbb{Z}$.

(iv) $f \in V_0 \Rightarrow f(x-k) \in V_0$ for all $k \in \mathbb{Z}$.

(v) There exists a function φ in V_0 such that $\{\varphi(x-k) : k \in \mathbb{Z}\}$ is an orthonormal basis of V_0.

The function φ is called a *scaling function*.

By conditions (v) and (iii) the functions $\varphi_{1,k}(x) = \sqrt{2}\varphi(2x-k)$ $(k \in \mathbb{Z})$ constitute an orthonormal basis in V_1. Since $V_0 \subset V_1$, the scaling function φ belongs to V_1 and has the Fourier expansion

$$\varphi(x) = \sum_k h_k\sqrt{2}\varphi(2x-k), \quad h_k = \int \varphi(x)\overline{\varphi_{1,k}(x)}dx. \tag{9.6}$$

Since V_j is a closed subspace of V_{j+1}, the space V_{j+1} has an orthogonal decompositon

$$V_{j+1} = V_j \oplus W_j,$$

which implies that W_j is defined as the orthogonal complement of V_j with respect to V_{j+1}. Let $P_j : L^2(\mathbb{R}) \mapsto V_j$ be the orthogonal projection.

The argument of the preceding section is stated in general as follows.

THEOREM 9.2. *Let $\psi \in W_0$ and $\psi_{j,k}(x) = 2^{j/2}\psi(2^j x - k)$. Suppose that $\{\psi_{0,k}\}$ is an orthonormal basis in W_0. Then:*

(i) *The system $\{\psi_{j,k}(x) : k \in \mathbb{Z}\}$ is an orthonormal basis of the space W_j. In particular, we have*

$$P_{j+1} = P_j + \sum_k (\cdot, \psi_{j,k})\psi_{j,k}.$$

(ii) *For $f \in L^2(\mathbb{R})$ we have the expression*

$$f = \sum_{j=-\infty}^{\infty} \sum_{k=-\infty}^{\infty} (f, \psi_{j,k})\psi_{j,k}, \quad (f, \psi_{j,k}) = \int f(x)\overline{\psi_{j,k}(x)}dx,$$

where the sum converges in L^2. Thus the space $L^2(\mathbb{R})$ can be decomposed as follows:

$$L^2 = \bigoplus_{j=-\infty}^{\infty} W_j.$$

PROOF. (i) The orthogonality and normality of $\{\psi_{j,k}\}$ follow by a simple direct computation. By condition (iii) of Definition 9.2 we have

$$f(x) \in W_j \iff f(x) \in V_{j+1} \text{ and } f(x) \perp V_j \iff f(2^{-j}x) \in W_0.$$

Therefore, for any function $f(x)$ in W_0 we have $f(2^{-j}x) = \sum_k c_k\psi(x-k)$, where $\sum |c_k|^2 < \infty$. This implies that $f(x) = \sum_k c_k 2^{-j/2}\psi_{j,k}(x)$, and thus $\{\psi_{j,k} : k \in \mathbb{Z}\}$ is a basis of W_j.

(ii) Let $f \in L^2$. By the orthogonal decomposition $V_j = V_{j-1} \oplus W_{j-1}$ we have

$$P_j f = P_{j-1} f + g_{j-1},$$

where $g_{j-1} \in W_{j-1}$. Applying this argument repeatedly, we get

$$P_j f = P_{j-n} f + (g_{j-1} + g_{j-2} + \cdots + g_{j-n}), \quad n = 1, 2, \ldots. \tag{9.7}$$

We have $\|P_{j-n}f\|_2 \to 0$ as $n \to \infty$. In fact, since $W_j \perp W_k$ $(j \neq k)$, for any fixed j the sequence $h_n = g_{j-1} + g_{j-2} + \cdots + g_{j-n}$ converges in L^2. Thus for each j the limit $\lim_{n\to\infty} P_{j-n}f$ exists in L^2, and it belongs to $\bigcap_{k=-\infty}^{\infty} V_k = \{0\}$.

Furthermore, we have $\|f - P_j f\|_2 \to 0$ as $j \to \infty$, since $\bigcup V_j$ is dense in L^2.

Thus from (9.7) it follows that $f = \sum_{j=-\infty}^{\infty} g_j$ in L^2 with $g_j \in W_j$, which implies (ii). Since $W_j \perp W_k$ $(j \neq k)$, we obtain (ii). □

Suppose that a multiresolution analysis $\{V_j\}$ is given and that a function $\psi \in W_0$ is constructed so that $\{\psi_{0,k}\}$ is an orthonormal basis in W_0. Then Theorem 9.2 enables us to express functions in L^2 by an orthonomal wavelet system.

In the following subsections we shall try to find scaling functions which are associated with multiresolution analysis and localized in the time-frequency coordinates, and then to construct wavelets which inherit the locality property from the scaling functions.

9.3.2. Construction of wavelets. Let $\{V_j\}$ be a given multiresolution analysis with scaling function φ. Then the functions $\varphi_{1,k}(x) = \sqrt{2}\varphi(2x-k)$ $(k \in \mathbb{Z})$ form an orthonormal basis of V_1.

LEMMA 9.1. *For $f \in V_1$ define*

$$m_f(\xi) = \frac{1}{\sqrt{2}} \sum_k (f, \varphi_{1,k}) e^{-2\pi i \xi x}. \tag{9.8}$$

Then the series converges in $L^2(\mathbb{T})$, and we have

$$\widehat{f}(\xi) = m_f\left(\frac{\xi}{2}\right) \widehat{\varphi}\left(\frac{\xi}{2}\right). \tag{9.9}$$

PROOF. Since $(f, \varphi_{1,k})$ are the Fourier coefficients of f with respect to the orthonormal basis $\{\varphi_{1,k}(x)\}$, $\|f\|_2^2 = \sum |(f, \varphi_{1,k})|^2 < \infty$. Thus the series (9.8) converges in $L^2(\mathbb{T})$.

Let $\widehat{V_1}$ be the set of Fourier transforms of functions in V_1. Then $\widehat{V_1}$ is a closed subspace of $L^2(\mathbb{R})$, since V_1 is a closed subspace, and the family $\{\widehat{\varphi}_{1,k}\}$ is an orthonormal basis in $\widehat{V_1}$ by the Plancherel theorem. Thus we have

$$\widehat{f}(\xi) = \sum (\widehat{f}, \widehat{\varphi}_{1,k}) \widehat{\varphi}_{1,k}(\xi).$$

Note that $(\widehat{f}, \widehat{\varphi}_{1,k}) = (f, \varphi_{1,k})$ again by the Plancherel theorem, and that $\widehat{\varphi}_{1,k}(\xi) = (1/\sqrt{2}) e^{-\pi i k \xi} \widehat{\varphi}(\xi/2)$ by a simple calculation. Therefore we get (9.9). □

LEMMA 9.2. *Let ρ be a function in $L^2(\mathbb{R})$. A necessary and sufficient condition for a system $\{\rho(x-k) : k \in \mathbb{Z}\}$ to be orthonormal in $L^2(\mathbb{R})$ is that*

$$\sum_{l=-\infty}^{\infty} |\widehat{\rho}(\xi + l)|^2 = 1 \ a.e. \tag{9.10}$$

Furthermore, if $\rho \in V_1$, then, under the condition (9.10), *we have*

$$|m_\rho(\xi)|^2 + \left| m_\rho\left(\xi + \frac{1}{2}\right)\right|^2 = 1 \ a.e., \tag{9.11}$$

where m_ρ is a function given by formula (9.8).

In particular, for the scaling function φ which belongs to V_0 we have

$$\sum_{l=-\infty}^{\infty} |\widehat{\varphi}(\xi + l)|^2 = 1 \text{ a.e.}, \tag{9.12}$$

and

$$|m_\varphi(\xi)|^2 + |m_\varphi(\xi + (1/2))|^2 = 1 \text{ a.e.} \tag{9.13}$$

PROOF. By the Plancherel theorem we have

$$\begin{aligned} \int_{-\infty}^{\infty} \rho(x)\overline{\rho(x-k)}dx &= \int_{-\infty}^{\infty} |\widehat{\rho}(\xi)|^2 e^{2\pi i k \xi} d\xi \\ &= \sum_{l=-\infty}^{\infty} \int_l^{l+1} |\widehat{\rho}(\xi)|^2 e^{2\pi i k \xi} d\xi \\ &= \int_0^1 \left(\sum_{l=-\infty}^{\infty} |\widehat{\rho}(\xi + l)|^2 \right) e^{2\pi i k \xi} d\xi. \end{aligned}$$

Therefore a necessary and sufficient condition for $\{\rho(x-k) : k \in \mathbb{Z}\}$ to be orthonormal in $L^2(\mathbb{R})$ is that

$$\int_0^1 \left(\sum_{l=-\infty}^{\infty} |\widehat{\rho}(\xi+l)|^2 \right) e^{2\pi i k \xi} d\xi = \delta_{0,k}, \tag{9.14}$$

which is equivalent to equation (9.10) by the uniqueness of Fourier coefficients.

Suppose that $\rho \in V_1$ and ρ satisfies (9.10). By the definition of m_ρ we have $\widehat{\rho}(\xi) = m_\rho(\xi/2)\widehat{\rho}(\xi/2)$. Thus from (9.14) we obtain

$$\delta_{0,k} = \int_0^1 \left(\sum_{l=-\infty}^{\infty} \left| m_\rho\left(\frac{\xi+l}{2}\right) \widehat{\rho}\left(\frac{\xi+l}{2}\right) \right|^2 \right) e^{2\pi i k \xi} d\xi.$$

Summing over even and odd l's separately and then applying (9.10), we get

$$\begin{aligned} \delta_{0,k} &= \int_0^1 \left(\sum_{l=-\infty}^{\infty} \left| m_\rho\left(\frac{\xi}{2}\right) \widehat{\rho}\left(\frac{\xi}{2}+l\right) \right|^2 \right. \\ &\qquad \left. + \sum_{l=-\infty}^{\infty} \left| m_\rho\left(\frac{\xi+1}{2}\right) \widehat{\rho}\left(\frac{\xi+1}{2}+l\right) \right|^2 \right) e^{2\pi i k \xi} d\xi \\ &= \int_0^1 \left(\left| m_\rho\left(\frac{\xi}{2}\right) \right|^2 + \left| m_\rho\left(\frac{\xi+1}{2}\right) \right|^2 \right) e^{2\pi i k \xi} d\xi. \end{aligned}$$

Since this equation holds for all k in $\mathbb{Z}$, we obtain (9.11). □

LEMMA 9.3. *Let $f \in V_1$. A necessary and sufficinent condition for f to belong to W_0 is that*

$$m_f(\xi)\overline{m_\varphi(\xi)} + m_f\left(\xi + \frac{1}{2}\right) \overline{m_\varphi\left(\xi+\frac{1}{2}\right)} = 0 \quad a.e. \tag{9.15}$$

PROOF. Suppose that $f \in V_1$. Then we have

$$\begin{aligned} f \in W_0 &\Leftrightarrow f \perp V_0 \Leftrightarrow (f, \varphi_{0,k}) = 0 \text{ for all } k \in \mathbb{Z} && \text{(by the definition of } W_0) \\ &\Leftrightarrow \int_{-\infty}^{\infty} \widehat{f}(\xi)\overline{\widehat{\varphi}(\xi)} e^{2\pi i k \xi} d\xi = 0 \text{ for all } k \in \mathbb{Z} && \text{(by the Plancherel theorem)} \\ &\Leftrightarrow \int_0^1 \left(\sum_l \widehat{f}(\xi+l)\overline{\widehat{\varphi}(\xi+l)} \right) e^{2\pi i k \xi} d\xi = 0 \text{ for all } k \in \mathbb{Z} \\ &\Leftrightarrow \sum_l \widehat{f}(\xi+l)\overline{\widehat{\varphi}(\xi+l)} = 0 \text{ a.e.} && \text{(by unicity of the Fourier coefficients).} \end{aligned}$$

If we substitute $\widehat{f}(\xi) = m_f(\xi/2)\widehat{\varphi}(\xi/2)$ and $\widehat{\varphi}(\xi) = m_\varphi(\xi/2)\widehat{\varphi}(\xi/2)$ in the last equation, and divide the sum into the parts over even integers and odd integers, then the last sum is transformed to

$$\begin{aligned} 0 &= \sum_l m_f\left(\frac{\xi+l}{2}\right) \overline{m_\varphi\left(\frac{\xi+l}{2}\right)} \left| \widehat{\varphi}\left(\frac{\xi+l}{2}\right) \right|^2 \\ &= m_f\left(\frac{\xi}{2}\right) \overline{m_\varphi\left(\frac{\xi}{2}\right)} \sum_l \left| \widehat{\varphi}\left(\frac{\xi}{2}+l\right) \right|^2 \\ &\quad + m_f\left(\frac{\xi+1}{2}\right) \overline{m_\varphi\left(\frac{\xi+1}{2}\right)} \sum_l \left| \widehat{\varphi}\left(\frac{\xi+1}{2}+l\right) \right|^2 \quad \text{a.e.} \end{aligned}$$

Therefore by (9.12) we obtain

$$0 = m_f\left(\frac{\xi}{2}\right)\overline{m_\phi\left(\frac{\xi}{2}\right)} + m_f\left(\frac{\xi+1}{2}\right)\overline{m_\phi\left(\frac{\xi+1}{2}\right)} \quad \text{a.e.}$$

□

LEMMA 9.4. *Every function f in W_0 can be written as*

$$\widehat{f}(\xi) = e^{-\pi i\xi}\nu(\xi)\overline{m_\varphi\left(\frac{\xi+1}{2}\right)}\widehat{\varphi}\left(\frac{\xi}{2}\right), \tag{9.16}$$

where ν is a function in $L^2(\mathbb{T})$ and satisfies

$$\|\nu\|_{L^2(\mathbb{T})} = \|f\|_{L^2(\mathbb{R})}.$$

PROOF. Let $f \in W_0$. Then there exists a periodic function λ with period 1 which satisfies

$$m_f(\xi) = \lambda(\xi)\overline{m_\varphi\left(\xi+\frac{1}{2}\right)} \quad \text{a.e.} \tag{9.17}$$

and

$$\lambda(\xi) + \lambda\left(\xi+\frac{1}{2}\right) = 0 \quad \text{a.e.} \tag{9.18}$$

In fact, by (9.13) $m_\varphi(\xi)$ and $m_\varphi(\xi+(1/2))$ do not vanish simultaneously. Suppose that $m_\varphi(\xi+(1/2)) \neq 0$. Define $\lambda(\xi)$ by (9.17) and then $\lambda(\xi+(1/2))$ by (9.15). Then the function $\lambda(\xi)$ satisfies (9.17) and (9.18). When $m_\varphi(\xi) \neq 0$ the argument is similar.

Substituting (9.17) into (9.9), we get

$$\widehat{f}(\xi) = m_f\left(\frac{\xi}{2}\right)\widehat{\varphi}\left(\frac{\xi}{2}\right) = \lambda\left(\frac{\xi}{2}\right)\overline{m_\varphi\left(\frac{\xi+1}{2}\right)}\widehat{\varphi}\left(\frac{\xi}{2}\right).$$

By (9.18) the Fourier coefficients of λ vanish on the even integers. Thus it can be written as

$$\lambda(\xi) = e^{-2\pi i\xi}\nu(2\xi),$$

where $\nu(\xi) = \sum \widehat{\lambda}_{2k-1}e^{2\pi ik\xi}$. Thus (9.16) is obtained.

The Plancherel theorem and (9.16) give

$$\begin{aligned}\int_{-\infty}^{\infty} |f|^2 dx &= \int_{-\infty}^{\infty} |\widehat{f}|^2 d\xi \\ &= \int_{-\infty}^{\infty} |\nu(\xi)|^2 \left|m_\varphi\left(\frac{\xi+1}{2}\right)\right|^2 \left|\widehat{\varphi}\left(\frac{\xi}{2}\right)\right|^2 d\xi \\ &= \int_0^1 |\nu(\xi)|^2 \left(\sum_{l=-\infty}^{\infty} \left|m_\varphi\left(\frac{\xi+l+1}{2}\right)\right|^2 \left|\widehat{\varphi}\left(\frac{\xi+l}{2}\right)\right|^2\right) d\xi.\end{aligned}$$

If the sum in the integrand is divided into two parts, the sum over odd integers and the sum over even integers, then it becomes

$$\sum_{l=-\infty}^{\infty} \left|m_\varphi\left(\frac{\xi+1}{2}\right)\right|^2 \left|\widehat{\varphi}\left(\frac{\xi}{2}+l\right)\right|^2 + \sum_{l=-\infty}^{\infty} \left|m_\varphi\left(\frac{\xi}{2}\right)\right|^2 \left|\widehat{\varphi}\left(\frac{\xi+1}{2}+l\right)\right|^2$$

$$(9.19) \qquad = \left|m_\varphi\left(\frac{\xi+1}{2}\right)\right|^2 + \left|m_\varphi\left(\frac{\xi}{2}\right)\right|^2 = 1 \text{ a.e.}$$

Therefore

$$\int_{-\infty}^{\infty} |f|^2 dx = \int_0^1 |\nu|^2 d\xi.$$

□

Now, define a function ψ by the Fourier transform

$$\widehat{\psi}(\xi) = e^{-\pi i(\xi+1)}\overline{m_\varphi\left(\frac{\xi+1}{2}\right)}\widehat{\varphi}\left(\frac{\xi}{2}\right).$$

It may be a good candidate for wavelet, thanks to the following properties.

(I) The function ψ belongs to $L^2(\mathbb{R})$, since m_φ is bounded and thus $\widehat{\psi} \in L^2$.

(II) More precisely, $\psi \in W_0$.

In fact, by equation (9.9) we have $m_\psi(\xi) = \widehat{\psi}(2\xi)/\widehat{\varphi}(\xi)$, which is equal, by the definition of ψ, to $e^{-2\pi i\xi+\pi i}\overline{m_\varphi(\xi + (1/2))}$. Therefore ψ satisfies condition (9.15) of Lemma 9.3 by a simple computation.

(III) The locality property of φ is reflected in that of ψ, if the function m_φ is smooth enough, for example, a trigonometric function.

LEMMA 9.5. *Let ψ be the function defined above. Then the family $\{\psi_{0,k} : k \in \mathbb{Z}\}$ is an orthonormal basis in W_0.*

PROOF. Equations (9.19) and (9.12) imply that

$$\sum_{l=-\infty}^{\infty} |\widehat{\psi}(\xi+l)|^2 = \sum_{l=-\infty}^{\infty} \left|m_\varphi\left(\frac{\xi+l+1}{2}\right)\right|^2 \left|\widehat{\varphi}\left(\frac{\xi+l}{2}\right)\right|^2 = 1 \text{ a.e.}$$

Thus the system $\{\psi_{0,k}\}$ is orthonormal by Lemma 9.2.

Next we shall prove that $\{\psi_{0,k}\}$ is total. Let f be a function in W_0. By Lemma 9.4 there exists a periodic function ν such that

$$\widehat{f}(\xi) = \nu(\xi)\widehat{\psi}(\xi)$$

and

$$\|\nu\|_{L^2(\mathbb{T})} = \|f\|_{L^2(\mathbb{R})}.$$

It suffices to show that

$$f(x) = \sum_k \widehat{\nu}_{-k}\psi(x-k),$$

and the sum converges in $L^2(\mathbb{R})$, where $\widehat{\nu}_k$ are Fourier coefficients of ν.

In fact, let

$$f_N(x) = \sum_{k=-N}^{N} \widehat{\nu}_{-k}\psi(x-k) \quad \text{and} \quad s_N(\nu)(\xi) = \sum_{k=-N}^{N} \widehat{\nu}_{-k}e^{2\pi ik\xi}$$

be Nth partial sums of f and ν respectively. Then, since $\widehat{f_N}(\xi) = s_N(\nu)(\xi)\widehat{\psi}(\xi)$, an application of the argument in the proof of Lemma 9.4 to $f - f_N$ leads to

$$\|f - f_N\|_{L^2(\mathbb{R})} = \|\nu - s_N(\nu)\|_{L^2(\mathbb{T})}.$$

The last term converges to 0 as $N \to \infty$. □

THEOREM 9.3. *Let $\{V_j\}$ be a multiresolition analysis. Then there exists a function ψ in L^2 such that $\{\psi_{j,k} : k \in \mathbb{Z}\}$ is an orthonormal basis in W_j.*

In particular, we can choose a function ψ given by the formula

$$\widehat{\psi}(\xi) = e^{-\pi i(\xi+1)} \overline{m_\varphi\left(\frac{\xi+1}{2}\right)} \widehat{\varphi}\left(\frac{\xi}{2}\right). \tag{9.20}$$

REMARK 9.1. We have

$$\psi(x) = \sum_{k=-\infty}^{\infty} (-1)^k \overline{h}_{-k+1} \varphi_{1,k}(x), \tag{9.21}$$

where $h_k = (\varphi, \varphi_{1,k})$. This formula implies that the wavelet ψ is computable as a linear combination of $\varphi_{1,k}$ if the Fourier coefficients of φ with respect to $\{\varphi_{1,k}\}$ are given.

In fact, $m_\varphi(\xi) = (1/\sqrt{2}) \sum_k h_k e^{-2\pi i k \xi}$ by (9.6) and (9.8). Substituting this into (9.20), we get

$$\widehat{\psi}(\xi) = \frac{1}{\sqrt{2}} \sum_{k=-\infty}^{\infty} \overline{h}_k e^{\pi i(k-1)} e^{\pi i(k-1)\xi} \widehat{\varphi}\left(\frac{\xi}{2}\right),$$

which implies (9.21) by the Fourier inversion formula.

REMARK 9.2. Let ρ be any periodic function with period 1 such that $|\rho(\xi)| = 1$ a.e., and let ψ^* be a function defined by

$$\widehat{\psi^*}(\xi) = e^{-\pi i \xi} \rho(\xi) \overline{m_\varphi\left(\frac{\xi+1}{2}\right)} \widehat{\varphi}\left(\frac{\xi}{2}\right).$$

Then ψ^* generates an orthonormal basis in W_0.

The proof is quite similar to that of Lemma 9.4, and is left to the reader.

9.3.3. Orthogonalization of a scaling function. We have imposed, in the definition of multiresolution analysis, the condition that the translations of the scaling function make up an orthonormal system, but it is more practical to relax the orthogonality condition. In fact, we shall consider the following more convenient assumption in place of condition (v).

(v)′ There exists a function φ in V_0 such that $V_0 = \overline{\operatorname{span}}^{L^2}\{\varphi_{0,k} : k \in \mathbb{Z}\}$, and, furthermore, there exist constants $A, B > 0$ such that

$$A \sum_k |c_k|^2 \le \|\sum_k c_k \varphi_{0,k}\|_2^2 \le B \sum_k |c_k|^2 \tag{9.22}$$

for all sequences $\{c_k\}$ with $\sum_k |c_k|^2 < \infty$.

In this case, a system $\{\varphi_{0,k}\}$ is called a *Riesz basis* of the space V_0 with bounds A and B.

REMARK 9.3. Under condition (v)′ a function f belongs to V_0 if and only if there exists $\{c_k\} \in l^2$ such that

$$f = \sum_k c_k \varphi_{0,k} \quad \text{in } L^2. \tag{9.23}$$

In fact, if f is given by (9.23), then obviously f belongs to V_0.

Suppose $f \in V_0$. Then by assumption there exists a sequence $c^j = \{c_k^j\}$ such that $\|f - \sum_k c_k^j \varphi_{0,k}\|_{L^2(\mathbb{R})} \to 0$ as $j \to 0$. Then we have, by (9.22), $A\|c^i - c^j\|_{l^2}^2 \le$

$\|\sum_k c_k^j \varphi_{0,k} - \sum_k c_k^j \varphi_{0,k}\|^2_{L^2(\mathbb{R})} \to 0$. Let $c = \{c_k\}$ be the l^2-limit of c^j, $c^j = \{c_k^j : k \in \mathbb{Z}\}$. Then, again by (9.22), $\|\sum_k c_k \varphi_{0,k} - \sum_k c_k^j \varphi_{0,k}\|^2_{L^2} \leq B\|c - c^j\|^2_{l^2} \to 0$ as $j \to \infty$, and thus $f = \sum_k c_k \varphi_{0,k} \in V_0$.

LEMMA 9.6. *A function φ satisfies condition* (9.22) *if and only if*

$$A \leq \sum_{j=-\infty}^{\infty} |\widehat{\varphi}(\xi + j)|^2 \leq B \quad a.e. \tag{9.24}$$

PROOF. For any finite sequence $\{c_k\}$ we have

$$\|\sum_k c_k \varphi_{0,k}\|_2^2 = \int_{-\infty}^{\infty} |\sum_k c_k e^{-2\pi i k \xi} \widehat{\varphi}(\xi)|^2 d\xi$$
$$= \int_0^1 |\sum_k c_k e^{-2\pi i k \xi}|^2 \sum_{j=-\infty}^{\infty} |\widehat{\varphi}(\xi + j)|^2 d\xi.$$

Thus (9.22) is equivalent to

$$A\|C\|^2_{L^2(\mathbb{R})} \leq \int_0^1 |C(\xi)|^2 \sum_{j=-\infty}^{\infty} |\widehat{\varphi}(\xi + j)|^2 d\xi \leq B\|C\|^2_{L^2(\mathbb{R})} \text{ a.e.}$$

for all $C(\xi) = \sum_k c_k e^{2\pi i k x}$ in $L^2(\mathbb{T})$. Taking the supremum for $\int_0^1 |C(\xi)|^2 d\xi = 1$, (9.24) is obtained. The converse is obvious. □

Now, put

$$\widehat{\varphi^*}(\xi) = (\sum_j |\widehat{\varphi}(\xi + j)|^2)^{-1/2} \widehat{\varphi}(\xi).$$

Then obviously $\sum_l |\widehat{\varphi^*}(\xi + l)|^2 = 1$. Therefore, by Lemma 9.2, $\{\varphi^*_{0,l}\}$ is an orthonormal system in V_0.

LEMMA 9.7. *If $\{\varphi_{0,k}\}$ is a Riesz basis in V_0, then the system $\{\varphi^*_{0,l}\}$ defined above is an orthonormal basis in V_0.*

PROOF. It suffices to prove that $\varphi^*_{0,k}$ is total in V_0, which is done by the following synopsis.

$f \in V_0 \Leftrightarrow$ there exists $\{c_k\} \in l^2$ such that $f = \sum_k c_k \varphi_{0,k}$ (by Remark 9.3)

$\Leftrightarrow$ there exists $\{c_k\} \in l^2$ such that $\widehat{f}(\xi) = (\sum_k c_{-k} e^{-2\pi i k \xi}) \widehat{\varphi}(\xi)$ (by Fourier transform)

$\Leftrightarrow$ there exists $C(\xi) \in L^2(\mathbb{T})$ sucht that $\widehat{f}(\xi) = C(\xi)\widehat{\varphi}(\xi)$

$\Leftrightarrow$ there exists $D(\xi) \in L^2(\mathbb{T})$ such that $\widehat{f}(\xi) = \dfrac{D(\xi)}{(\sum_j |\widehat{\varphi}(\xi + j)|^2)^{1/2}} \widehat{\varphi}(\xi)$ (by Lemma 9.6)

$\Leftrightarrow$ there exists $D(\xi) = \sum_k d_k e^{-2\pi i k \xi} \in L^2(\mathbb{T})$ such that $\widehat{f}(\xi) = D(\xi)\widehat{\varphi^*}(\xi)$

$\Leftrightarrow$ there exists $\{d_k\} \in l^2$ such that $f = \sum_k d_k \varphi^*_{0,k}$.

□

LEMMA 9.8. *Assume that* $\{\varphi_{0,k}\}$ *is a Riesz basis in* V_0 *with bounds* $A, B > 0$. *Then*

$$A\|f\|_2^2 \le \sum_k |(f, \varphi_{0,k})|^2 \le B\|f\|_2^2 \quad (f \in V_0). \tag{9.25}$$

PROOF. For a function $f \in V_0$ define a function f^* by the Fourier transform

$$\widehat{f^*}(\xi) = (\sum_j |\widehat{\varphi}(\xi + j)|^2)^{1/2} \widehat{f}(\xi).$$

Then by Lemma 9.6

$$A\|f\|_2^2 \le \|f^*\|_2^2 \le B\|f\|_2^2.$$

On the other hand, by the Plancherel theorem

$$\begin{aligned}(f^*, \varphi_{0,k}^*) &= \int \widehat{f^*}(\xi)\overline{\widehat{\varphi^*}(\xi)} e^{-2\pi i k\xi} d\xi = \int \widehat{f}(\xi)\overline{\widehat{\varphi}(\xi)} e^{-2\pi i k\xi} d\xi \\ &= (f, \varphi_{0,k}).\end{aligned}$$

Thus $\|f^*\|_2^2 = \sum |(f^*, \varphi_{0,k}^*)|^2 = \sum |(f, \varphi_{0,k})|^2$. Therefore (9.25) is obtained. □

9.4. Examples of wavelets

EXAMPLE 9.4 (The Haar wavelet). Recall the Haar function $\varphi_H(x)$, which is the characteristic function of the interval $[0, 1)$. Put $\varphi_{j,k}(x) = 2^{j/2}\varphi_H(2^j x - k)$ and $V_j = \overline{\text{span}}^{L^2}\{\varphi_{j,k} : k \in \mathbb{Z}\}$. V_j is the space of functions in $L^2(\mathbb{R})$ which are constant in each interval $[k2^{-j}, (k+1)2^{-j})$, $k \in \mathbb{Z}$. Since

$$h_k = (\varphi_{0,0}, \varphi_{1,k}) = \begin{cases} 1/\sqrt{2} & (k = 0, 1), \\ 0 & (k \ne 0, 1), \end{cases}$$

the wavelet function defined by formula (9.20) (see also (9.21)) is written as

$$\psi(x) = \overline{h}_1\varphi_{1,0}(x) - \overline{h}_0\varphi_{1,1}(x) = \begin{cases} 1 & (0 \le x < 1/2), \\ -1 & (1/2 \le x < 1), \\ 0 & (\text{otherwise}), \end{cases}$$

which is idetical with the Haar wavelet given in § 9.2.2.

EXAMPLE 9.5 (A spline wavelet). Let χ be the characteristic function of the interval $(-1/2, 1/2)$. The function $\varphi(x) = \chi * \chi(x)$ is called the *spline function* of order 1. The function $\varphi(x)$ is piecewise linear, continuous, $\varphi(0) = 1$ and $\operatorname{supp}\varphi = [-1, 1]$. Thus $\varphi(x) = (1 - |x|)_+$. The space

$$V_j = \overline{\text{span}}^{L^2}\{\varphi_{j,k} : k \in \mathbb{Z}\}$$

consists of functions in $L^2(\mathbb{R})$ which are continuous and linear in each interval $[k2^{-j}, (k+1)2^{-j})$ $(k \in \mathbb{Z})$.

The family $\{\varphi_{0,k}\}$ is a Riesz basis in V_j but not orthogonal. Now, we shall give an orthogonalization of it following the preceeding process.

First, we show that

$$\sum_j |\widehat{\varphi}(\xi + j)|^2 = \frac{2}{3} + \frac{1}{3}\cos 2\pi\xi = \frac{1}{3}(1 + 2\cos^2 \pi\xi).$$

In fact, the Fourier coefficients of each side coincide by the following simple calculation:

$$\int_0^1 \Big(\sum_{j=-\infty}^{\infty} |\widehat{\varphi}(\xi+j)|^2\Big) e^{-2\pi i k\xi} d\xi = \int_{-\infty}^{\infty} |\widehat{\varphi}(\xi)|^2 e^{-2\pi i k\xi} d\xi$$

$$= \int_{-\infty}^{\infty} \varphi(x-k)\varphi(x)dx = \begin{cases} 2/3 & (k=0), \\ 1/6 & (|k|=1), \\ 0 & (|k|\geq 2). \end{cases}$$

Thus $1/3 \leq \sum_j |\widehat{\varphi}(\xi+j)|^2 \leq 1$.

Since $\widehat{\varphi}(\xi) = (\sin \pi\xi/\pi\xi)^2$, an orthogonalization is given by

$$\widehat{\varphi^*}(\xi) = \frac{\widehat{\varphi}(\xi)}{(\sum_j |\widehat{\varphi}(\xi+j)|^2)^{1/2}} = \frac{\sqrt{3}\sin^2 \pi\xi}{\pi^2\xi^2(1+2\cos^2\pi\xi)^{1/2}}.$$

Thus

$$m_{\varphi^*}(\xi) = \frac{\widehat{\varphi^*}(2\xi)}{\widehat{\varphi^*}(\xi)} = \left[\frac{1+2\cos^2\pi\xi}{1+2\cos^2 2\pi\xi}\right]^{1/2} \cos^2\pi\xi.$$

Therefore the formula (9.20) which defines a wavelet is given as

$$\widehat{\psi}(\xi) = e^{-\pi i(\xi+1)} \left[\frac{1+2\sin^2\frac{\pi\xi}{2}}{1+2\cos^2\pi\xi}\right]^{1/2} \sin^2\frac{\pi\xi}{2}\widehat{\varphi^*}\left(\frac{\xi}{2}\right).$$

Substituting $\widehat{\varphi^*} = \widehat{\varphi}(\xi/2)/(\sum |\widehat{\varphi}((\xi/2+j)|^2)^{1/2}$, we get

$$\widehat{\psi}(\xi) = \sqrt{3}e^{-\pi i(\xi+1)} \left[\frac{1+2\sin^2\frac{\pi\xi}{2}}{(1+2\cos^2\pi\xi)\left(1+2\cos^2\frac{\pi\xi}{2}\right)}\right]^{1/2} \sin^2\frac{\pi\xi}{2}\widehat{\varphi}\left(\frac{\xi}{2}\right)$$

$$= -\frac{\sqrt{3}}{4}\left[\frac{1+2\sin^2\frac{\pi\xi}{2}}{(1+2\cos^2\pi\xi)\left(1+2\cos^2\frac{\pi\xi}{2}\right)}\right]^{1/2} (e^{-2\pi i\xi} - 2e^{-\pi i\xi}+1)\widehat{\varphi}\left(\frac{\xi}{2}\right).$$

If $[\cdots]^{1/2} = \sum_k d_k e^{-\pi i k\xi}$ is the Fourier expansion, then from the last equation it follows that

$$\psi(x) = -\frac{\sqrt{3}}{2}\sum_k (d_k - 2d_{k-1} + d_{k-2})\varphi(2x-k).$$

The graph of the function ψ is a broken line with knots at $\frac{1}{2}\mathbb{Z}$. Figure 9.6 is obtained by a numerical computation of d_k.

EXAMPLE 9.6 (Meyer's wavelet). Let φ be a function in $\mathcal{S}(\mathbb{R})$ such that

$$\widehat{\varphi}(\xi) = 1\ (|\xi| < 1/3), \quad \operatorname{supp}\widehat{\varphi} \subset (-2/3, 2/3),$$

and

$$|\widehat{\varphi}(\xi)|^2 + |\widehat{\varphi}(\xi-1)|^2 = 1\ (0 \leq \xi < 1)$$

(see Figure 9.7).

Since $\sum_k |\widehat{\varphi}(\xi+k)|^2 = 1$, the system $\{\varphi_{0,k}\}$ is orthonormal.

Let $V_j = \overline{\operatorname{span}}^{L^2}\{\varphi_{j,k} : k \in \mathbb{Z}\}$. Assuming temporarily that $\{V_j\}$ is a multiresolution analysis with the scaling function φ, we construct a wavelet associated with $\{V_j\}$.

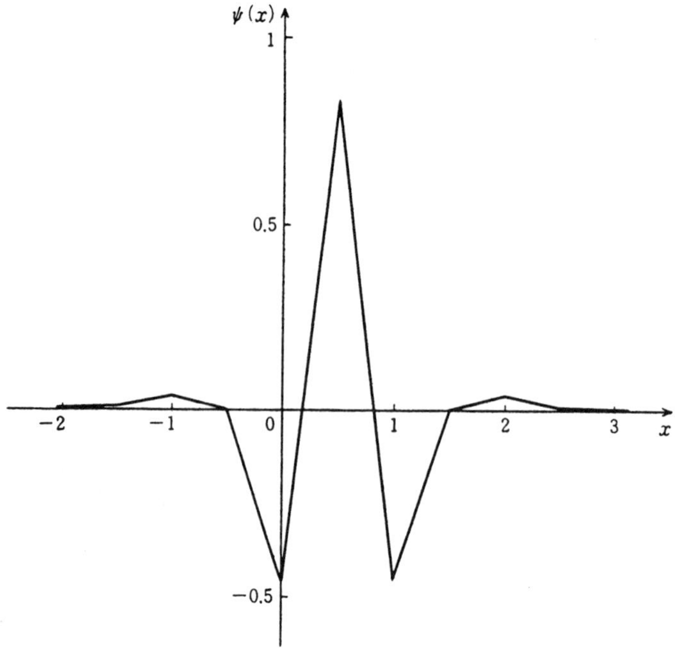

FIGURE 9.6. A spline wavelet

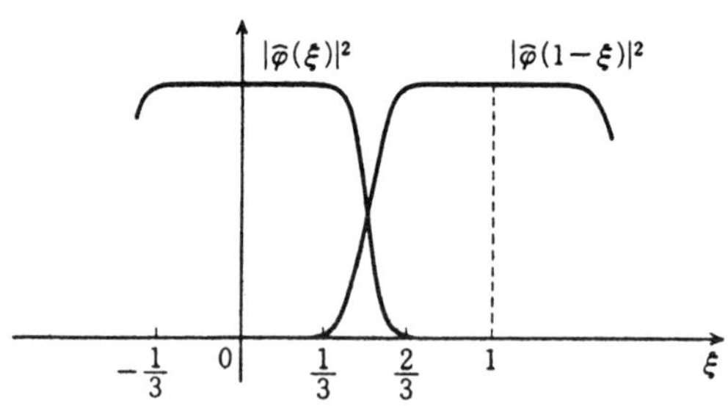

FIGURE 9.7. Graph of $|\widehat{\varphi}(\xi)|^2$

The function $m_\varphi(\xi)$ is given by

$$m_\varphi(\xi) = \sum_l \widehat{\varphi}(2(\xi + l)). \tag{9.26}$$

In fact, the right-hand side is periodic with period 1, and the kth Fourier coefficient equals $\int \widehat{\varphi}(2\xi)e^{-2\pi i k\xi}d\xi = \varphi(k/2)/2$. By definition (9.8) the kth Fourier coefficient of $m_\varphi(\xi)$ is $\sqrt{2}^{-1}(\varphi, \varphi_{1,k}) = \int \varphi(x)\varphi(2x-k)dx$, which is equal to $2^{-1}\int \widehat{\varphi}(\xi)\overline{\widehat{\varphi}(\xi/2)}e^{-\pi i k\xi}d\xi = 2^{-1}\int \widehat{\varphi}(\xi)e^{-\pi i k\xi}d\xi = 2^{-1}\varphi(k/2)$ due to the Plancherel theorem.

Since $m_\varphi(\xi/2)\widehat{\varphi}(\xi/2) = \widehat{\varphi}(\xi)$, a wavelet is given by

$$\begin{aligned}\widehat{\psi}(\xi) &= e^{-\pi i(\xi+1)}\overline{m_\varphi\left(\frac{\xi+1}{2}\right)}\widehat{\varphi}\left(\frac{\xi}{2}\right)\\ &= e^{-\pi i(\xi+1)}[\widehat{\varphi}(\xi+1)+\widehat{\varphi}(\xi-1)]\widehat{\varphi}\left(\frac{\xi}{2}\right).\end{aligned}$$

Now we shall show that $V_j = \overline{\text{span}}^{L^2}\{\varphi_{j,k}\}$ satisfies condition (i) of multiresolution analysis. In fact, by (9.26) m_φ is a C^∞-function. More precisely, we have $m_\varphi(\xi) = \sum_k 2^{-1}\varphi(k/2)e^{-2\pi ik\xi}$. Since $m_\varphi(\xi) = \widehat{\varphi}(\xi)/\widehat{\varphi}(\xi/2)$, we have $\varphi(x) = \sum_k \varphi(k/2)\varphi(2x-k)$ by Fourier transform, which implies that $V_0 \subset V_1$.

To check condition (ii) we shall prove a more general lemma:

LEMMA 9.9. *Let φ be a function in $L^2(\mathbb{R})$ and $V_j = \overline{\text{span}}^{L^2}\{\varphi_{j,k} : k \in \mathbb{Z}\}$. If $\{\varphi_{0,k}\}$ is a Riesz basis in V_0, then $\bigcap_{j=-\infty}^{\infty} V_j = \{0\}$.*

PROOF. Let $f \in \bigcap V_j$. For a given $\varepsilon > 0$ choose $u \in C_c^\infty(\mathbb{R})$ so that $\|f-u\|_2 < \varepsilon$. Let P_j be the orthogonal projection of L^2 to V_j. Then we have

$$\|f - P_j u\|_2 = \|P_j(f-u)\|_2 \le \|f-u\|_2 < \varepsilon$$

for every $j \in \mathbb{Z}$. Thus

$$\|f\|_2 \le \|P_j u\|_2 + \varepsilon.$$

Since $\{\varphi_{j,k}\}$ is a Riesz basis in V_j,

$$\|P_j u\|_2^2 \le B\sum\nolimits_k |(P_j u, \varphi_{j,k})|^2,$$

where the constant B is a Riesz bound of $\{\varphi_{0,k}\}$. Observe that $(P_j u, \varphi_{j,k}) = (u, \varphi_{j,k}) = 2^{j/2}\int u(x)\varphi(2^j x - k)dx$. Then we have, by Schwarz's inequality,

$$\|P_j u\|_2^2 \le B\|u\|_2^2 \sum_k 2^j \int_{|x|<R} |\varphi(2^j x - k)|^2 dx,$$

where $R > 0$ is chosen so that $\operatorname{supp} u \subset [-R, R]$. Put $S_j = \bigcup_k(-2^j R - k, 2^j R - k)$. Then we have $\sum_k 2^j \int_{|x|<R} |\varphi(2^j x - k)|^2 dx = \int_{S_j} |\varphi(x)|^2 dx \to 0 \ (j \to -\infty)$. Thus $\|f\|_2^2 \le \varepsilon$ for any $\varepsilon > 0$, which implies that $f = 0$. □

LEMMA 9.10. *Under the same assumption as in Lemma 9.9, if there exists $\delta > 0$ such that $|\widehat{\varphi}(\xi)| > \delta$ in a neighborhood of 0, then $\overline{\bigcup V_j}^{L^2} = L^2(\mathbb{R})$.*

PROOF. Let $f \in (\bigcup_j V_j)^\perp$. We shall prove that $f = 0$.

Since the space $C_c^\infty(\mathbb{R})$ is dense in $L^2(\mathbb{R})$, $\widehat{C_c^\infty}(\mathbb{R})$, the set of Fourier transforms of functions in $C_c^\infty(\mathbb{R})$, is also dense in $L^2(\mathbb{R})$. For a given $\varepsilon > 0$ choose a function u in $\widehat{C_c^\infty}(\mathbb{R})$ such that $\|f-u\|_2 < \varepsilon$. Note that $P_j f = 0$, since $(P_j f, P_j f) = (f, P_j f) = 0$. Thus

$$\|P_j u\|_2 = \|P_j u - P_j f\|_2 \le \|u - f\|_2 < \varepsilon \quad (j \in \mathbb{Z}).$$

Since $\{\varphi_{j,k}\}$ is a Riesz basis of the space V_j, we have

$$\|P_j u\|_2^2 \ge A\sum_k |(P_j u, \varphi_{j,k})|^2 = A\sum_k |(u, \varphi_{j,k})|^2.$$

Let j be any sufficiently large integer. Then $\operatorname{supp}\widehat{u} \subset (-2^{j-1}, 2^{j-1})$ and $|\widehat{\varphi}(2^{-j}\xi)| > \delta\ (\xi \in \operatorname{supp}\widehat{u})$.

By the Plancherel theorem we have

$$2^{-j/2}(u,\varphi_{j,k}) = 2^{-j}\int \widehat{u}(\xi)\widehat{\varphi}(2^{-j}\xi)e^{-2\pi i k 2^{-j}\xi}d\xi,$$

which is considered as the kth Fourier coefficient of the function $\widehat{u}(\xi)\widehat{\varphi}(2^{-j}\xi)$ on the interval $[-2^{j-1}, 2^{j-1})$. Therefore the Parseval relation implies that

$$\begin{aligned}\sum_{k=-\infty}^{\infty} |2^{-j/2}(u,\varphi_{j,k})|^2 &= 2^{-j}\int_{-2^{j-1}}^{2^{j-1}} |\widehat{u}(\xi)\widehat{\varphi}(2^{-j}\xi)|^2 d\xi \\ &\geq 2^{-j}\delta^2 \int |\widehat{u}(\xi)|^2 d\xi = 2^{-j}\delta^2\|u\|_2^2.\end{aligned}$$

Thus

$$\varepsilon \geq \|P_j u\|_2 \geq A^{1/2}\delta\|u\|_2.$$

Therefore we get

$$\|f\|_2 \leq \varepsilon + \|u\|_2 \leq \varepsilon + \varepsilon A^{-1/2}\delta^{-1},$$

which implies that $\|f\|_2 = 0$ and thus $f = 0$ a.e.

□

9.5. Compactly supported wavelets

The Haar wavelet is an example of a wavelet whose support is compact. There exists a wavelet which has given smoothness and whose support is compact. Such a wavelet has important applications, both mathematically and practically. We shall give a construction following I. Daubechies' idea.

9.5.1. Construction of a scaling function and main theorem. Suppose that the trigonometric polynomial

$$m_0(\xi) = \frac{1}{\sqrt{2}}\sum_{n=L}^{M} h_n e^{-2\pi i n\xi} \tag{9.27}$$

satisfies the conditions

$$m_0(0) = 1 \tag{9.28}$$

and

$$|m_0(\xi)|^2 + |m_0(\xi + (1/2))|^2 = 1. \tag{9.29}$$

Our first object is to show that under some conditions on m_0 there exists a scaling function φ such that $m_\varphi(\xi) = m_0(\xi)$.

To construct such a scaling function, put formally

$$\widehat{\varphi}(\xi) = \prod_{k=1}^{\infty} m_0(2^{-k}\xi). \tag{9.30}$$

(I) Assume that the infinite product converges. Then equation (9.9) holds:

$$\widehat{\varphi}(\xi) = m_0(\xi/2)\widehat{\varphi}(\xi/2).$$

Furthermore assume the following:

(II) The function $\widehat{\varphi}$ belongs to $L^2(\mathbb{R})$.

(III) Its inverse Fourier transform φ has compact support.

If a function ψ is defined by

$$\widehat{\psi}(\xi) = e^{-\pi i(\xi+1)}\overline{m_0\left(\frac{\xi+1}{2}\right)}\widehat{\varphi}\left(\frac{\xi}{2}\right),$$

then by the definition (9.27) we have

$$\psi(x) = \sum_{n=-M+1}^{-L+1} (-1)^n \overline{h}_{-n+1}\varphi_{1,n}(x),$$

which has compact support by our assumption (III).

In the following we shall show that (I) holds good, (III) holds if we assume (II), and (II) is valid under some condition on the trigonometric polynomial m_0. Finally we shall give a trigonometric polynomial which satisfies all these conditions.

Let m_0 be a trigonometric polynomial given by (9.27). For a complex number ζ we have

$$\begin{aligned} m_0(\zeta) &= e^{-\pi i(M+L)\zeta}\frac{e^{\pi i(M-L)\zeta}}{\sqrt{2}}\sum_{n=0}^{M-L} h_{n+L}e^{-2\pi i n\zeta} \\ &= e^{-\pi i(M+L)\zeta}m_1(\zeta), \quad \text{say.} \end{aligned}$$

Define $\varphi_1(\zeta)$ formally by

$$\widehat{\varphi}_1(\zeta) = \prod_{k=1}^{\infty} m_1(2^{-k}\zeta).$$

Then we have

$$\widehat{\varphi}(\zeta) = \prod_{k=1}^{\infty} m_0(2^{-k}\zeta) = e^{-\pi i(M+L)\zeta}\widehat{\varphi}_1(\zeta).$$

In the following lemma we shall show that the function φ satisfies the assumptions (I) and (III).

LEMMA 9.11. *Let m_0 be a trigonometric polynomial given by* (9.27) *which satisfies* (9.28) *and* (9.29).

Then the infinite product $\widehat{\varphi}_1(\zeta)$ converges uniformly on any compact set. Thus $\widehat{\varphi}(\zeta)$ is an entire function. If $\widehat{\varphi}(\xi) \in L^2(\mathbb{R})$, then the inverse Fourier transform φ has support contained in the interval $[L, M]$.

PROOF. By definition we have

$$\begin{aligned} |m_1(\zeta) - m_1(0)| &\le \frac{1}{\sqrt{2}}\sum_{n=0}^{M-L} |h_{n+L}||e^{\pi i(M-L)\zeta}e^{-2\pi i n\zeta} - 1| \\ &\le C_1|\zeta|, \end{aligned}$$

where C_1 is a constant independent of ζ (L, M and h_n are fixed). Since $m_1(0) = 1$ by assumption (9.28), we have the estimate

$$|m_1(\zeta)| \le 1 + C_1|\zeta| \le e^{C_1|\zeta|}, \quad |\zeta| \le 1.$$

Thus the product $\widehat{\varphi}_1$ converges uniformly in $|\zeta| \le 1$, and

$$\left|\prod_{k=1}^{\infty} m_1(2^{-k}\zeta)\right| \le \prod_{k=1}^{\infty} e^{C_1 2^{-k}|\zeta|} \le e^{C_1}. \tag{9.31}$$

Let $2^{K-1} \le |\zeta| < 2^K$, $K \ge 1$. We have

$$\left|\prod_{k=1}^{\infty} m_1(2^{-k}\zeta)\right| = \left|\prod_{k=1}^{K} m_1(2^{-k}\zeta)\right| \left|\prod_{k=1}^{\infty} m_1(2^{-k}(2^{-K}\zeta))\right|.$$

Apply the trivial estimate $|m_1(\zeta)| \le C_2 e^{\pi(M-L)|\zeta|}$ to the first product on the right-hand side, where C_2 is a constant not depending on ζ . Then we obtain

$$\left|\prod_{k=1}^{K} m_1(2^{-k}\zeta)\right| \le C_2^K e^{\pi(M-L)|\xi|}.$$

For the second product we apply estimate (9.31). Then we get

$$\left|\prod_{k=1}^{\infty} m_1(2^{-k}\zeta)\right| \le e^{C_1} e^{C_3 \log|\zeta| + \pi(M-L)|\zeta| + 1},$$

where $C_3 = \log C_2/\log 2$. Therefore the product converges uniformly in $2^{K-1} \le |\zeta| < 2^K$ for each $K \ge 1$. Therefore the function $\widehat{\varphi}_1(\zeta)$ is analytic in the whole plane. Furthermore the last estimate shows that $\widehat{\varphi}_1$ is of type A for any $A > (M-L)/2$. Thus by the Paley-Wiener theorem $\varphi_1(x) = 0$ a.e. for $|x| > A$; that is, we can suppose that $\operatorname{supp}\varphi_1 \subset [-(M-L)/2, (M-L)/2]$. Thus $\operatorname{supp}\varphi \subset [L, M]$. □

By the above lemma the function φ is legitimately defined by (9.30). The following lemma gives a condition under which assumption (II) is valid.

LEMMA 9.12. *Let K be a compact set in $\mathbb{R}$ such that* $\operatorname{int}(K) \ni 0$, $m(K) = 1$, *and* $K \equiv [0,1] \pmod{\mathbb{Z}}$. *Put*

$$\widehat{\varphi^n}(\xi) = \prod_{k=1}^{n} m_0(2^{-k}\xi)\chi_K(2^{-n}\xi),$$

where $m_0(\xi)$ is the function given in Lemma 9.11. If

$$C = \inf_{k>0} \inf_{\xi\in K} |m_0(2^{-k}\xi)| > 0, \tag{9.32}$$

then $\widehat{\varphi^n}$ converges in L^2 to $\widehat{\varphi}$.

Furthermore, the system $\{\varphi(x-n) : n \in \mathbb{Z}\}$ is orthonormal.

PROOF. First step. *We have*

$$\int_{-\infty}^{\infty} |\widehat{\varphi^n}(\xi)|^2 e^{2\pi i m\xi} d\xi = \delta_{0,m}. \tag{9.33}$$

In fact, if a function f is periodic with period 1, then obviously $\int_K f(\xi)d\xi = \int_0^1 f(\xi)d\xi$. Therefore it follows that

$$\begin{aligned}\int_{-\infty}^{\infty} |\widehat{\varphi^n}(\xi)|^2 e^{2\pi i m\xi} d\xi &= 2^n \int_{-\infty}^{\infty} \prod_{k=1}^{n} |m_0(2^{n-k}\xi)|^2 \chi_K(\xi) e^{2\pi i m 2^n \xi} d\xi \\ &= 2^n \int_0^1 \prod_{k=0}^{n-1} |m_0(2^k\xi)|^2 |m_0(\xi)|^2 e^{2\pi i m 2^n \xi} d\xi.\end{aligned}$$

Divide the integral domain into two intervals $(0, 1/2)$ and $[1/2, 1)$. Then the last term equals

$$2^n \int_0^{1/2} \prod_{k=1}^{n-1} |m_0(2^k\xi)|^2 \left[|m_0(\xi)|^2 + |m_0(\xi + (1/2))|^2 \right] e^{2\pi i m 2^n \xi} d\xi.$$

By the assumption (9.29) and changing the variable, this becomes

$$\begin{aligned} 2^n \int_0^{1/2} \prod_{k=1}^{n-1} |m_0(2^k\xi)|^2 e^{2\pi i m 2^n \xi} d\xi &= 2^{n-1} \int_0^1 \prod_{k=0}^{n-2} |m_0(2^k\xi)|^2 e^{2\pi i m 2^{n-1}\xi} d\xi \\ &= \int |\widehat{\varphi^{n-1}}(\xi)|^2 e^{2\pi i m \xi} d\xi. \end{aligned}$$

Iterating this process, we get

$$\begin{aligned} \int |\widehat{\varphi^n}(\xi)|^2 e^{2\pi i m\xi} d\xi = \cdots &= \int |\widehat{\varphi^1}(\xi)|^2 e^{2\pi i m \xi} d\xi \\ &= 2\int_0^{1/2} e^{-2\pi i 2 m \xi} d\xi = \delta_{0,m}. \end{aligned}$$

Since K contains a neighborhood of 0, $\widehat{\varphi^n}(\xi)$ converges uniformly on every compact set to $\widehat{\varphi}(\xi)$ by the last lemma.

Second Step. *There exists a constant $C_1 > 0$ such that*

$$|\widehat{\varphi^n}(\xi)| \leq C_1 |\widehat{\varphi}(\xi)|. \tag{9.34}$$

In fact, since $m_0(0) = 1$, there exists a constant $C_2 > 0$ such that $|m_0(\xi)| \geq 1 - C_2|\xi|$. Choose an integer k_0 so that $2^{-k}C_2|\xi| < 1/2$ for all $\xi \in K$ and $k \geq k_0$, which is possible, since K is compact. By the inequality $1 - x \geq e^{-2x}$ $(0 \leq x \leq 1/2)$ we see that $|m_0(2^{-k}\xi)| \geq \exp[-2C_2 2^{-k}|\xi|]$ for all $\xi \in K$ and $k \geq k_0$. On the other hand, $|m_0(2^{-k}\xi)| \geq C$ for $\xi \in K$ by assumption (9.32).

Thus for $\xi \in K$ we have

$$\begin{aligned} |\widehat{\varphi}(\xi)| &= \prod_{k=1}^{k_0} |m_0(2^{-k}\xi)| \prod_{k=k_0+1}^{\infty} |m_0(2^{-k}\xi)| \\ &\geq C^{k_0} \prod_{k=k_0+1}^{\infty} \exp[-2C_2 2^{-k}|\xi|] \\ &\geq C^{k_0} \exp[-2C_2 2^{-k_0+1} \max_{\xi\in K}|\xi|] = C_1^{-1} > 0, \quad \text{say}. \end{aligned}$$

Now we have

$$\begin{aligned} |\widehat{\varphi^n}(\xi)| &= \prod_{k=1}^{n} |m_0(2^{-k}\xi)| \chi_K(2^{-n}\xi) \\ &\leq C_1 \prod_{k=1}^{n} |m_0(2^{-k}\xi)||\widehat{\varphi}(2^{-n}\xi)| = C_1|\widehat{\varphi}(\xi)|. \end{aligned}$$

Third Step. *Proof of the Lemma.* Since $\int |\widehat{\varphi^n}(\xi)|^2 d\xi = 1$ and since $\widehat{\varphi^n}$ converges uniformly to $\widehat{\varphi}$ on any compact set, we get $\int |\widehat{\varphi}(\xi)|^2 d\xi \leq 1$ by Fatou's lemma. Thus by (9.34) $\widehat{\varphi^n}$ converges in L^2 to $\widehat{\varphi}$.

Let $n \to \infty$ in (9.33). Then we obtain

$$\int |\widehat{\varphi}(\xi)|^2 e^{2\pi i m\xi} d\xi = \int \varphi(\xi)\overline{\varphi(x-m)} d\xi = \delta_{0,m}.$$

Therefore the family $\{\varphi_{0,n}\}$ is orthonormal. □

Define

$$V_j = \overline{\operatorname{span}}^{L^2}\{\varphi_{j,k}\}.$$

LEMMA 9.13. *Assume $m_0(\xi)$ satisfies conditions* (9.28), (9.29) *and* (9.32). *Then $\{V_j\}$ is a multiresolution analysis with scaling function φ.*

PROOF. Conditions (iv) and (v) of multiresolution analysis are obviously satisfied. (i) and (iii) follow from the relation $\widehat{\varphi}(\xi) = m_0(\xi/2)\widehat{\varphi}(\xi/2)$.

By Lemma 9.9 we have $\bigcap_{j=-\infty}^{\infty} V_j = \{0\}$.

Since the function $\widehat{\varphi}$ is continuous and $\widehat{\varphi}(0) = 1$, Lemma 9.10 implies that $\overline{\bigcup V_j}^{L^2} = L^2$. Thus (ii) is satisfied. Therefore $\{V_j\}$ is a multiresolution analysis. □

Summarizing the above lemmas, we obtain the following:

THEOREM 9.4. *Let $m_0(\xi) = (1/\sqrt{2})\sum_{n=L}^{M} h_n e^{-2\pi i n\xi}$ be a trigonometric polynomial satisfying the following conditions:*

(a) $|m_0(\xi)|^2 + |m_0(\xi + 1/2)|^2 = 1$ *and* $m_0(0) = 1$.
(b) *There exists a compact set K such that* $\operatorname{int}(K) \ni 0$, $m(K) = 1$, $K \equiv [0,1] \pmod{\mathbb{Z}}$, *and*

$$\inf_{k>0} \inf_{\xi \in K} |m_0(2^{-k}\xi)| > 0.$$

If functions φ and ψ are defined by Fourier transforms

$$\widehat{\varphi}(\xi) = \prod_{j=1}^{\infty} m_0(2^{-j}\xi),$$

and

$$\widehat{\psi}(\xi) = e^{-\pi i(\xi+1)}\overline{m_0((\xi+1)/2)}\widehat{\varphi}(\xi/2),$$

then

(i) *there exists a multiresolution analysis with scaling function φ,*
(ii) $\{\psi_{j,k} : j, k \in \mathbb{Z}\}$ *is an orthonormal system in L^2,*
(iii) $\operatorname{supp}\psi \subset [-(M-L)/2 + 1/2, (M-L)/2 + 1/2]$,
(iv) *we have the expression*

$$\psi(x) = \sum_{n=-M+1}^{-L+1} (-1)^n \overline{h}_{-n+1}\varphi_{1,n}(x).$$

9.5.2. Construction of trigonometric polynomials. In this subsection we shall construct trigonometric polynomials m_0 which satisfy condition (a) of Theorem 9.4.

LEMMA 9.14. *Let N be a positive integer and put*

$$p_N(x) = \sum_{n=0}^{N-1} \binom{N-1+n}{n} x^n.$$

Then we have

$$(1-x)^N p_N(x) + x^N p_N(1-x) = 1. \tag{9.35}$$

PROOF. By Euclid's algorithm there exist unique polynomials $p(x)$ and $q(x)$ of order at most $N-1$ such that

$$(1-x)^N p(x) + x^N q(x) = 1$$

(see Appendix B.6). Substituting $1-x$ for x, we get $x^N p(1-x) + (1-x)^N q(1-x) = 1$. Therefore we have $q(x) = p(1-x)$ by the uniqueness.

This implies that

$$\begin{aligned} p(x) &= (1-x)^{-N}[1 - x^N p(1-x)] \\ &= \sum_{n=0}^{N-1} \binom{N-1+n}{n} x^n + x^N r(x) \quad (|x| < 1), \end{aligned}$$

where $r(x)$ is an infinite series $\sum_{k=0}^{\infty} a_k x^k$; but since $\deg p < N$, $r(x)$ is necessarily zero. □

REMARK 9.4. $p_N(x) \geq 0$ for $x \geq 0$.

If $r(x)$ is a polynomial such that $r(x) = -r(-x)$, then the polynomial

$$p_N(x) + x^N r(1/2 - x)$$

satisfies (9.35), too.

LEMMA 9.15. *Let $a_0, a_1, \ldots, a_M$ be a sequence of real numbers with $a_M \neq 0$. If*

$$A(e^{2\pi i\xi}) = \frac{a_0}{2} + \sum_{n=1}^{M} a_n \cos 2\pi n\xi \geq 0,$$

then there exists a polynomial $B(z) = \sum_{n=0}^{M} b_n z^n$ of order M such that

$$|B(e^{2\pi i\xi})|^2 = A(e^{2\pi i\xi}).$$

PROOF. Put $p(z) = (1/2) z^M \sum_{n=-M}^{M} a_{|n|} z^n$. Then $p(e^{2\pi i\xi}) = e^{2\pi i M\xi} A(e^{2\pi i\xi})$.

We observe that any solution on the unit circle of the equation $p(z) = 0$ is a multiple root of even order, because $A(e^{2\pi i\xi}) \geq 0$. Furthermore, if $p(z_j) = 0$, then $\overline{z}_j, z_j^{-1}$, and $\overline{z}_j^{-1}$ are all sollutions, since $p(z) = z^{2M} p(z^{-1})$ and all coefficients of $p(z)$ are real. Note that $p(0) = a_M \neq 0$. Let $r_j, j = 1, \ldots, J$, be real solutions to $p(z) = 0$, and $z_l, \overline{z}_l, l = 1, \cdots, L$, be complex solutions in the open unit disk, respectively, and let $e^{2\pi i\theta_k}, k = 1, \ldots, 2K$, be complex solutions on the unit circle, such that $\theta_k \equiv \theta_{k+K} \pmod 1$. Then $2(J + K + 2L) = 2M$, and

$$\begin{aligned} p(z) = \frac{a_M}{2} &\left\{ \prod_{j=1}^{J} (z - r_j)(z - r_j^{-1}) \right\} \left\{ \prod_{k=1}^{K} (z - e^{2\pi i\theta_k})^2 \right\} \\ &\times \left\{ \prod_{l=1}^{L} (z - z_l)(z - \overline{z}_l)(z - z_l^{-1})(z - \overline{z}_l^{-1}) \right\}. \end{aligned}$$

Observe that if $|z| = 1$, then

$$|(z - z_l)(z - \overline{z}_l)(z - z_l^{-1})(z - \overline{z}_l^{-1})| = |z_l|^{-2} |z - z_l|^2 |z - \overline{z}_l|^2.$$

Define

$$B(z) = \left\{\frac{|a_M|}{2}\right\}^{1/2} \left\{\prod_{j=1}^{J} |r_j| \prod_{l=1}^{L} |z_l|^2\right\}^{-1/2}$$
$$\times \left\{\prod_{j=1}^{J}(z - r_j) \prod_{k=1}^{K}(z - e^{2\pi i\theta_k}) \prod_{l=1}^{L}(z - z_l)(z - \overline{z}_l)\right\}.$$

Then $B(z)$ is a polynomial of order M, and

$$A(e^{2\pi i\xi}) = |p(e^{2\pi i\xi})| = |B(e^{2\pi i\xi})|^2.$$

□

Let $r(x)$ be a polynomial such that $r(x) = -r(-x)$, and suppose that

$$P(x) = p_N(x) + x^N r(1/2 - x)$$

is a polynomial nonnegative in $0 \le x \le 1$. Note that $p_N(0) = 1$.

Put

$$A(\xi) = (\cos^2 \pi\xi)^N P(\sin^2 \pi\xi). \tag{9.36}$$

Then by Lemma 9.15 there exists a trigonometric polynomial $m_0(\xi)$ such that $A(\xi) = |m_0(\xi)|^2$. Thus $m_0(\xi)$ satisfies condition (a).

It is easily checked that the function $m_0(\xi)$ satisfies (b) with $K = [-1/2, 1/2]$.

PROPOSITION 9.2. *There exists a trigonometric polynomial m_0 which satisfies conditions* (a) *and* (b) *of Theorem 9.4.*

9.5.3. An example of a compactly supported wavelet. We consider the polynomial of order $N = 2$ given in Lemma 9.14.

By a simple computation, $p_2(x) = 1 + 2x$. We suppose that $r(x) = 0$. Then by (9.36) we have

$$|m_0(\xi)|^2 = A(\xi) = (\cos^2 \pi\xi)^2(1 + 2\sin^2 \pi\xi)$$
$$= \frac{1}{2}\left|\left(\frac{e^{2\pi i\xi} + 1}{2}\right)^4 (e^{2\pi i\xi} - 2 - \sqrt{3})(e^{2\pi i\xi} - 2 + \sqrt{3})\right|.$$

Therefore, by a computation,

$$m_0(\xi) = \frac{1 - \sqrt{3}}{2}\left(\frac{e^{-2\pi i\xi} + 1}{2}\right)^2 (e^{-2\pi i\xi} - 2 - \sqrt{3})$$
$$= \frac{1 - \sqrt{3}}{8}\left\{e^{-6\pi i\xi} - \sqrt{3}e^{-4\pi i\xi} - \sqrt{3}(2 + \sqrt{3})e^{-2\pi i\xi} - (2 + \sqrt{3})\right\}.$$

Put

$$\widehat{\varphi}_2(\xi) = \prod_{j=1}^{\infty} m_0(2^{-j}\xi),$$

and let ψ_2 be the corresponding wavelet. Then by Theorem 9.4 we obtain

$$\psi_2(x) = \frac{1}{8}\Big\{ - (\sqrt{3} - 1)(\varphi_2)_{1,-2}(x) - \sqrt{3}(\sqrt{3} - 1)(\varphi_2)_{1,-1}(x)$$
$$+ \sqrt{3}(\sqrt{3} + 1)(\varphi_2)_{1,0}(x) - (1 + \sqrt{3})(\varphi_2)_{1,1}(x)\Big\}.$$

From Lemma 9.11 and Theorem 9.4 it follows that

$$\operatorname{supp}\varphi_2 \subset [0,3], \quad \text{and} \quad \operatorname{supp}\psi_2 \subset [-1,2]$$

(see Figure 9.8).

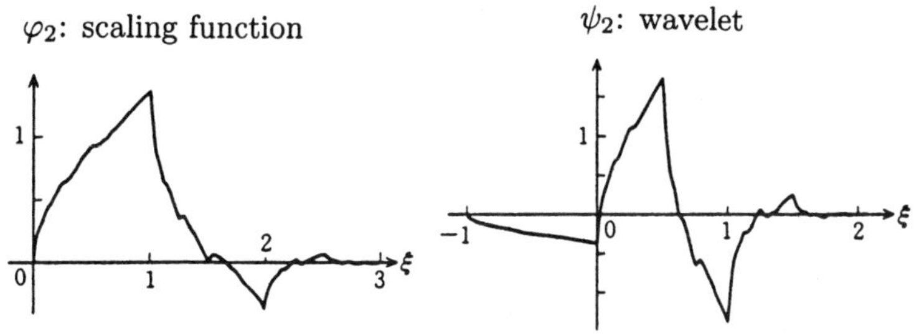

FIGURE 9.8. Daubechies' wavelet with $N = 2$

The function φ_2 is continuous. In fact, we shall show that φ_2 satisfies a Lipschitz condition of order $\gamma = 1 - [\log 3/(2\log 2)] = 0.2075\ldots$.

Since $m_0(\xi) = ((e^{2\pi i\xi}+1)/2)^2 L(\xi)$ with a trigonometric polynomial $L(\xi)$ such that $L(0) = 1$, there exists a positive constant C such that $|L(\xi)| \le 1 + C|\xi|$. Thus

$$\prod_{j=1}^{\infty} |L(2^{-j}\xi)| \le \prod_{j=1}^{\infty} e^{C2^{-j}|\xi|} \le e^C \quad (|\xi| \le 1).$$

Furthermore, we have

$$\max|L(\xi)|^2 = \max_{0\le x\le 1} p_2(x) = \sum_{n=0}^{1} \binom{1+n}{n} = 3.$$

Therefore when $2^k < |\xi| \le 2^{k+1}$, we have the estimate

$$\begin{aligned}\prod_{j=1}^{\infty} |L(2^{-j}\xi)| &= \prod_{j=1}^{k} |L(2^{-j}\xi)| \prod_{j=1}^{\infty} |L(2^{-k-j}\xi)| \\ &\le \sqrt{3}^k e^C \le C'|\xi|^\alpha,\end{aligned}$$

where $\alpha = \log 3/(2\log 2)$ and C' is a constant independent of ξ.

The above estimate combined with Euler's formula

$$\prod_{j=1}^{\infty} \cos \pi 2^{-j}\xi = \frac{\sin \pi\xi}{\pi\xi}$$

(see Appendix B.5) gives

$$|\widehat{\varphi}_2(\xi)| = \prod_{k=1}^{\infty} \cos^2(\pi 2^{-k}\xi)|L(2^{-k}\xi)| \le C'(1+|\xi|)^{-2}|\xi|^\alpha \quad (\xi \in \mathbb{R}).$$

Therefore our assertion follows from the next lemma.

LEMMA 9.16. *Let $0 < \gamma < 1$. If $f \in L^2(\mathbb{R})$ and*

$$|\widehat{f}(\xi)| \le C(1+|\xi|)^{-\gamma-1},$$

then f is a Lipschitz function of order γ.

PROOF. By definition we have

$$f(x+h)-f(x)=\int \widehat{f}(\xi)[e^{2\pi ih\xi}-1]e^{2\pi ix\xi}d\xi$$
$$=\left(\int_{|h\xi|<1}+\int_{|h\xi|>1}\right)\widehat{f}(\xi)[e^{2\pi ih\xi}-1]e^{2\pi ix\xi}d\xi=I+J,$$

say. Let C_1, C_2 and C_3 be appropriate constants independent of ξ. Using the assumption $|\widehat{f}(\xi)|\leq C(1+|\xi|)^{-\gamma-1}$, we get

$$|I|\leq\int_{|h\xi|<1}2\pi|h\xi||\widehat{f}(\xi)|d\xi\leq\int_{|\xi|<1/|h|}2\pi C|h\xi|(1+|\xi|)^{-\gamma-1}d\xi\leq C_1|h|^{\gamma},$$

and

$$|J|\leq C\int_{|h\xi|>1}2|\xi|^{-\gamma-1}d\xi\leq C_2|h|^{\gamma}.$$

Therefore we obtain

$$|f(x+h)-f(x)|\leq C_3|h|^{\gamma}.$$

□

Corresponding to each polynomial p_N, $N=3,4,\ldots$, we can construct a compactly supported wavelet. It is known that the order of continuity of wavelets increases with the growth of the number N. For the details the reader may consult Daubechies' book [Da].

Appendix A

A.1. Zorn's lemma

A.1.1. Relation and ordering. Let X and Y be sets. The *Cartesian product* $X \times Y$ of X and Y is the collection of all orderd pairs (x, y) such that $x \in X$ and $y \in Y$.

A *relation* is any set of ordered pairs. Thus a relation is a subset of the Cartesian product.

Let X be a set. For a relation $R \subset X \times X$ denote $(x, y) \in R$ by xRy.

DEFINITION A.1. Let X be a set. An *equivalence relation* in X is a relation $\sim \subset X \times X$ such that for $x, y, z \in X$ we have:

(i) *Reflexive relation*: $x \sim x$.
(ii) *Symmetric relation*: $x \sim y$ implies $y \sim x$.
(iii) *Transitive relation*: $x \sim y$ and $y \sim z$ implies $x \sim z$.

DEFINITION A.2. Let X be a set. A *partial ordering* in X is a relation $\prec \subset X \times X$ such that, for $x, y, z \in X$ we have:

(i) *Reflexive relation*: $x \prec x$.
(ii) *Antisymmetric relation*: $x \prec y$ and $y \prec x$ imply $x = y$.
(iii) *Transitive relation*: $x \prec y$ and $y \prec z$ imply $x \prec z$.

Furthermore, if the relation $\prec$ satisfies

(iv) *Triconomy relation*: $x, y \in X$ implies $x \prec y$ or $y \prec x$,

then $\prec$ is called a *total* (*linear*) *ordering*.

A linearly ordered subset of a partially ordered set is called a *chain*.

DEFINITION A.3. Let A be a subset of a partially ordered set $(X, \prec)$. An element u of X is called an *upper bound* for A, if $a \prec u$ for all $a \in A$. An upper bound u_0 of A is called the *least upper bound* or the *supremum* of A if $u_0 \prec u$ for all upper bounds u of A.

An element $m \in X$ is called *maximal* if $x \in X$ and $m \prec x$ imply $x = m$.

By the same way an element l of X is called a lower bound for A, if $l \prec a$ for all $a \in A$, and a lower bound l_0 of A is called the *greatest lower bound* or the *infimum* of A if $l \prec l_0$ for all lower bounds l of A.

A.1.2. Zorn's lemma.

DEFINITION A.4. A *choice function* for a set X is a function f which associates to each nonempty subset A of X an element $f(A)$ of A.

The following three axioms are equivalent:

AXIOM A.1 (Axiom of choice). For any nonempty set there exists a choice function.

AXIOM A.2 (Zorn's lemma). If each chain in a nonempty partially ordered set has an upper bound, then there exists a maximal element of the set.

AXIOM A.3 (The Hausdorff maximality principle). Every nonempty partially ordered set contains a maximal chain.

To prove the equivalence of these three axioms we use the following lemma

LEMMA A.1. *Let $(X, \prec)$ be a nonempty partially ordered set such that every chain contained in X has an upper bound. Let f be a function from X to X such that $x \prec f(x)$ for all $x \in X$. For $a \in X$ let $X_a = \{x \in X : a \prec x\}$. Then for every $a \in X$ there exists an element $a^* \in X_a$ such that $f(a^*) = a^*$.*

PROOF. First Step. Let $a \in X$. Let $\mathcal{A}$ be the collection of all subsets A of X_a such that

(a) $a \in A$,
(b) $f(x) \in A$ for all $x \in A$,
(c) if $B \subset A$ and if the least upper bound of B exists, then it belongs to A.

Remark that $X_a \in \mathcal{A}$, and any finite intersection of sets of $\mathcal{A}$ belongs to $\mathcal{A}$.

Let I be the intersection of all sets in $\mathcal{A}$. Assume that I is a chain of X. Then by the assumption of the lemma the set I has an upper bound, say a^*. Then by property (c) $a^* \in A$ for all $A \in \mathcal{A}$. Thus $a^* \in I$, and we have $f(a^*) = a^*$ by (b).

In the following we shall prove that I is a chain.

Second Step. Let N be the set of n in I such that if $x \in I$ and $x \precneqq n$, then $f(x) \prec n$. Observe that $a \in N$, because there is no element x in I such that $x \precneqq a$. For $n \in N$ put $B_n = \{b \in I : b \prec n \text{ or } f(n) \prec b\}$. Then $B_n \in \mathcal{A}$.

In fact, (a): Since $a \prec n$, the element a is contained in B_n. (b): Let $b \in B_n$. If $f(n) \prec b$, then $f(n) \prec b \prec f(b)$, which implies $f(b) \in B_n$. If $b = n$, then $f(n) = f(b)$, and if $b \precneqq n$, then $f(b) \prec n$ by the definition of N. In any case these imply that $f(b) \in B_n$. (c): Let $B \subset B_n$ and assume that the least upper bound b_0 of B exists. Obviously b_0 belongs to I. If there exists $b \in B$ such that $f(n) \prec b$, then $f(n) \prec b_0$. Thus $b_0 \in B_n$. Otherwise, $b \prec n$ for all $b \in B_n$. Thus $b_0 \prec n$, which implies $b_0 \in B_n$.

Therefore $I = B_n$ for all $n \in N$. This means that if $x \in I$, then $x \prec n$ or $n \prec f(n) \prec x$ for all $n \in N$.

Third Step. $N \in \mathcal{A}$.

In fact, condition (a) is observed in the second step. (b): Let $n \in N$. To prove that $f(n) \in N$, it suffices to show that if $x \in I$ and $x \precneqq f(n)$, then $f(x) \prec f(n)$. Since $x \in B_n$, we see that $x \prec n$ or $f(n) \prec x$. The last case is impossible, so that $x \precneqq n$ or $x = n$. In either case we have $f(x) \prec f(n)$ by the property of N. (c): Let $C \subset N$ and assume that the least upper bound c_0 of C exists. To prove that $c_0 \in N$, suppose that $x \in I$ and $x \precneqq c_0$. Our aim is to show that $f(x) \prec c_0$. Now, we have shown in the second step that $x \prec c$ or $c \prec x$ for $x \in I = B_n$ and $c \in C$. By our assumption we can choose an element $c \in C$ such that $x \precneqq c$. Thus $f(x) \prec c \prec c_0$.

Therefore we have $I \subset N$, and thus $N = I = B_n$ for all $n \in N$.

Fourth Step. Let $x, y \in I$. Then, since $y \in B_x$, we have $y \prec x$ or $x \prec f(x) \prec y$. Therefore I is a chain. □

AXIOM A.1 $\Rightarrow$ AXIOM A.2. Assume that every chain in the set $(X, \prec)$ has an upper bound and X has no maximal element. Then the set $F_a = \{x \in X : a \not\preceq x\}$ is not empty for all $a \in X$. Let f be a function which associates to F_a a point in F_a, that is, $f(a) \in F_a$. Thus by definition $a \not\preceq f(a)$ for all $a \in X$. On the other hand by Lemma A.1 for any $a \in X$ there exists $a^* \in F_a$ such that $f(a^*) = a^*$, which is absurd. □

AXIOM A.2 $\Rightarrow$ AXIOM A.3. Let $\mathcal{C}$ be the family of all chains in $(X, \prec)$, which is partially ordered by inclusion. If $\mathcal{C}_0$ is a chain of $(\mathcal{C}, \subset)$, then the union $C_0 = \bigcup_{C \in \mathcal{C}_0} C$ is a chain in X and C_0 is un upper bound of $\mathcal{C}_0$. Thus by our hypothesis $\mathcal{C}$ has a maximal element. □

AXIOM A.3 $\Rightarrow$ AXIOM A.1. Let $\mathcal{P}$ be a family of nonempty subsets of a set X. Let $\mathcal{A}$ be a subfamily of $\mathcal{P}$ and suppose that there exists a function f such that $f(A) \in A$. Then we denote by $(\mathcal{A}, f)$ the set $\{(A, f(A)) : A \in \mathcal{A}\}$.

Let us consider the set $\mathcal{F} = \{(\mathcal{A}, f) : \mathcal{A} \subset \mathcal{P}$ and f is a choice function of $\mathcal{A}\}$. The set $\mathcal{F}$ is nonempty. For $(\mathcal{A}, f)$ and $(\mathcal{B}, g) \in \mathcal{F}$ define an order $(\mathcal{A}, f) \prec (\mathcal{B}, g)$ by

$$\mathcal{A} \subset \mathcal{B} \quad \text{and} \quad f = g \text{ on } \mathcal{A}.$$

Then $(\mathcal{F}, \prec)$ is a partially ordered set. Thus by the hypothesis the set $\mathcal{F}$ has a maximal chain $\mathcal{C}$.

Let us consider the set $F = \bigcup_{\mathcal{A}} (\mathcal{A}, f)$, where $\mathcal{A}$ runs over all $(\mathcal{A}, f) \in \mathcal{C}$. If $(A, f(A)), (A, f'(A)) \in F$, then, as is easily seen, we have $f(A) = f'(A)$. Thus $F \in \mathcal{F}$. From the maximality of $\mathcal{C}$ it follows that the domain of F coincides with $\mathcal{P}$. □

A.2. Urysohn's lemma

THEOREM A.1 (Urysohn's lemma). *Let X be a locally compact Hausdorff space. Let K be a compact subspace and O be an open subset. If $K \subset O$, then there exists $u \in C_c(X)$ such that $K \prec u \prec O$, that is,*

$$u(x) = 1\,(x \in K), \quad u(x) = 0\,(x \notin O), \quad \text{and} \quad 0 \leq u(x) \leq 1.$$

PROOF. Let $\{r_j\}$ be an enumeration of all rational numbers in $[0, 1]$ such that $r_0 = 0$ and $r_1 = 1$.

By Theorem 6.16 there exist open sets O_0 and O_1 with compact closures such that

$$K \subset O_1 \subset \overline{O_1} \subset O_0 \subset \overline{O_0} \subset O.$$

Let $n \geq 1$ and suppose that open sets $O_0, O_1, \ldots, O_n$ are chosen so that if $r_k < r_j$, then $O_j \subset \overline{O_j} \subset O_k$. Let r_p, $1 \leq p \leq n$, be the largest rational number such that $r_p < r_{n+1}$, and let r_q, $1 \leq q \leq n$, be the smallest one such that $r_{n+1} < r_q$. Then, since $\overline{O_q} \subset O_p$, we can find, again by Theorem 6.16, an open set O_{n+1} such that

$$\overline{O_q} \subset O_{n+1} \subset \overline{O_{n+1}} \subset O_p.$$

Thus we get a sequence $\{O_j\}$ of open sets with the properties that $\overline{O_j}$ is compact for each j, and $K \subset O_j \subset \overline{O_j} \subset O_k \subset O\,(r_k < r_j)$.

Define

$$f(x) = \sup\{r_j : x \in O_j\}$$

for $x \in O_0$ and $f(x) = 0$ outside O_0. It is obvious that $f(x) = 1$ on K and $= 0$ outside O, and $0 \leq f(x) \leq 1$. Suppose $0 < f(x_0) < 1$. For any $1 > \varepsilon > 0$ there exist rational numbers r and s such that $f(x_0) - \varepsilon < r < f(x_0) < s < f(x_0) + \varepsilon$. Put $U = O_r - \overline{O_s}$. Then U is an open set containing x_0, and

$$|f(x_0) - f(x)| < \varepsilon, \quad x \in U.$$

Thus f is continuous at x_0. If $f(x_0) = 0$, then $x_0 \notin O_r$ for any positive small rational number r. Thus $x_0 \notin \overline{O_{2r}}$. Thus there exists a neighborhood U of x_0 such that $U \subset \overline{O_{2r}}^c$, which implies that $f(x) \leq 2r$ on U. In the same way, if $f(x_0) = 1$ and $0 < s < 1$, then we can find a neighborhood U of x_0 such that $f(x) > s$ on U. Thus $f(x)$ is continuous on X. □

Appendix B

B.1. Gamma function

The gamma function

$$\Gamma(\alpha) = \int_0^\infty t^{\alpha-1} e^{-t} dt$$

converges in the sense of Lebesgue and is continuous if $\Re\alpha > 0$.

Integration by parts gives

$$\Gamma(\alpha+1) = \left[-t^\alpha e^{-t}\right]_0^\infty + \alpha \int_0^\infty t^{\alpha-1} e^{-t} dt = \alpha\Gamma(\alpha).$$

Thus

$$\text{(B.1)} \qquad \Gamma(\alpha+1) = \alpha\Gamma(\alpha), \quad \Re\alpha > 0.$$

In particular, since $\Gamma(1) = 1$, we have $\Gamma(n+1) = n!$.

The integral

$$B(\alpha, \beta) = \int_0^1 t^{\alpha-1}(1-t)^{\beta-1} dt$$

is called the *beta function*. It converges if $\Re\alpha > 0$ and $\Re\beta > 0$.

We have the following formula:

$$\text{(B.2)} \qquad B(\alpha, \beta) = \frac{\Gamma(\alpha+\beta)}{\Gamma(\alpha)\Gamma(\beta)}, \quad \Re\alpha > 0 \text{ and } \Re\beta > 0.$$

In fact, by definition we have

$$(u+1)^{-\alpha-\beta} = \frac{1}{\Gamma(\alpha+\beta)} \int_0^\infty t^{\alpha+\beta-1} e^{-(u+1)t} dt.$$

Multiplying by $u^{\alpha-1}$ and integrating with respect to u, we get

$$\int_0^\infty u^{\alpha-1}(u+1)^{-\alpha-\beta} du = \frac{1}{\Gamma(\alpha+\beta)} \int_0^\infty u^{\alpha-1} du \int_0^\infty t^{\alpha+\beta-1} e^{-(u+1)t} dt.$$

If we reverse the integrals, the last term equals

$$\begin{aligned} &\frac{1}{\Gamma(\alpha+\beta)} \int_0^\infty t^{\alpha+\beta-1} e^{-t} dt \int_0^\infty u^{\alpha-1} e^{-ut} du \\ &= \frac{1}{\Gamma(\alpha+\beta)} \int_0^\infty t^{\beta-1} e^{-t} dt \int_0^\infty u^{\alpha-1} e^{-u} du = \frac{\Gamma(\alpha)\Gamma(\beta)}{\Gamma(\alpha+\beta)}, \end{aligned}$$

that is,

$$\frac{\Gamma(\alpha)\Gamma(\beta)}{\Gamma(\alpha+\beta)} = \int_0^\infty u^{\alpha-1}(u+1)^{-\alpha-\beta} du.$$

If we put $u = v(1-v)^{-1}$, then we obtain (B.2).

In particular, $\Gamma(\frac{1}{2})^2 = B(\frac{1}{2}, \frac{1}{2})$ and

$$B(\frac{1}{2}, \frac{1}{2}) = \int_0^1 t^{-1/2}(1-t)^{-1/2}dt = 2\int_0^{\pi/2} d\theta = \pi$$

by a change of variable, $t = \sin\theta$. Therefore we have

$$\Gamma(\frac{1}{2}) = \sqrt{\pi}. \tag{B.3}$$

Furthermore, since $\Gamma(\frac{1}{2}) = \int_0^\infty t^{-1/2}e^{-t}dt = 2\int_0^\infty e^{-u^2}du$, we get

$$\int_{-\infty}^{\infty} e^{-u^2}du = \sqrt{\pi}. \tag{B.4}$$

We shall prove the duplication formula

$$\Gamma(2\alpha+1) = 2^{2\alpha}\pi^{-1/2}\Gamma(\alpha+1)\Gamma(\alpha+\frac{1}{2}), \quad \alpha > -1/2. \tag{B.5}$$

In fact, since the function $t(1-t)$ is symmetric about $1/2$,

$$\frac{[\Gamma(\alpha+\frac{1}{2})]^2}{\Gamma(2\alpha+1)} = \int_0^1 [t(1-t)]^{\alpha-(1/2)}dt = 2\int_0^{1/2} [t(1-t)]^{\alpha-(1/2)}dt.$$

Put $u = 4t(1-t)$. Then we have $du = 4(1-u)^{1/2}dt$. Thus the last term equals

$$2^{-2\alpha}\int_0^1 u^{\alpha-\frac{1}{2}}(1-u)^{-1/2}du = 2^{-2\alpha}\frac{\Gamma(\alpha+\frac{1}{2})\Gamma(\frac{1}{2})}{\Gamma(\alpha+1)},$$

which implies formula (B.5).

B.2. Unit ball

Let $B(0,1)$ be the unit ball in $\mathbb{R}^d$ and S^{d-1} the unit sphere $\{x \in \mathbb{R}^d : |x| = 1\}$. Let Ω_d and ω_{d-1} be the volume of $B(0,1)$ and the area of S^{d-1}, respectively.

For $x \in \mathbb{R}^d$ denote $x = rw$, where $r \geq 0$ and $w \in S^{d-1}$. Then

$$\int_{\mathbb{R}^d} e^{-\pi|x|^2}dx = \int_0^\infty e^{-\pi r^2}r^{d-1}dr \int_{S^{d-1}} d\sigma(w),$$

where $d\sigma(w)$ is the area element of S^{d-1}. The right-hand side of the above equation equals

$$\omega_{d-1}\frac{1}{2\pi^{d/2}}\int_0^\infty t^{(d-2)/2}e^{-t}dt = \omega_{d-1}\frac{1}{2\pi^{d/2}}\Gamma(\frac{d}{2}).$$

Since the left-hand side equals 1, we get

$$\omega_{d-1} = 2\pi^{d/2}/\Gamma(\frac{d}{2}).$$

Since $\Omega_d = \int_{|x|<1} dx = \int_0^1 \int_{S^{d-1}} r^{d-1}drd\sigma(w)$, we obtain

$$\Omega_d = \pi^{d/2}/\Gamma(\frac{d}{2}+1).$$

B.3. The Bessel function

For $\Re\mu > -1$ the *Bessel function* of the first kind is defined by the series

$$J_\mu(t) = \left(\frac{t}{2}\right)^\mu \sum_{k=0}^{\infty} \frac{(-1)^k}{2^{2k}\Gamma(k+1)\Gamma(\mu+k+1)} t^{2k}. \tag{B.6}$$

Since $a^{k/2} \leq k! = \Gamma(k+1)$ for $k > 2a$, the series converges in the whole complex plain.

Term by term differentiation gives the following formulas:

$$-\frac{1}{t}\frac{d}{dt}\frac{J_\mu(t)}{t^\mu} = \frac{J_{\mu+1}(t)}{t^{\mu+1}}, \tag{B.7}$$

$$\frac{1}{t}\frac{d}{dt}t^{\mu+1}J_{\mu+1}(t) = t^\mu J_{\mu+1}(t). \tag{B.8}$$

B.4. General form of Bochner's formula

We shall give a generalization of Theorem 8.19.

Let us define a differential operator L by

$$L = -\frac{1}{t}\frac{d}{dt}.$$

Put $V_\mu(t) = J_\mu(t)/t^\mu$. Then (B.7) implies

$$LV_\mu(t) = V_{\mu+1}(t).$$

LEMMA B.1. *Let $u \in C^\infty(\mathbb{R})$. For $\xi \in \mathbb{R}^d$ we have*

$$\frac{\partial^n}{\partial\xi_j^n}u(|\xi|) = \sum_{k=0}^{\infty} \frac{(-1)^{n-k}}{2^k k!}(L^{n-k}u)(|\xi|)\frac{\partial^{2k}}{\partial\xi_j^{2k}}\xi_j^n, \tag{B.9}$$

whre the kth summand is assumed to be 0 if $k > n/2$.

PROOF. Formula (B.9) is trivial when $n = 0$. Suppose it holds for n. Then, differentiating both sides with respect to ξ_j, we get

$$\begin{aligned}\frac{\partial^{n+1}}{\partial\xi_j^{n+1}}u &= \sum_{k=0}^{\infty} \frac{(-1)^{n-k}}{2^k k!}\left\{(L^{n-k+1}u)\left(\frac{\partial^{2k}}{\partial\xi_j^{2k}}\xi_j^n\right)(-\xi_j) + (L^{n-k}u)\frac{\partial^{2k+1}}{\partial\xi_j^{2k+1}}\xi_j^n\right\}\\ &= \sum_{k=1}^{\infty} \frac{(-1)^{n-k}}{2^k k!}(L^{n-k+1}u)\left\{(-\xi_j)\frac{\partial^{2k}}{\partial\xi_j^{2k}}\xi_j^n - 2k\frac{\partial^{2k-1}}{\partial\xi_j^{2k-1}}\xi_j^n\right\}\\ &\quad + (-1)^n L^{n+1}u)(-\xi_j),\end{aligned}$$

where $u = u(|\xi|)$. The term in braces is, by a simple computation, equal to $-(\partial/\partial\xi_j)^{2k}\xi_j^{n+1}$. Therefore formula (B.9) is valid for $n+1$. Thus the lemma is proved by induction. □

Let $\Re\mu > -1$. By successive application of Lemma B.1 we get for $\alpha \in \mathbb{Z}_+^d$

$$D^\alpha V_\mu(2\pi|\xi|) = \sum_k \frac{(-1)^{\|\alpha\|-\|k\|}(2\pi)^{2\|\alpha\|-2\|k\|}}{2^{\|k\|}k!}V_{\mu+\|\alpha\|-\|k\|}(2\pi\xi)D^{2k}\xi^\alpha, \tag{B.10}$$

where k runs over $\mathbb{Z}_+^d$.

Let $p(x), x \in \mathbb{R}^d$, be a harmonic homogeneous polynomial of order N, that is, $p(x) = \sum_{\|\alpha\|=N} a_\alpha x^\alpha$ and $\Delta p \equiv 0$. Then by (B.10) we have

$$p(D)V_\mu(2\pi|\xi|) = \sum_{n=0}^{\infty} \sum_{\{k:\|k\|=n\}} \frac{(-1)^{N-n}(2\pi)^{2N-2n}}{2^n k!} V_{\mu+N-n}(2\pi|\xi|) D^{2k} p(\xi).$$

The sum of $n!(k!)^{-1} D^{2k} p(\xi)$ over $\{k \in \mathbb{Z}_+^d : \|k\| = n\}$ equals $\Delta^n p(\xi)$, which vanishes except when $n = 0$. Therefore we get

$$p(D)V_\mu(2\pi|\xi|) = (2\pi)^N p(-2\pi\xi) V_{\mu+N}(2\pi|\xi|). \tag{B.11}$$

As an easy consequence of (8.31) and (B.11) we obtain

LEMMA B.2. *If $p(x)$ is a harmonic homogeneous polynomial of order N, then*

$$\int_{S^{d-1}} p(w) e^{-2\pi i \xi w} d\sigma(w) = (2\pi)^{d/2} p(-2\pi i \xi) V_{(d/2)+N-1}(2\pi|\xi|). \tag{B.12}$$

THEOREM B.1 (Bochner's formula). *Let $p(x), x \in \mathbb{R}^d$, be a harmonic homogeneous polynomial of order N, and f^* be a radial function in $L^1(\mathbb{R}^d)$. If $f(x) = p(w) f^*(r)$, where $x = rw, r \geq 0$ and $w \in S^{d-1}$, then*

$$\widehat{f}(\xi) = p(-2\pi i \xi) F^*(\rho),$$

where $\xi = \rho\omega$ with $\rho \geq 0$ and $\omega \in S^{d-1}$, and

$$F^*(\rho) = (2\pi)^{d/2} \int_0^\infty f^*(r) V_{(d/2)+N-1}(2\pi\rho r) r^{d+N-1} dr.$$

PROOF. By polar coordinates we have

$$\widehat{f}(\xi) = \int_0^\infty f^*(r) r^{d-1} dr \int_{S^{d-1}} p(w) e^{-2\pi i \rho r \omega w} d\sigma(w),$$

The last integral implies Theorem B.1 by Lemma B.2. □

B.5. Euler's formula

$$\prod_{j=1}^{\infty} \cos \pi 2^{-j} \xi = \frac{\sin \pi\xi}{\pi\xi} \quad (\xi \in \mathbb{R}). \tag{B.13}$$

PROOF. By the formula $\sin \pi 2^{-j}\xi = 2 \sin \pi 2^{-j-1}\xi \cos \pi 2^{-j-1}\xi$, we have

$$\begin{aligned}\prod_{j=1}^{k} \cos \pi 2^{-j}\xi &= \frac{\sin \pi\xi}{2 \sin \pi 2^{-1}\xi} \frac{\sin \pi 2^{-1}\xi}{2 \sin \pi 2^{-2}\xi} \cdots \frac{\sin \pi 2^{-k+1}\xi}{2 \sin \pi 2^{-k}\xi} \\ &= \frac{\sin \pi\xi}{2^k \sin \pi 2^{-k}\xi}.\end{aligned}$$

The last term converges to $\frac{\sin \pi\xi}{\pi\xi}$ as $k \to \infty$. □

B.6. Euclid's algorithm

THEOREM B.2. *Let $p_0(x)$ and $p_1(x)$ be nonzero polynomials which are relatively prime. Then there exist unique polynomials $q_0(x)$ and $q_1(x)$ such that*

$$p_0(x)q_0(x) + p_1(x)q_1(x) = 1, \quad \deg q_j < \deg p_j \ (j = 0, 1).$$

PROOF. We can assume that $\deg p_0 \geq \deg p_1$. Then we have

$$p_0(x) = a_1(x)p_1(x) + p_2(x), \quad \deg p_2 < \deg p_1,$$

and

$$p_1(x) = a_2(x)p_2(x) + p_3(x), \quad \deg p_3 < \deg p_2.$$

In the same way we get

$$p_{n-1}(x) = a_n(x)p_n(x) + p_{n+1}(x), \quad \deg p_{n+1} < \deg p_n.$$

Since $\deg p_n$ is strictly decreasing, this process stops at a finite step, say the Nth; that is, $p_N \neq 0$ and $p_{N+1} = 0$.

We observe that p_N divides p_{N-1}, since $p_{N-1} = a_N p_N$, and thus p_N divides p_{N-2}, since $p_{N-2} = a_{N-1}p_{N-1} + p_N$. And finally we find that p_N divides both p_0 and p_1. By assumption p_0 and p_1 are relatively prime. Thus p_N must be a nonzero constant, which we can suppose 1 without loss of generality. Therefore we have

$$1 = p_{N-2} - a_{N-1}p_{N-1}.$$

Substitute successively $p_n = p_{n-2} - a_{n-1}p_{n-1}$, $n = N+1, \ldots, 2$, for the last formula. Then we get

$$\begin{aligned} 1 &= (1 + a_{N-1}a_{N-2})p_{N-2} - a_{N-1}p_{N-3} \\ &= -(a_{N-1} + a_{N-3} + a_{N-1}a_{N-2}a_{N-3})p_{N-3} + (1 + a_{N-1}a_{N-2})p_{N-4} \\ &= \cdots. \end{aligned}$$

Finally we can express this as $1 = q_1p_1 + q_0p_0$ with some polynomials q_1 and q_0.

Next, we show that $\deg q_1 < \deg p_1$ and $\deg q_0 < \deg p_0$. In fact, we have

$$\begin{aligned} \deg q_1 &= \deg a_{N-1} + \deg a_{N-2} + \cdots + \deg a_1 \\ &= (\deg p_{N-2} - \deg p_{N-1}) + (\deg p_{N-3} - \deg p_{N-2}) + \cdots \\ &+ (\deg p_1 - \deg p_2) \\ &= \deg p_1 - \deg p_{N-1} < \deg p_1. \end{aligned}$$

In the same way

$$\begin{aligned} \deg q_0 &= \deg a_{N-1} + \deg a_{N-2} + \cdots + \deg a_2 \\ &= \deg p_2 - \deg p_{N-1} < \deg p_0. \end{aligned}$$

Finally, to prove the uniqueness of polynimials q_0 and q_1, suppose that $1 = p_0q_0 + p_1q_1 = p_0q_0' + p_1q_1'$, where $\deg q_j' < \deg p_j \ (j = 0, 1)$. Then $p_1(q_1 - q_1') = p_0(q_0' - q_0)$. Since p_0 and p_1 are relatively prime, it follows that $q_j = q_j' \ (j = 0, 1)$. □

Solutions to Problems

Chapter 1

1.1. Condition (O1) is trivial. Let $A^i_\alpha = A_\alpha$, and $A = \bigcup A_\alpha$. If $x \in A$, then $x \in A_\alpha$ for some α. Thus we can choose $\varepsilon > 0$ so small that $B(x,\varepsilon) \subset A_\alpha$. Thus $B(x,\varepsilon) \subset A$, which implies (O2). We have $(A_\alpha \cap A_\beta)^i \subset A^i_\alpha, A^i_\beta$, from which it follows that $(A_\alpha \cap A_\beta)^i \subset A^i_\alpha \cap A^i_\beta$. On the other hand, if $x \in A^i_\alpha \cap A^i_\beta$, there exists $\varepsilon > 0$ such that $B(x,\varepsilon) \subset A_\alpha$ and $B(x,\varepsilon) \subset A_\beta$. This implies that x is an inner point of $A_\alpha \cap A_\beta$. Thus $A^i_\alpha \cap A^i_\beta \subset (A_\alpha \cap A_\beta)^i$.

1.2. Suppose that the unit disc $B(0,1)$ is a countable union of mutually disjoint open intervals $\{I_j\}$. Denote $\{(a_1,0) : -1 < a_1 < 1\} = (-1,1)$ for simplicity. Then we have $(-1,1) \cap \bigcup_{j=1}^\infty I_j = (-1,1)$. The right-hand side is the countable union of mutually disjoint open intervals in $\mathbb{R}$. Thus $(-1,1) \cap I_j = (-1,1)$ for some I_j, which is absurd. In fact, pick a rectangle I_j so that $(a_1,b_1) = (-1,1) \cap I_j \neq \emptyset$. Since $a = (a_1,0)$ is not contained in the open interval I_j, a is an inner point of I_k for some $k \neq j$. Then $I_j \cap I_k \neq \emptyset$.

1.3. (i) $y \in \varphi(\bigcup A_\alpha) \Leftrightarrow \exists x \in \bigcup A_\alpha; y = \varphi(x) \Leftrightarrow \exists x \& \alpha_0; x \in A_{\alpha_0}$ and $y = \varphi(x) \Leftrightarrow y \in \bigcup \varphi(A_\alpha)$. The other cases are proved similarly.

1.4. Assume that $A' = A \neq \emptyset$. Pick a point a in A. For every $\varepsilon > 0$ the set $A \cap B(a,\varepsilon)$ contains infinitely many points. Fix an $\varepsilon > 0$ and choose two distinct points a_0 and a_1 in $A \cap B(a,\varepsilon)$. Take $\varepsilon/2 > \varepsilon' > 0$, so that $B(a_0,\varepsilon')$ and $B(a_1,\varepsilon')$ are mutually disjoint. Next pick two distinct points $a_{0,0}, a_{0,1}$ in $A \cap B(a_0,\varepsilon')$ and two distinct points $a_{1,0}, a_{1,1}$ in $A \cap B(a_1,\varepsilon')$. In the same way we get sequences $\{a_{i_1,i_2\dots,i_k}; i_k = 0,1\}$. Let $\lim_{k\to\infty} a_{i_1,\dots,i_k} = a_{i_1,i_2,\dots}$. Then $a_{i_1,i_2,\dots}$ ($i_j = 0$ or 1) are distinct points in A. Remark that $a_{i_1,i_2,\dots} \mapsto \sum_{k=1}^\infty i_k/2^k$ is a mapping of a subset of A onto $[0,1]$.

1.5. We can assume that F_1 is a compact set. By Theorem 1.14 the intersection $\bigcap F_j$ is nonempty and does not contain any distinct points by assumption.

1.6. Let I be a bounded closed cube of $\mathbb{R}^d$ which contains $\{a_j\}$. Divide I into 2^d congruent closed cubes. One of these cubes, say I_1, contains infinitely many points of $\{a_j\}$, which we denote by $\{a_{1,k}\}$. Next, divide I_1 into 2^d congruent cubes. One of these cubes, say I_2, contains infinitely many points of $\{a_{1,j}\}$, which we denote by $\{a_{2,l}\}$. Continuing this process, we get a decreasing sequence of closed cubes I_k and an infinite subsequence $\{a_{k,n}\}$ of $\{a_j\}$. Observe that $a_{k,k} \in I_k$ and $\{a_{k,k} : k = 1,2,\dots\}$ is a convergent sequence.

1.7. Let $\varepsilon > 0$ be given. Take $R > 0$ so that $|f(x)| < \varepsilon/2$ if $|x| > R-1$. Then we have $|f(x) - f(y)| < \varepsilon$ if $|x| > R-1$. Since a continuous function on a compact set is uniformly continuous, there exists $1 > \delta > 0$ such that $|f(x) - f(y)| < \varepsilon$ if $|x-y| < \delta$ and $|x|, |y| \le R$. We remark that if $|x-y| < \delta$, then both $|x|$ and $|y|$ are larger than $R-1$ or smaller than R.

1.8. Assume that f is nondecreasing in $[a,b]$. Let $\Delta = \{a = a_0 < a_1 < \cdots < a_N = b\}$ be a partition of $[a,b]$, and put $I_j = [a_{j-1}, a_j]$. Since $\overline{f}_j = f(a_j), \underline{f}_j = f(a_{j-1})$, we have

$$\sum_{j=1}^{N} (\overline{f}_j - \underline{f}_j) = \sum_j [f(b_j) - f(a_j)] = f(b) - f(a). \qquad (*)$$

We have $0 \leq \overline{s}(f,\Delta) - \underline{s}(f,\Delta) = \sum_{j=1}^{N} [\overline{f}_j - \underline{f}_j]|I_j| \leq \sum_j [\overline{f}_j - \underline{f}_j] \sup_j |I_j|$. The last term is not greater than $[f(b) - f(a)] \sup_j |I_j|$ by (*).

1.9. (i) It is obvious that $1 < m_{J*}(O)$. Let $J_1, \ldots, J_k$ be mutually disjoint right half-open intervals contained in O. Let $J'_j \subset J_j$ be an arbitrary closed subinterval and put $J' = \bigcup_{j=1}^{k} J'_j$. Since J' is compact, it is covered by a finite number of open intervals $I_1, \ldots, I_l$. Thus $\sum_{j=1}^{k} |J'_j| \leq \sum_{i=1}^{l} |I_i| < \sum_{i=1}^{\infty} 2/2^i = 2$. Thus $m_{J*}(O) \leq 2$.

(ii) Since $\{r_j\}$ is dense in $\mathbb{R}$, the set O is dense in $\mathbb{R}$. This implies that if a finite number of open intervals cover O, then these open intervals cover the whole axis.

Chapter 2

2.1. $\chi_{\limsup E_j}(x) = 1 \Leftrightarrow x \in \limsup E_j \Leftrightarrow$ for any k we have $x \in E_j$ for infinitely many $j \geq k \Leftrightarrow$ for any k we have $\chi_{E_j}(x) = 1$ for infinitely many $j \geq k \Leftrightarrow \limsup \chi_{E_j}(x) = 1$.

2.2. Since $F = \limsup_{j\to\infty} E_j \subset \bigcup_{j=k}^{\infty} E_j$, we have $m(F) \leq \sum_{j=k}^{\infty} m(E_j) \to 0$ as $k \to 0$.

2.3. If E is an interval, (i) and (ii) are obvious. Since open sets are countable unions of mutually disjoint right half-open intervals, they hold good for open sets. When E is a measurable set, observe that $m(E) = \inf\{m(O) : E \subset O \in \mathcal{O}\}$.

2.4. Note that $m(E) = m(E \cap F) + m(E - F)$ and $m(E - F) \leq m([0,1] - F) = 1 - 1 = 0$.

2.5. When E is bounded, the proof follows from Theorem 2.8 and the definition of the inner measure. For a general set E put $E_R = E \cap B(O,R)$. Then $m(E) \geq \sup\{m(C) : C \subset E\} \geq \sup\{m(C) : C \subset E_R\} = m(E_R) \to m(E)$ as $R \to \infty$.

2.6. From the square $[0,1] \times [0,1]$ remove a cross-type open set $\{(x,y) : 1/3 < x < 2/3\} \cup \{(x,y) : 1/3 < y < 2/3\}$, getting the closed set $\tilde{C}_1$. Thus $\tilde{C}_1$ consists of 4 closed squares. In the next step remove 1/3 cross-type open sets from each of the 4 squares. The remainder consists of 4^2 closed squares. Denote it by $\tilde{C}_2$. In the same way we get the closed set $\tilde{C}_n$ at the nth step.

Define $\tilde{C} = \bigcap_{n=1}^{\infty} \tilde{C}_n$. Then $(x,y) \in \tilde{C} \Leftrightarrow x \in C$ and $y \in C$, where C denotes the Cantor ternary set. If $0 \leq a \leq 2$, the line $a = x + y$ intersects $\tilde{C}_1, \tilde{C}_2, \ldots$. Thus by the finite intersection property it intersects $\tilde{C}$, which means that $a = x + y$ for some $(x,y) \in \tilde{C}$.

2.7. If f is continuous at a point a, then f is continuous at any point x, because $f(x+h) = f(x+a) - f(a-h)$ by the hypothesis. For positive integers n and m we have $f(n) = f(1 + \cdots + 1) = nf(1)$ and $f(1) = f(1/m + \cdots + 1/m)(n \text{ sum}) = mf(1/m)$. Thus $f(n/m) = f(1/m + \cdots + 1/m)(n \text{ sum}) = nf(1/m) = (n/m)f(1)$. If $x \geq 0$, choose positive integers m and n so that $n/m \to x$. Then we get $f(x) = xf(1)$ by the continuity. For a negative x use $f(-x) + f(x) = 0$.

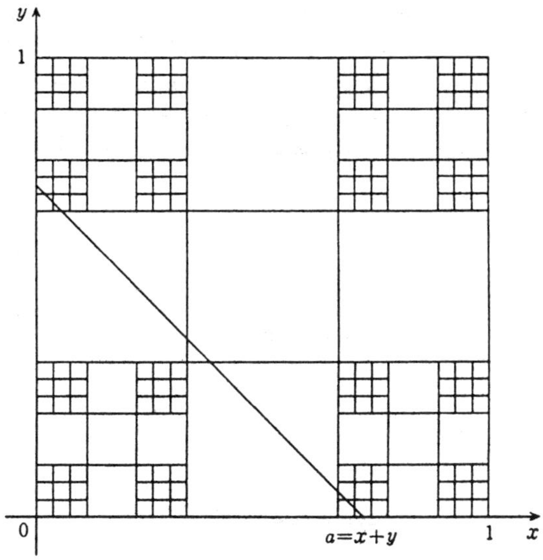

Chapter 3

3.1. Let $f = u + iv$ and $g = u' + iv'$. (i) is obvious. (ii) Observe that $f/g = (uu' + vv')/(u'^2 + v'^2) + i(uv' + vu')/(u'^2 + v'^2)$ and apply the real valued function case. To prove (iii), apply (iii) for real valued functions to the function $f \cdot \overline{f}$.

3.2. Put $f_n(x) = (nx)/[1 + (nx)^2]$. Then $f_n \in C[0,1]$. For each $x \in [0,1]$, $f_n(x) \to 0$ as $n \to \infty$, but $f_n(1/n) = 1/2$.

3.3. Let E be a nonmeasurable set. Then $f = \chi_E - \chi_{E^c}$ is not measurable, but $|f| \equiv 1$.

3.4. $\{x : \sup f_\alpha(x) > a\} = \bigcup_\alpha \{x : f_\alpha(x) > a\}$. The right-hand side is an open set.

3.5. Let $E_n = \{x : f(x) > 1/n\}$. By Chebyshev's inequality

$$m(E_n) \leq n \int f dm = 0, \quad n = 1, 2, \dots .$$

Now observe that $\{x : f(x) \neq 0\} = \bigcup_{n=1}^\infty E_n$.

3.6. Apply Fatou's lemma to the sequence $F_j = |f| - f_j$.

3.7. Consider the functions $-f_j$, where the f_j are the functions given in Remark 3.2.

3.8. (i) Put $f_j(x) = \chi_{E_j}(x)$ and apply Fatou's lemma. To prove (ii), apply (i) to $E - E_j$. A counterexample for (iii) is given by the set $E_j = (j, j+1)$. In fact, we have $\limsup E_j = \emptyset$.

3.9. Observe that

$$\sum_{k=-\infty}^{\infty} 2^k \chi_{E_k}(x) \leq \sum_{k=-\infty}^{\infty} |f(x)| \chi_{E_k}(x) \leq \sum_{k=-\infty}^{\infty} 2^{k+1} \chi_{E_k}(x).$$

3.10. (i) For a given $\varepsilon > 0$ choose a simple function $s = \sum_{n=1}^N c_n \chi_{F_n}$ so that $\int |f - s| dm < \varepsilon$. Thus $\left| \int_{E_j} (f - s) dm \right| < \varepsilon$ for all j. On the other hand, $\int_{E_j} s dm = \sum_{n=1}^N c_n m(F_n \cap E_j) \to 0 \, (j \to \infty)$.

(ii) Suppose $t_j \to t$. Let $E_j = (\min(t, t_j), \max(t, t_j))$. Then $|F(t_j) - F(t)| \leq \int_{E_j} |f| dm \to 0$. A proof for the case where $t \to -\infty$ is similar.

3.11. For a given $\varepsilon > 0$ there exists a simple function $s(x) = \sum_{j=1}^{N} a_j \chi_{E_j}(x)$ such that $\int |f - s| dm < \varepsilon$, where the I_j are bounded intervals. Note that if $|y|$ is large enough, $\int |s(x+y) - s(x)| dm(x) = \int |s(x+y)| dm(x) + \int |s(x)| dm(x)$.

3.12. Apply Lebesgue's dominated convergence theorem to $F_n(x) = x^n f(x)$.

3.13. We have $|f(y)\varphi(x-y)| \le |f(y)| \|\varphi\|_\infty \in L^1(\mathbb{R})$. Thus

$$|F(x)| \le \int |f(y)| dm \|\varphi\|_\infty.$$

By the dominated convergence theorem and by the continuity of φ we have

$$\lim_{x\to a} \int f(y)\varphi(x-y) dm(y) = \int \lim_{x\to a} f(y)\varphi(x-y) dm(y) = \int f(y)\varphi(a-y) dm(y).$$

3.14. We have

$$[F(x+h) - F(x)]/h = \int_{-\infty}^{\infty} f(y)[\varphi(x+h-y) - \varphi(x-y)]/h dm(y). \qquad (*)$$

By the mean value theorem $[\varphi(x+h-y) - \varphi(x-y)]/h = \partial\varphi(x+\theta h - y)/\partial x$, where $0 < \theta < 1$. Since $d\varphi(x)/dx$ is bounded and continuous, the dominated convergence theorem can be applied to (*).

3.15. We have

$$\int f(x) \frac{e^{-2\pi i \eta x} - 1}{\eta} e^{2\pi i \xi x} dm(x) = -2\pi i \int f(x) \frac{\sin \pi \eta x}{\pi \eta} e^{-\pi i \eta x} e^{-2\pi i \xi x} dm(x).$$

Observe that the absolute value of the integrand of the last integral is $\le |xf(x)|$, which is integrable, and apply the dominated convergence theorem.

3.16. By the Taylor expansion, $x^p/(1-x)\log(1/x) = \sum_{n=0}^{\infty} x^{n+p} \log(1/x)$. Since each term on the right-hand side is positive in $(0,1)$, we can integrate the sum term by term by B. Levi's theorem. Furthermore, integration by parts yields

$$\int_0^1 x^{p+n} \log(1/x) dx = [x^{p+n+1} \log(1/x)/(p+n+1)]_0^1 + \int_0^1 x^{p+n} dx/(p+n+1).$$

3.17. $\int_0^\infty u_n dm = 0$, and $\int_0^\infty [e^{-x}/(1+e^{-x})] dx = \log 2$.

Chapter 4

4.1. Let E be a nonmeasurable set. Put $f(x) = 1\ (x \in E)$, and $= -1\ (x \notin E)$. If $\varphi(x) = |x|$, then $\varphi(f) \equiv 1$.

4.2. $f^{-1}(A)$ consists of countable middle third intervals and a subset of the Cantor ternary set which is null.

4.3. First check that the series coincides with the Cantor singular function in each middle third interval, and then observe that both the sum and the Cantor singular function are nondecreasing.

4.4. For $\delta > 0$ the function $F_\delta(x) = \sup_{0<h<\delta}[f(x+h) - f(x)]/h$ is lower semicontinuous. Thus $\{x : D^+ f(x) \le a\} = \bigcap_{n=1}^{\infty} \{x : F_{1/n}(x) \le a\}$ is measurable.

4.5. For any partition of $[a,b]$ with points of division $\{a_j\}$ we have

$$\sum |f(a_j) - f(a_{j-1})| \le C \sum |a_j - a_{j-1}| = C(b-a).$$

Thus f is of bounded variation. Furthermore, every function of bounded variation is a linear combination of four nondecreasing functions (see Corollary 4.1).

Chapter 5

5.1. See Problem 1.2 for (i) and (ii). (iii) Let $\mathcal{X} = \{\emptyset, [-1,1), [-1,0), [0,1)\}$ and $f(x) = x^2$. Then $f(\mathcal{X})$ is not a ring of sets.

5.2. Let $E \in \mathcal{M}(\mathbb{R})$. Put $B = \{y \in Y : g(y) \in E\}$. Then $B \in \mathcal{Y}$. Therefore $\{x \in X : g(T(x)) \in E\} = \{x \in X : T(x) \in B\} \in \mathcal{X}$.

5.3. Check the definition of measure.

5.4. Observe that $\rho(E,F) = \int_X |\chi_E(x) - \chi_F(x)| d\mu(x)$.

5.5. When f is a simple function on $\mathbb{N}$, the result is a direct consequence of the definition. The general case is obtained by passing to the limit.

5.6. f_k is uniquely given by $f_k(x) = 2^k \int_{I_j} f(t)dt \quad (x \in I_j)$.

5.7. The set of points x such that $\mu(\{x\}) \neq 0$ is countable.

5.8. We can assume that $\mu_j\,(j = 1,2,3)$ are nonnegative without losing generality. If N is a μ_3-null set, then it is μ_2-null, so also μ_1-null. Put $\frac{d\mu_1}{d\mu_2} = f$ and $\frac{d\mu_2}{d\mu_3} = g$. Then $f, g \geq 0$. For a simple function $s = \sum a_j \chi_{E_j}$, we have $\int_E s d\mu_2 = \sum_j a_j \mu_2(E \cap E_j) = \sum_j a_j \int_{E\cap E_j} g d\mu_3 = \int_E (\sum_j a_j \chi_{E_j}) g d\mu_3$. Replace s by simple functions s_n such that $0 \leq s_1 \leq s_2 \leq \cdots \to f$ and let $n \to \infty$. Then $\int_E f d\mu_2 = \lim \int_E s_n d\mu_2 = \lim \int_E s_n g d\mu_3 = \int_E f g d\mu_3$. Since $\mu_1(E) = \int_E f d\mu_2$, it follows that $\mu_1(E) = \int_E f g d\mu_3$.

5.9. Observe that $\operatorname{diam} f(B(a,r)) \leq C(2r)^\alpha$ and apply the definition of the Hausdorff dimension.

5.10. Cf. Example 5.11. The Hausdorff dimension of D is 2.

Chapter 6

6.1. Apply Hölder's inequality with the indices $1/(q/p) + 1/[q/(q-p)] = 1$. Then $\int |f|^p d\mu \leq \left(\int |f|^q d\mu\right)^{p/q} \left(\int 1 d\mu\right)^{(q-p)/q}$. Put $f(x) = x^{-1/p}$ for $x > 1, = 0$ otherwise.

6.2. Let $p \neq 2$. Check the parallelogram law for the functions $f = \chi_{[0,1]}$ and $g = \chi_{[2,3]}$.

6.3. $\left|\int (f - f_n) g d\mu\right| \leq \left(\int |f - f_n|^p d\mu\right)^{1/p} \left(\int |g|^{p'} d\mu\right)^{1/p'}$.

6.4. The first part follows from the inequality $|\|f\|_p - \|g\|_p| \leq \|f - g\|_p$.

To prove the second part, notice that $\int_F |f|^p d\mu \to 0$ if $\mu(F) \to 0$. Put $E(n) = \{x : 2|f(x)| \geq |f_n(x)|\}$ and $g_n = f_n \chi_{E(n)}$. Then $g_n \to f$ a.e. and $|g_n| \leq 2|f|$. Thus $\int |g_n - f|^p \to 0$, and $\int |f_n - g_n|^p = \int |f_n|^p - \int |f_n|^p \chi_{E(n)} = \int |f_n|^p - \int |g_n|^p \to \int |f|^p - \int |f|^p = 0$.

6.5. Put $E = \{x : |f(x)| > M\}$. By the assumption that $\|f\|_q < \infty$, $\mu(E) < \infty$. If $\mu(E) > 0$, then $\|f\|_p \geq \left(\int_E M^p d\mu\right)^{1/p} \to M$ as $p \to \infty$. On the other hand, $\|f\|_p \leq \left(\int_X |f(x)|^q \|f\|_\infty^{p-q} d\mu\right)^{1/p} \to \|f\|_\infty$.

6.6. We observe that

$$\int_Y \left|\int_X f(x,y) d\mu(x)\right|^p d\nu(y) \leq \int_Y \int_X \left(\int_X |f(x',y)| d\mu(x')\right)^{p-1} |f(x,y)| d\mu(x) d\nu(y).$$

Change the order of integration and apply Hörder's inequality to the integral over Y with indices $1/[p/(p-1)] + 1/p = 1$.

6.7. The case $n = 3$. We may suppose that $f_1, f_2, f_3 \geq 0$. If

$$1/q_1 = 1/[p_1(1 - 1/p_3)] \text{ and } 1/q_2 = 1/[p_2(1 - 1/p_3)],$$

then $1/q_1 + 1/q_2 = 1$. Apply Hölder's inequality with indices $1/q_1 + 1/q_2 = 1$ to

$$\int f_1 f_2 f_3 d\mu = \int (f_1 f_3^{1/q_1})(f_2 f_3^{1/q_2}) d\mu,$$

and then, with indices $1/p_3 + 1/(p_j/q_j) = 1$, to $\int f_3 f_j^{q_j} d\mu\, (j = 1, 2)$.

6.8. We may suppose that $f, g \geq 0$. Observe that

$$f * g(x) = \int f(y)^{p/r} g(x-y)^{q/r} f(y)^{p[(1/p)-(1/r)]} g(x-y)^{q[(1/q)-(1/r)]} dy.$$

If $1/s = 1/p - 1/r$ and $1/t = 1/q - 1/r$, then $1/r + 1/s + 1/t = 1$. Apply Hölder's inequality.

6.9. (i) Consider the family of simple functions $s = \sum_{j=1}^{n} a_j \chi_{I_j}$, where the real and imaginary parts of a_j and the end points of the interval I_j are all rational. The family of such simple functions is dense in L^p and countable. (ii) Let D be a dense subset of $L^\infty(0,1)$. Put $f_t(x) = \chi_{[0,t]}(x)$ for $0 < t < 1$. For each $0 < t < 1$ pick a function $u^t \in L^\infty(0,1)$ such that $\|f_t - u^t\|_\infty < 1/2$. Since $u^s \neq u^t$ if $s \neq t$, D is not countable.

Chapter 7

7.1. We have $|\langle T_\mu, u\rangle| \leq p_{0,K}(u)|\mu|(K)$, where $K = \operatorname{supp} u$.

7.2. See Theorem 6.13.

7.3. $\langle \Delta U, u\rangle = \langle T, \langle -(4\pi)^{-1}\Delta|x|^{-1}, \tau_{(-\cdot)}u\rangle\rangle$.

7.4. (i) δ. (ii) $-|\sin t| + 2\sum_{n=-\infty}^{\infty} \delta_{2\pi n}$.

7.5. By the definition of differentiation, $\langle D^\alpha(\varphi T), u\rangle = (-1)^{\|\alpha\|}\langle T, \varphi D^\alpha u\rangle$ and $\left\langle \sum_{\beta\leq\alpha} \binom{\alpha}{\beta}(D^{\alpha-\beta}\varphi)(D^\beta T), u\right\rangle = \sum_{\beta\leq\alpha}\binom{\alpha}{\beta}(-1)^{\|\beta\|}\left\langle T, D^\beta((D^{\alpha-\beta}\varphi)u)\right\rangle$. Observe that $\sum_{\beta\leq\alpha}(-1)^{\|\beta\|}\binom{\alpha}{\beta}\sum_{\gamma\leq\beta}\binom{\beta}{\gamma}(D^{\alpha-\gamma}\varphi)D^\gamma u = \varphi D^\alpha_u$.

7.6. Apply the formula (7.3).

Chapter 8

8.1. The residue of $e^{-itz}/(1+z^2)$ at the point $z = -i$ is $e^{-t}/2i$. Integrate $e^{-itz}/(1+z^2)$ on the curve $\{x : -R \leq x \leq R\} \cup \{Re^{i\theta} : \pi \leq \theta < 2\pi\}$ to get $ie^{-t}/2i = (2\pi i)^{-1}\int_R^{-R} e^{-itx}/(1+x^2)dx + (2\pi i)^{-1}\int_\pi^{2\pi} e^{-itR[\cos\theta + i\sin\theta]} \times iRe^{i\theta}/(1+R^2e^{2i\theta})d\theta$. The second term on the right-hand side converges to 0 as $R \to \infty$.

8.2. (i) $\widehat{x^\alpha} = (2\pi\xi)^{-\|\alpha\|}D^\alpha\delta$. (ii) $\widehat{Y}(\xi) = (1/2)\delta + (1/2\pi i)\mathrm{P.V.}(1/\xi)$. (iii) $1/(2\pi i\xi + 1)$.

8.3. A direct computation yields

$$\sum_{j,k} c_j\overline{c_k}\widehat{\mu}(\xi_j - \xi_k) = \int_{\mathbb{R}} \sum_{j,k} c_j\overline{c_k} e^{-2\pi i\xi_j x + 2\pi i\xi_k x} d\mu(x)$$
$$= \int_{\mathbb{R}} |\sum_j c_j e^{-2\pi i\xi_j x}|^2 d\mu(x) \geq 0.$$

8.4. Suppose that there exists a unit e in L^1. Then $\widehat{e * P_t}(\xi) = \widehat{e}(\xi)\widehat{P_t}(\xi) = \widehat{P_t}(\xi)$. Thus $\widehat{e}(\xi) \equiv 1$. On the other hand, by the Riemann-Lebesgue theorem we have $\widehat{e}(\xi) \to 0\ (\xi \to \infty)$.

8.5. We have $\tilde{x} = m$. Thus by translation we can assume $m = 0$. We have $\int_{-\infty}^{\infty} f(x)^2 dx = c^2(\pi/2a)^{1/2}$ and $\int_{-\infty}^{\infty} x^2 f(x)^2 dx = c^2(1/2)(d/da)\int e^{-2ax^2}dx = c^2(\pi/2a)^{1/2} \times (1/4a)$. Thus $\Delta_f^2 = 1/4a$. Since $\widehat{f}(\xi) = c^2\sqrt{\pi/a}e^{-(\xi-b)^2\pi^2/a}$, we get $\Delta_{\widehat{f}}^2 = a/4\pi^2$.

8.6. Let $0 < t < 1$ and $|x| < 1/2$. (s1): Integrate w_t term by term. (s2)$'$: Let $n \neq 0$. Since $(x+n)^2/4t > n^2/16t$, we have $e^{-(x+n)^2/4t} < 16t/n^2$, and thus $w_t(x) \leq (4\pi t)^{-1/2}(1 + \sum_{n\neq 0} t/n^2) < (\text{constant}) \times t^{-1/2}$. (s3)$'$: If $x^2/t > 1$, then

$e^{-x^2/4t} < 4t/x^2$. Thus $w_t(x) \leq (4\pi t)^{-1/2}[(4t/x^2) + \sum_{n\neq 0} t/n^2] <$ (constant) $\times$ $t^{1/2}/x^2$. Therefore we get (s2)$'$ and (s3)$'$ with $t^{1/2}$ in place of t.

Bibliography

[Da] I.Daubechies, *Ten Lectures on Wavelets*, SIAM, Philadelphia, 1992.

[Fa] K. J. Falconer, *The Geometry of Fractal Sets*, Cambridge University Press, 1985.

[Fo] G.B.Folland, *Real Analysis, Modern Techniques and Their Applications*, Wiley, 1984.

[GR] J.Garcia-Cuerva and J.Rubio de Francia, *Weighted Norm Inequalities and Related Topics*, North-Holland, 1985.

[Ha] P.R.Halmos, *Measure Theory*, Van Nostrand, 1950.

[HS] E.Hewitt and K.Stromberg, *Real and Abstract Analysis*, Springer-Verlag, Berlin, 1965.

[HW] E.Hernández and G.Weiss, *A First Course on Wavelets*, CRC Press,1996.

[Ig] S.Igari, *Fourier Series* (Japanese), Iwanami Shoten, 1975.

[Je] T.J.Jech, *The Axiom of Choice*, North-Holland, 1973.

[Ka] Y.Katznelson, *An Introduction to Harmonic Analysis*, Wiley, 1968.

[Ma] L.Mattila, *Geometry of Sets and Measures in Euclidean Spaces, Fractals and Rectifiability*, Cambridge University Press, 1995.

[Me] Y.Meyer, *Ondeletts et Opérateurs*, I,II,III(with R.R.Coifman), Hermann, Paris, 1990. English transl. I; *Wavelets and Operators*, Cambridge University Press, 1993. II, III; *Wavelets: Calderón-Zygmund and Multilinear Operators*, ibidem, 1997.

[Na] L.Nachbin, *The Haar Integral*, Van Nostrand, 1964.

[Ro] H.L.Royden, *Real Analysis*, Macmillan, 1968.

[Ru] W.Rudin, *Real and Complex Analysis*, McGraw-Hill, 1966.

[Sa] S.Sakakibara, *Wavelets : A Guide for Beginners* (Japanese), Tokyo Denkidai Press, Tokyo, 1995.

[Sk] S.Saks, *Theory of the Integral,* 2nd ed., Hafner, New York, 1937.

[Sc] L.Schwartz, *Méthodes Mathématiques pour les Sciences Physiques*, Hermann, Paris, 1961.

[St] E.M.Stein, *Harmonic Analysis: Real-Variable Methods, Orthogonality, and Oscillatory Integrals*, Princeton University Press, 1994.

[SW] E.M.Stein and G.Weiss, *Introduction to Fourier Analysis on Euclidean Spaces*, Princeton University Press, 1971.

[Tr] F.Trèves, *Topological Vector Spaces, Distributions and Kernels*, Academic Press, 1967.

[Wo] P.Wojtaszczyk, *A Mathematical Introduction to Wavelets*, London Math. Soc. Student Texts, vol. 37, Cambridge University Press, 1997.

[WZ] R.L.Wheeden and A.Zygmund, *Measure and Integral, An Introduction to Real Analysis*, Marcel Dekker, 1977.

[Zy] A.Zygmund, *Trigonometric Series*, Cambridge University Press, 1959.

Index

Selected Titles in This Series

(Continued from the front of this publication)

142 **N. V. Krylov,** Introduction to the theory of diffusion processes, 1995
141 **A. A. Davydov,** Qualitative theory of control systems, 1994
140 **Aizik I. Volpert, Vitaly A. Volpert, and Vladimir A. Volpert,** Traveling wave solutions of parabolic systems, 1994
139 **I. V. Skrypnik,** Methods for analysis of nonlinear elliptic boundary value problems, 1994
138 **Yu. P. Razmyslov,** Identities of algebras and their representations, 1994
137 **F. I. Karpelevich and A. Ya. Kreinin,** Heavy traffic limits for multiphase queues, 1994
136 **Masayoshi Miyanishi,** Algebraic geometry, 1994
135 **Masaru Takeuchi,** Modern spherical functions, 1994
134 **V. V. Prasolov,** Problems and theorems in linear algebra, 1994
133 **P. I. Naumkin and I. A. Shishmarev,** Nonlinear nonlocal equations in the theory of waves, 1994
132 **Hajime Urakawa,** Calculus of variations and harmonic maps, 1993
131 **V. V. Sharko,** Functions on manifolds: Algebraic and topological aspects, 1993
130 **V. V. Vershinin,** Cobordisms and spectral sequences, 1993
129 **Mitsuo Morimoto,** An introduction to Sato's hyperfunctions, 1993
128 **V. P. Orevkov,** Complexity of proofs and their transformations in axiomatic theories, 1993
127 **F. L. Zak,** Tangents and secants of algebraic varieties, 1993
126 **M. L. Agranovskiĭ,** Invariant function spaces on homogeneous manifolds of Lie groups and applications, 1993
125 **Masayoshi Nagata,** Theory of commutative fields, 1993
124 **Masahisa Adachi,** Embeddings and immersions, 1993
123 **M. A. Akivis and B. A. Rosenfeld,** Élie Cartan (1869–1951), 1993
122 **Zhang Guan-Hou,** Theory of entire and meromorphic functions: deficient and asymptotic values and singular directions, 1993
121 **I. B. Fesenko and S. V. Vostokov,** Local fields and their extensions: A constructive approach, 1993
120 **Takeyuki Hida and Masuyuki Hitsuda,** Gaussian processes, 1993
119 **M. V. Karasev and V. P. Maslov,** Nonlinear Poisson brackets. Geometry and quantization, 1993
118 **Kenkichi Iwasawa,** Algebraic functions, 1993
117 **Boris Zilber,** Uncountably categorical theories, 1993
116 **G. M. Fel′dman,** Arithmetic of probability distributions, and characterization problems on abelian groups, 1993
115 **Nikolai V. Ivanov,** Subgroups of Teichmüller modular groups, 1992
114 **Seizô Itô,** Diffusion equations, 1992
113 **Michail Zhitomirskiĭ,** Typical singularities of differential 1-forms and Pfaffian equations, 1992
112 **S. A. Lomov,** Introduction to the general theory of singular perturbations, 1992
111 **Simon Gindikin,** Tube domains and the Cauchy problem, 1992
110 **B. V. Shabat,** Introduction to complex analysis Part II. Functions of several variables, 1992
109 **Isao Miyadera,** Nonlinear semigroups, 1992
108 **Takeo Yokonuma,** Tensor spaces and exterior algebra, 1992
107 **B. M. Makarov, M. G. Goluzina, A. A. Lodkin, and A. N. Podkorytov,** Selected problems in real analysis, 1992
106 **G.-C. Wen,** Conformal mappings and boundary value problems, 1992

(See the AMS catalog for earlier titles)